Monitoring and Diagnosis of Turbine-Driven Generators

Monitoring and Diagnosis of Turbine-Driven Generators

Avelino J. Gonzalez
M. Stanley Baldwin
J. Stein
N. E. Nilsson

Prentice Hall, Englewood Cliffs, New Jersey 07632

Monitoring and diagnosis of turbine-driven generators / by Avelino J. Gonzalez . . . [et al.].
p. cm.
Includes bibliographical references and index.
ISBN 0-13-606096-X
1. Gas-turbine power-plants—Management. 2. Turbogenerators—Maintenance and repair. 3. Electric power failures. I. Gonzalez, Avelino J.
TK1076.M65 1995
621.31'34–dc20 94-36848
CIP

Publisher: Alan Apt
Project Manager: Mona Pompili
Cover Designer: Wendy Alling Judy
Copy Editor: Shirley Michaels
Manufacturing Buyer: Lori Bulwin
Editorial Assistant: Shirley McGuire

Printed in the United States of America

10 9 8 7 6 5 4 3 2 1

ISBN 0-13-606096-X

Prentice-Hall International (UK) Limited, *London*
Prentice-Hall of Australia Pty. Limited, *Sydney*
Prentice-Hall Canada, Inc., *Toronto*
Prentice-Hall Hispanoamericana, S.A., *Mexico*
Prentice-Hall of India Private Limited, *New Delhi*
Prentice-Hall of Japan, Inc., *Tokyo*
Simon & Schuster Asia Pte. Ltd, *Singapore*
Editora Prentice-Hall do Brasil, Ltda., *Rio de Janeiro*

Preface

For many turbine generator failures, corrective action could have been taken to reduce the magnitude of the damage, or possibly prevent it altogether, if the incipient fault or abnormal operating condition had been detected earlier. Recent advances in instrumentation have helped eliminate this problem by providing an opportunity to better understand the internal state of electrical generators during their operation. With such understanding, new knowledge about the internal processes in operating units that can lead to a number of serious malfunctions has been developed over the last decade of field experience.

In some instances, however, the number of new sensors monitoring a turbine generator have had the effect of an "information overload" to unit operators and plant engineers. Without knowledge of what to look for (i.e., what to consider and what to disregard), the additional information available in the control room can be confusing and often useless. As a result, incorrect interpretation of significant readings have at times resulted in the disastrous neglect of serious developing faults, as well as in unnecessary outages. Proper training of plant personnel therefore can lead to more correct identification of incipient faults in a turbine generator and what corrective course of action to undertake to minimize or prevent such outages.

The main objective of this book is to teach the participants to recognize, in a timely fashion, the symptoms of an on-line steam turbine generator whenever abnormal conditions exist that can lead to a significant unit outage. These symptoms of incipient faults are reflected through on-line sensor readings as well as other means. Readers will also learn what corrective action to take to mitigate the circumstances or minimize the damage. The emphasis of the course is on both on-line and off-line monitoring and diagnostics.

The concept of this book came about from a three-day course of the same title offered since 1989 at the Electric Power Research Institute Monitoring and Diagnostic Center in Eddystone, Pennsylvania. The original suggestion for the book was made by Murthy Divakaruni of EPRI and it was identified as a formidable challenge by the first author. The book came to fruition as a result of Murthy's persistence and support.

The authors would like to thank the following individuals for their help in ensuring the highest quality in the book.

Murthy Divakaruni, *EPRI*
John Niemkiewicz, *EPRI M&D Center*
Russ Pflasterer, *EPRI*
Dick Boggia, *Consolidated Edison*
Arnold Friedeck, *Union Electric*
Geoff Klempner, *Ontario Hydro*
Ron Corkins, *Detroit Edison*
Jim Folda and Kevin Becker, *Siemens*
Don Ramey, Mike Twerdochlib, Ali Moradian, and Lyndol Michael, *Westinghouse*
Blake Lloyd and Greg Stone, *IRIS Power Engineering*
Don Albright, *Generator Tech, Inc.*
Elton Floyd, *TU Electric*
Lowell Brothers, *Southern Company Services*
Jack White, *Tampa Electric Co.*
Jim Bothwell, *Keystone Speciality Services*
John Rink, Dean Harrington, and Kurt Steinebronn

Lastly we acknowledge and thank Westinghouse Electric Corp., General Electric Co., Generatortech, IRIS Power Engineering, Siemens, and Ohio Edison for allowing the use of their illustrations for the descriptive figures.

Avelino Gonzalez
M. Stanley Baldwin
J. Stein
N. E. Nilsson

DISCLAIMER OF WARRANTIES AND LIMITATION OF LIABILITIES

B) Assumes responsibility for any damages or other liability whatsoever (including any consequential damages, even if EPRI or any EPRI representative has been advised of the possibility of such damages) resulting from your selection or use of this publication or any information, apparatus, method, process, or similar item disclosed in this publication.

Contents

CHAPTER 1

Introduction

Early in the afternoon of a hot summer day, Unit 2 of a power station was being brought to full load after a successful synchronization. As the load on the unit began to increase, numerous alarm windows on the annunciator panel became lit, as is typical during normal start-ups. In normal cases, these alarms will disappear as the unit begins to stabilize in load and in temperature. In this particular case, however, the alarms indicated a real problem within the generator. Overloaded with information and concentrating on the load ramping process, the operators did not notice these danger signals and allowed the malfunction to worsen. Several minutes later, an internal fault in the generator windings caused the unit to be tripped automatically by its protective relaying scheme.

Upon internal inspection of the unit, it was determined that significant damage had been sustained by the generator, and the unit had to be sent to the manufacturer for extensive repairs lasting several months. The cost of the forced outage was several million dollars, which included removal of the generator from its foundation, shipment to the manufacturer, cost of repair, and most importantly, cost of replacement power.

Analysis of the failure and the indications leading up to it revealed that the temperature and pressure indicators of the generator had provided an early indication of the incipient failure. Had the readings been noticed and correctly interpreted by a knowledgeable individual, the unit could have been brought off-line prior to its full failure, and the result might have instead been a three-day outage to repair the problem.

1.1 THE IMPORTANCE OF DIAGNOSTICS

The fictitious but realistic case described above indicates the magnitude of the problem that misdiagnosis (or, in this case, nondiagnosis) of an abnormal condition can cost the plant owner.

Early diagnosis of abnormal conditions in systems, whether natural or man-made, has always been a significant factor in their continued and reliable operation. This is quite evident in the medical field, where the early detection and treatment of certain conditions (e.g., cancer, coronary disease) significantly increases the patient's chance for survival. Conversely, the cost of not recognizing in a timely fashion the symptoms of a treatable disease can be loss of life in many cases.

Man-made systems or machines are similar to biological ones. Their reliable operation is based on continuous monitoring of their vital signs and in the diagnosis of any problems indicated, but the task of diagnosing problems with complex engineered systems, such as an electric generator, is not a simple one. It is typically performed by a person with a significant amount of expertise in the design and/or operation of the system or equipment. Much of the knowledge is based on procedures learned from an analysis of the system's design, but a significant portion of it is based on heuristic knowledge compiled over the many years of experience in confronting and solving such problems.

The steadily decreasing margins of reserve generation capacity in the United States have placed increased pressure on the utilities to maximize the reliability and availability of generating stations. Because the loss of a generator results in the forced outage of an entire generating unit, careful attention to the generator's operating condition can result in increased unit availability and lower forced outage rates.

Recent advances in generator monitoring systems and techniques have provided an excellent opportunity to better determine incipient faults within the generators. Over the last decade, new knowledge about their internal processes has been developed through experience with these sensors, which has led to a better understanding of serious malfunctions in operating units. This, in turn, has resulted in better diagnostic techniques.

On the negative side, the growth in number of new sensors monitoring a generator has had the effect of an information overload to plant operators and engineers. Misinterpretation of significant readings and neglect in acknowledging others because of previous false alarms and misoperations have led to unnecessary outages and impacted the severity of equipment damage. Training of plant personnel can accomplish the objectives of identifying developing situations in a generator and taking corrective action to minimize outages.

Thus, the main objective of this book is to instruct the plant personnel in the diagnostics of turbine generators.

1.2 A CATASTROPHE THAT WASN'T

On the morning of January 21, 1985, diagnostic experts assigned by the manufacturer to monitor a generator (with the help of a diagnostic expert system) along with plant operations personnel, noted a slight discrepancy in the temperature readings of the exhaust gas coming from one of the stator winding conductor-cooled coils. While the discrepancy was not high enough to reach alarm level, it was sufficient evidence for an expert diagnostician to determine that a conductor problem existed. Upon discussion with the plant management, it was decided to take the unit off-line in order to inspect the situation. It was subsequently discovered that, as predicted, strands in the stator coil in question had suffered some cracks, which caused the current to be redirected to the other group transpositions within the same circuit. This redirection resulted in increased losses in the other transpositions within the same circuit, causing all coils within the affected circuit to operate slightly hotter than the coils in other circuits. This small but noticeable discrepancy provided the critical clue that led to the correct diagnosis of the problem. The repair took only a few hours, and the entire outage was limited to approximately three days.

The kind of incipient failure diagnosed in this factual case could have caused severe damage to the generator if it had been allowed to reach its natural conclusion. Downtimes for other such incidences have been in the order of months. The amount of money saved by the plant owner in this incident was quite significant.

This last case study points out the benefits to be had by implementing a good diagnostic program for electrical generators. This can be done by a) vigorously training the plant engineers responsible for the unit, and b) enhancing their capability by providing them with tools such as automated monitoring and diagnostic systems. The purpose of this book is to assist in the training of plant operators to recognize such problems in a timely manner.

1.3 THE GENERATOR SYSTEM

For the purposes of this book, the generator will be said to be composed of the following major components:

- The generator itself
- The excitation system
- The auxiliaries

Each of the above will be further divided into subsystems. The generator is composed of

- The stator
- The rotor
- The core and frame

The excitation system is similarly divided into

- The excitation supply system
- The excitation control system

Likewise, the auxiliaries are broken down into the following subsystems:

- The gland seal oil subsystem
- The hydrogen subsystem (if hydrogen-cooled)
- The stator coil cooling water subsystem (if water-cooled)
- The bearing lubrication subsystem

Although there are some air-cooled and oil-cooled generators in operation, these are few in number. Thus, these auxiliary systems will not be covered in this book.

1.4 THE PURPOSE AND ORGANIZATION OF THIS BOOK

From the preceding sections of this chapter, the reader should understand the significance of generator monitoring and diagnostics. Thus, the book has as its main objective the education of plant or headquarters personnel, whether engineers or operators, in the design and operation of large steam turbine-driven electrical generators so that a proper diagnosis of abnormal conditions can be made in a timely manner. Because different manufacturers frequently use different approaches in the construction and operation of turbine generators, an effort has been made to keep the discussion general. When that is not possible, however, the book incorporates as many of the variant features as reasonably possible in order to cover the important aspects in monitoring and diagnosis of machine behavior. However, it would impossible to properly cover all such variations in this book. Readers should consult the generator instruction manuals for their particular units, or the manufacturers, to determine whether these units fits any specific descriptions included herein.

Because of its general nature, therefore, **this book is not intended to be used as a troubleshooting guide.** The subtle differences between designs from manufacturer to manufacturer, and even between unit types from the same manufacturer, precludes the ability to provide specific diagnostic information herein. This is made worse by the unavailability to the authors of any one unit's operating conditions at any one time, as well as its history—two significant parameter in the diagnosis of any problem. Upon suspicion of a specific problem, the parties responsible for the operation of the generator in question should contact the manufacturer to confirm a diagnosis and/or to seek specific recommendations on what course of action to take.

The book describes the generator and its subsystems. The intent is to cover the major general designs for large turbine-driven generators used in the United States, regardless of the manufacturer. These descriptions include discussions of the instruments and sensors that monitor the performance of the relevant subsystem. The book also discusses some of the realistic failure modes of a generator system, as well as some of the sensor indications that help identify them. The book also discusses, in general terms, the proper action to take in order to prevent or minimize the damage.

The basic organization is along the lines of the generator subsystems described previously. The chapter titled "Generator Monitors" will introduce the more complex individual monitors commonly available for generators. The excitation control (the voltage regulator) is included in the same chapter as the other parts of the excitation system. The relatively small extent of this topic does not merit its own separate chapter.

Chapters 4 through 11 on generator subsystems are each composed of the following subsections:

- A description of the components or subsystems pertinent to that chapter. This section discusses the important aspects of the elements being described, their construction, and the variations in the different designs, if any.
- A description of the sensors and monitors that are typically used to detect any abnormal conditions within the subsystem described in that chapter.
- A discussion of some of the potential problems that can arise in the subject subsystem, as well as how to detect them and what, in general, to do about them once they are detected.

Additionally, such related topics as general concepts in power generation (Chapter 2), generator protective relaying (Chapter 12), and automated diagnostic systems (Chapter 13) are covered in separate chapters in order to enhance the usefulness of this book.

1.5 A CLARIFICATION TO USERS OF THIS BOOK

The information contained here is intended for instructional purposes only. It is general in nature and not specifically directed to any manufacturer's equipment, construction type, or rating. **Thus, it is not intended to be used as a troubleshooting guide.** Because generators vary in their capability and in their design and complexity, and because interpretation of sensor readings is often quite subjective in nature, the Electric Power Research Institute, its subcontractors, and the authors of this book shall not be held liable for any injury or damage resulting from a recommended course of action followed by a reader. Readers should always consult the manufacturers before allowing their particular units to operated with suspected incipient faults, regardless of the contents of this book. In case of doubt, the safest course of action should be followed.

1.6 DIAGNOSTIC BOXES

Chapters 4 through 11 of this text contain diagnostic boxes, which the authors have developed to help identify typical problems related to the particular subject matter under discussion. Each box contains a brief description of the pertinent problem, any observed sensor parameters, the diagnosis suggested, and the courses of action that should be pursued. It is hoped that they will serve as valuable guides to utility operations and maintenance personnel responsible for the generator.

CHAPTER 2

Electrical Generation Concepts

2.1 GENERATOR BASICS

Michael Faraday discovered in 1831 that when an electrical conductor (a wire) is moved through a magnetic field in a direction perpendicular to the direction of the field, a voltage is generated in the wire. The amount of voltage generated is directly proportional to the active length of the conductor, the strength of the magnetic field, and the velocity of the conductor. Refer to Figure 2.1. This became known as *Faraday's Law* and is the basis for electromagnetic induction, the fundamental principle of electric power generation.

Assume a uniform magnetic field of strength B, an active length of conductor L, and a velocity of v. The induced voltage, e, can thus be described by the following equation:

$$e = B * L * v \tag{2.1}$$

If B is in lines per square centimeter, L in centimeters, and v in centimeters per second, the voltage will be in *abvolts*, or 10^{-8} volts. The resulting equation is

$$e = B * L * v * 10^{-8} \text{ volts} \tag{2.2}$$

The active length of the conductor L is that portion of the conductor within the magnetic field or that component of the conductor perpendicular to the magnetic field.

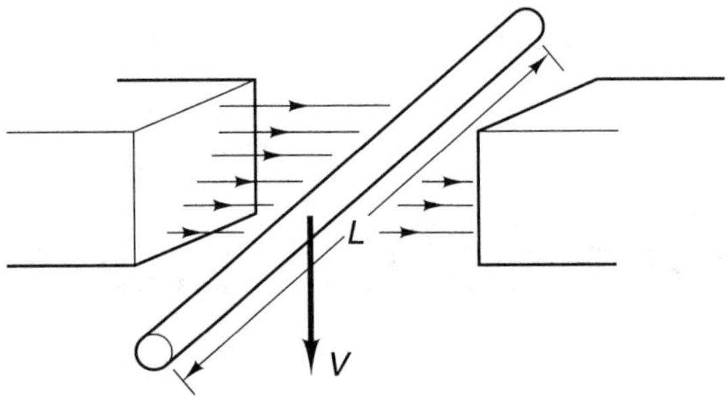

Figure 2.1 Faraday's Law.

Faraday's law implies that there must be either physical movement of the conductor relative to the magnetic field or movement of the magnetic field relative to the conductor. While this is perfectly satisfactory for a dynamo or generator, which consists of a set of coils rotating within a stationary magnetic field or a magnetic field rotating inside a stationary set of coils, it does not explain the voltage induced in a transformer winding when there is a changing voltage on another winding on the same magnetic core.

Lenz's Law expands Faraday's law to explain the operation of a transformer as well as that of a generator. The magnitude of voltage generated at any instant is proportional to the number of coils and the rate of change of magnetic flux threading these turns. The product of the number of turns and magnetic lines is called the *number of flux linkages*. So, the amount of voltage induced in a coil is proportional to the rate of change in the flux linkages. In equation form, Lenz's law is

$$e = N * \frac{d\phi}{dt} \tag{2.3}$$

In the above equation N is the number of turns of a coil or set of coils within the influence of the magnetic field, ϕ is the magnetic field strength, and $d\phi/dt$ is the time rate of change of magnetic lines of force threading the coil or coils. The voltage generated or induced is referred to as the *electromotive force* (emf).

Refer to Figure 2.2. The coil will have the most lines of magnetic flux threading it when the magnet is in a vertical position. However, the conductors of the coil will be cutting the maximum number of lines of force when the magnet is in the horizontal position. If the angle between the centerline of the magnet and the axis of the coil is designated as ωt, the amount of flux threading the coil will be proportional to the cosine of the angle ωt. The instantaneous value of flux linkages of the coil will therefore be

$$\phi = N * \phi_m * \cos \omega t \tag{2.4}$$

The emf generated in the coil is proportional to the rate of change of flux linkages and will be in a direction to oppose the change. In equation form, the instantaneous voltage can be expressed as

$$e = -N * \phi m * \frac{d(\cos \omega t)}{dt} \tag{2.5}$$

Because the derivative of the cosine is the negative sine, the above equation can be redefined as

$$e = N * \phi m * \sin \omega t \tag{2.6}$$

In Equations (2.5) and (2.6), ϕm is the maximum value of the magnetic flux or of the lines of force.

The resulting instantaneous voltage wave is shown in Figure 2.3.

2-1.1 The Elementary AC Generator

Figure 2.4 is an elementary alternating current (AC) generator with an electromagnet as the rotating field. Because of the large magnetic fields required, the mag-

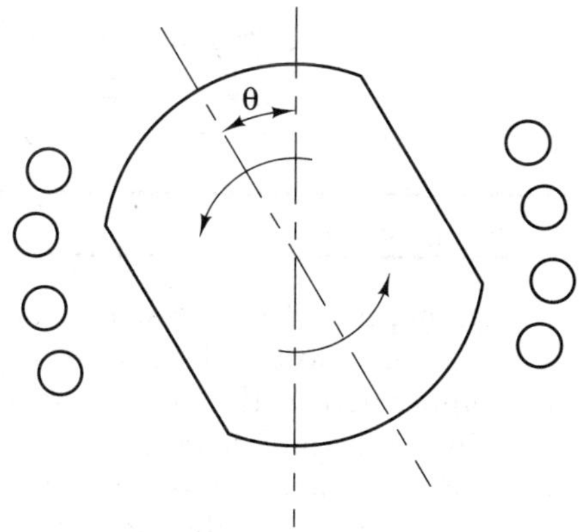

Figure 2.2 Alternating voltage generator.

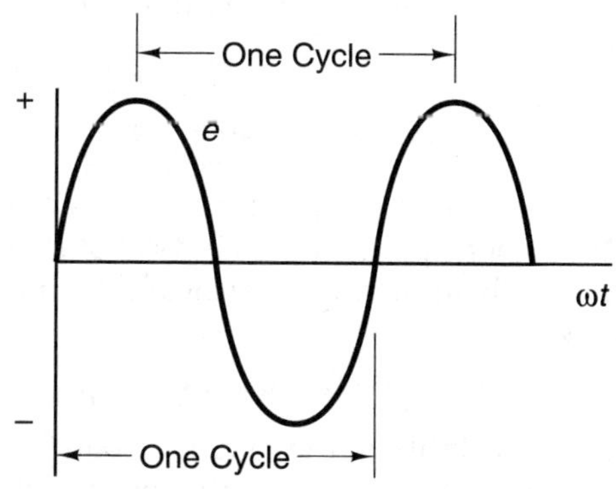

Figure 2.3 Sine wave of generated voltage.

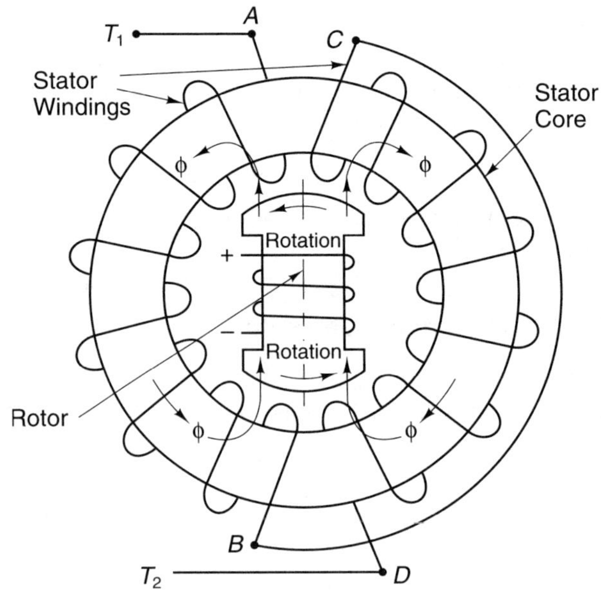

Figure 2.4 Elementary AC generator.

netic field in an actual medium- to large-size AC steam turbine generator (or alternator) typically consists of a slotted solid steel forged *rotor,* with a direct current (DC) field winding embedded in the slots to form the electromagnet. The field winding is supplied by a controllable DC *exciter.*

A current flowing in a coil or set of coils creates a magnetic field that is proportional to the number of coils and the magnitude of the current. This is referred to as a *magnetomotive force* (mmf). The actual amount of magnetic flux that will flow, or exist, in the magnetic path will depend also on the *reluctance* of the magnetic circuit. Reluctance of a magnetic circuit is analoguous to the resistance of an electrical circuit. Therefore, the flux that will exist in the circuit is defined by equation (2.7):

$$\phi = \frac{\text{mmf}}{\mathcal{R}} \tag{2.7}$$

In Equation (2.7), ϕ is the number of lines of magnetic flux and $\mathcal{R}$ is the reluctance of the magnetic path. The reciprocal of reluctance is the permeability of a magnetic circuit, with air having a permeability of one (1.0). The equation for flux, ϕ, can then be expressed as

$$\phi = \text{mmf} * \mu \tag{2.8}$$

where μ is the permeability of the magnetic path.

In an alternator, the magnetic path consists of the rotor, the air gap between the rotor and stator core, and the stator core. The stationary windings of the alternator where the power is being generated, called the *armature* winding, typically consists of a set of serially connected and suitably insulated electrical *coils* or *bars* that are embedded in axial slots in a laminated steel *core*. The laminations of this core are made from high permeability steel. The physical construction of these components will be described in Chapters 4, 5, and 6.

Refer to Equation (2.6), which is reprinted below for convenience.

$$e = N * \phi_m * \sin \omega t \tag{2.6}$$

In this equation, ω is the angular speed of the magnetic field. If there are more than one pair of poles on the generator rotor, there will be as many sine waves of voltage generated during each revolution of the rotor as there are pairs of poles on the rotor. The time for one cycle of voltage is the *period*. The reciprocal of the period is the *frequency* of the wave or the number of cycles per second of the wave. The relationship between generator speed, number of poles, and frequency is given in Equation (2.9):

$$f = \frac{p * n}{120} \tag{2.9}$$

In this equation, f is the frequency in cycles per second, or *hertz* (Hz), p is the number of poles, and n is the rotor speed in *revolutions per minute* (r/min).

Equation (2.6) could be rewritten as

$$\text{Voltage} = K * f * B \tag{2.10}$$

In this equation, K is a proportionality constant encompassing the parameters for a given generator, B is the magnetic flux density, and f is the frequency.

The electrical frequency in the United States is 60 Hz. All large steam turbine generators are therefore either 1800-r/min or 3600-r/min machines. This means that the 1800-r/min units have four rotor poles, and the 3600-r/min units have two poles. It should be obvious that, for a given magnetic field strength, the 3600-r/min generator will produce twice the emf per armature coil as the 1800-r/min unit. Practically, this means that the higher speed generator does not have to be physically as large as the lower speed one.

2-1.2 Real and Apparent Power in AC Generators

In a DC circuit, the product of voltage and current is the real power, measured in *watts*. In an AC circuit, the product of *instantaneous* voltage and *instantaneous* current is *instantaneous* real power; however, the instantaneous power can be either positive or negative! Positive power, in this case, is real power that the generator supplies to the load. Negative power is real power that the generator receives from the load. As shown in Figure 2.5, the instantaneous power curve is one for which the frequency is twice that of the generator frequency.

The current and the voltage waves are said to be *in phase* if they reach their maximum values or minimum values simultaneously. If, at all times in the cycle, the voltage and current are in phase, as shown in Figure 2.5, the instantaneous

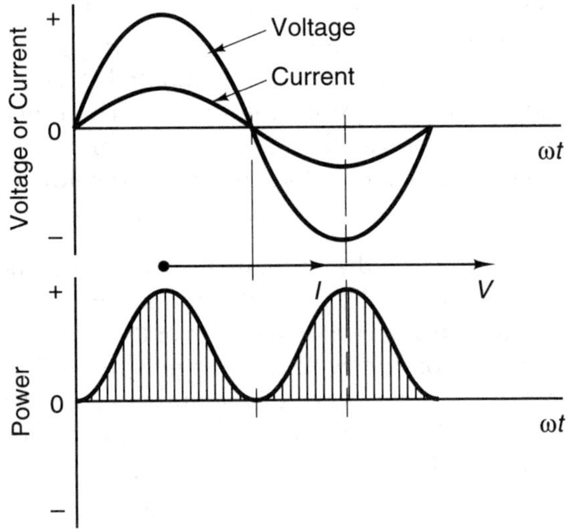

Figure 2.5 Power in an AC circuit.

power will be positive at all times. This condition occurs only when the load is a pure resistance. If the voltage and current are 90 electrical degrees out of phase, as shown in Figure 2.6, the power will be positive one-half of the time and negative the other half of the time, and the net power will be zero. This condition would occur with a purely inductive or purely capacitive load. In general, the current is somewhat out of phase with the voltage as shown in Figure 2.7. The power that is alternately absorbed by the load and returned to the generator (or source) is called *reactive power* or *volt-amperes-reactive* (vars).

In an AC circuit, the product of the *root mean square* (rms) voltage and current is referred to as *apparent power*, or *volt-amperes* (VA). The rms value of voltage or current is the peak value divided by the square root of two (2). The ratio of real power to apparent power in an AC circuit is defined as the *power factor* (PF) of the circuit, and is equal to the cosine of the angle between the instantaneous voltage and the instantaneous current.

$$\text{Power Factor} = \frac{\text{Real Power}}{\text{Apparent Power}} \tag{2.11}$$

The PF is always ≤ 1.0 and can be used to calculate the real power if the apparent power is known, as shown in Equation 2.12:

$$\text{Real Power (in watts)} = \text{Apparent Power (in volt-amperes)} * \text{PF} \tag{2.12}$$

The relationship between the apparent power and the reactive power is described in Equation 2.13, where θ is the angle between the instantaneous voltage and the instantaneous current:

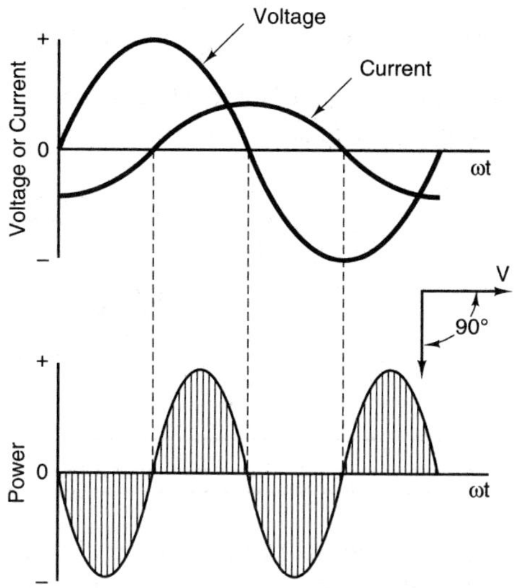

Figure 2.6 Power in AC circuit with voltage and current 90 degrees out of phase.

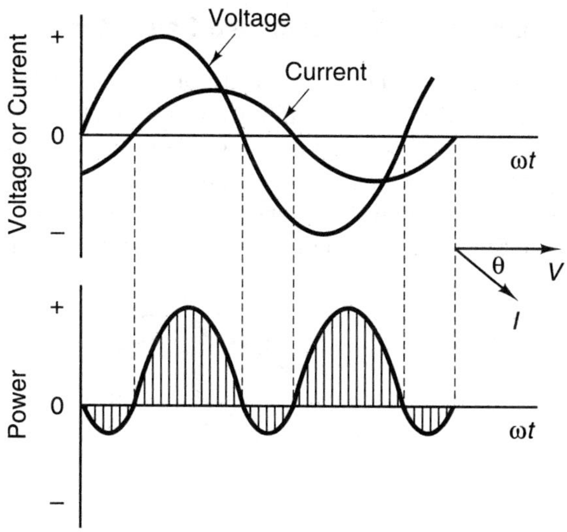

Figure 2.7 Typical phase relationship in an AC circuit.

Reactive Power (in volt-amperes) =

$$\text{Apparent Power (in volt-amperes)} * \text{Sin}\ \theta \quad (2.13)$$

Another term used is *reactive factor,* and it is defined as follows:

$$\text{Reactive Factor} = \frac{\text{Reactive Power}}{\text{Apparent Power}} \tag{2.14}$$

The phase relationships of currents and voltages as well as those of apparent, real, and reactive power are shown in phasor form in Figure 2.8.

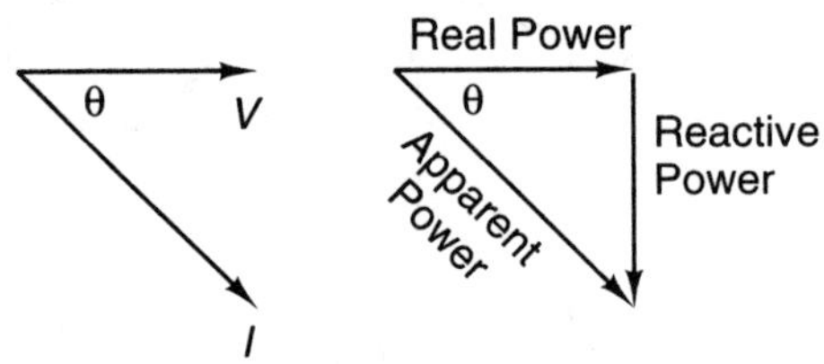

Figure 2.8 Phasor relationships of current and voltage and real and reactive power.

Of course, for a generator to provide real power output to a load, there must be a real power input to the generator. The generator converts mechanical power to electrical power. The mechanical power is supplied through the shaft connecting the generator to a steam turbine, which, in turn, converts the thermal energy of steam at high pressure into mechanical energy.

2-1.3 The Elementary AC Polyphase Generator

A *polyphase* generator is constructed by putting more than one set of windings on the armature, or stator. A simplified diagram of a basic three-phase generator is indicated in Figure 2.9. The individual windings can be connected in either a wye or delta configuration. Large generators always have the windings connected in wye. In a wye-connected generator there can be a fewer number of coils for a given terminal voltage than in a delta-connected machine. Also, the wye-connected genera-

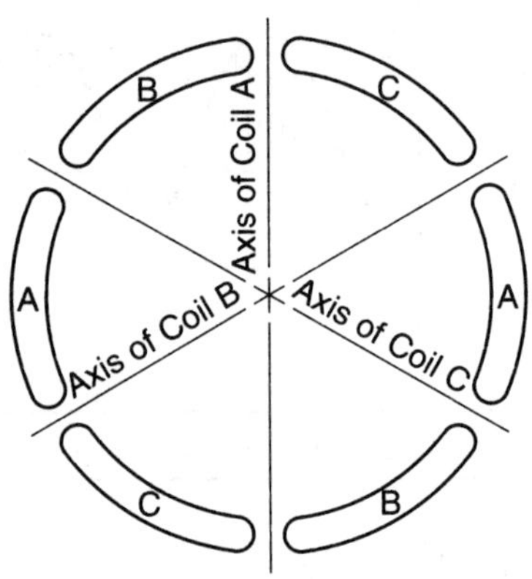

Figure 2.9 Basic three-phase generator.

tor has a neutral available for grounding. It is desirable to ground the generator to stabilize the power frequency voltages to ground and to minimize any transient overvoltages during faults on the generator terminal or in equipment connected to the generator.

The instantaneous voltages generated in the three-phase windings will be as shown in Figure 2.10. It should be noted that the algebraic sum of all three voltages will be equal to zero at any instant. Until now, the instantaneous values of voltage and current have been discussed, but the rms values of the voltage are actually used in calculations. A phasor representation of the three-phase rms voltages is shown in Figure 2.11. The phasor sum of the three rms voltages will also be zero.

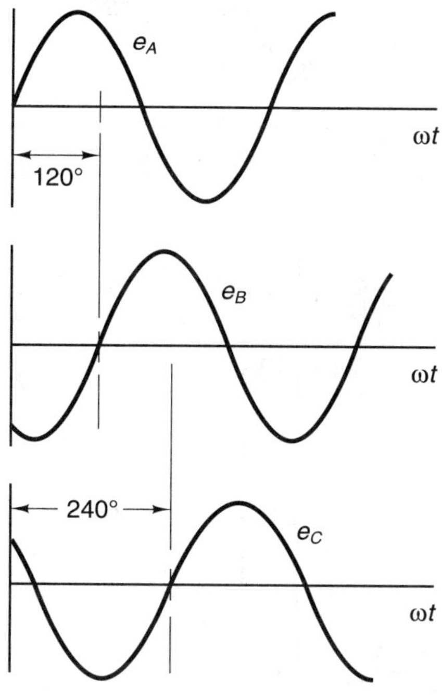

Figure 2.10 Instantaneous voltages of three-phase generator.

2.2 POWER SYSTEM BASICS

To be of any use, a generator must be connected to a transmission and distribution network that is capable of carrying its output to the load centers. Such a network is referred to as the *power system,* and it is described in the following sections.

2-2.1 The Power System

A simplified power system, or portion of one, is shown in Figure 2.12. In the United States, almost all of the power systems are interconnected at various transmis-

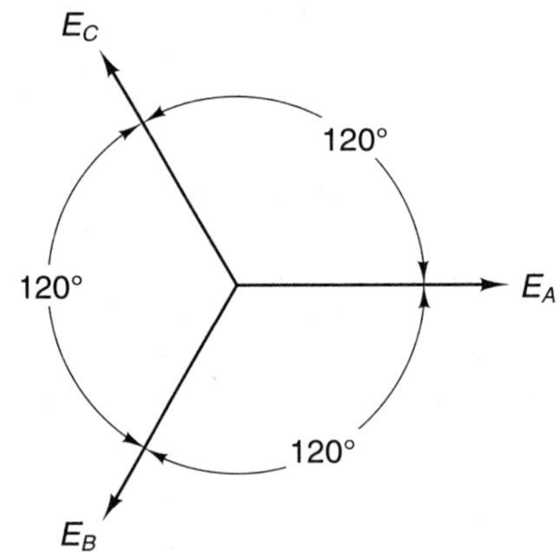

Figure 2.11 Phasor representation of AC voltages.

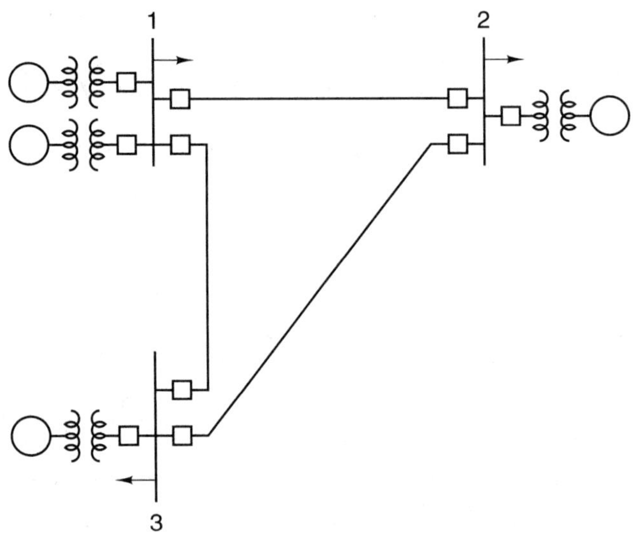

Figure 2.12 A simple power system.

sion voltage levels. Because a three-phase circuit, in which all three phase currents are equal (*balanced system*), can be treated as a single-phase system, it is customary to show a power system as a single-line representation for ease of analysis. Figure 2.12 shows a single-line representation of a three-phase system. A power system consists of generating units, transformers, transmission and distribution circuits, power circuit breakers, loads, lightning or surge protection devices, protective relay systems, metering devices, grounding equipment, and other devices.

2-2.2 Power Transfer Principles

The normal mode of operation of power systems is with all the generating stations and loads connected electrically in some way. It is necessary to understand the way in which the real and reactive power flow is controlled or "dispatched" as well as the interaction between generating units and the power systems.

Refer to Figure 2.13, which is a simple system consisting of two generating stations connected by a transmission circuit with loads on each of the station busses. For current to flow from one *bus* to another, there must be a difference in the voltages at locations *A* and *B*. If the voltages are of the same magnitude and in phase with each other, there can be no current flow in the line. Thus, a current will flow only if there is a difference in magnitude, phase angle, or both.

Refer to Figure 2.13. Assume that the transmission line consists of inductive impedance only. Actually there will be some resistance in a transmission line, but it is usually very low compared with the inductive reactance. If there is a difference only in voltage magnitude, the current between the *A* and *B* busses will lag the voltages, and the difference between them, by 90°, as shown in Figure 2.13. In this case only reactive power will flow between the two locations.

On the other hand, if there is a difference of phase angle only, as shown in the upper phasor diagram of Figure 2.14, there will be a real power component in the current between the two locations. Typically, there will be a difference in both phase angle and magnitude between the two voltages. There could be real power flow from bus *A* to bus *B* and reactive flow from *B* to *A* depending on how the two voltages differ.

It should be apparent from the preceding explanation that the real and reactive power flows can be controlled on a power system by controlling the magnitudes and phase angles of the voltages at the generating stations. While not strictly true, in general, real power (or megawatts) is controlled by manipulation of the

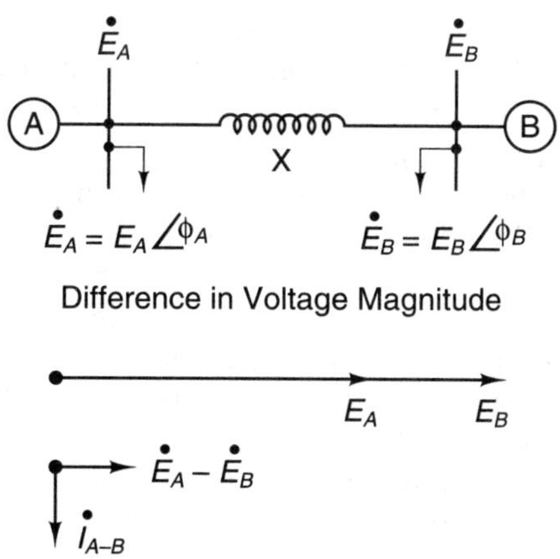

Figure 2.13 Reactive power transfer.

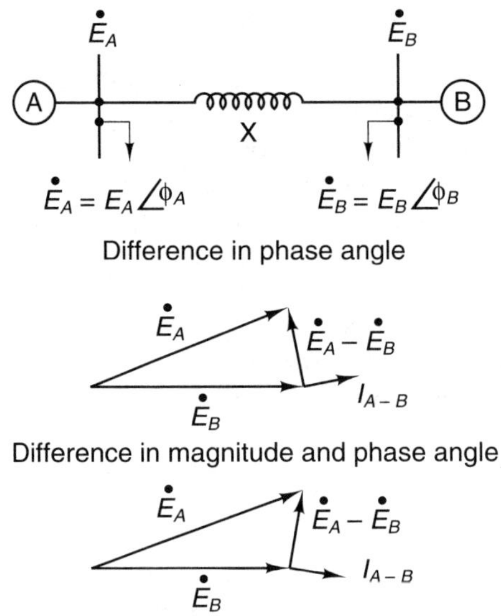

Figure 2.14 Real and reactive power transfer.

phase angle of the voltage; and, the reactive power (vars) is controlled by adjusting the voltage magnitude at the generating stations. How these can be accomplished is discussed below.

When a generator is electrically connected to a power system and delivering load, it must be in *synchronism* with the other generators connected to the power system. The current in the generator armature creates a rotating magnetic field or mmf in the armature that is magnetically "locked" with the magnetic field of the rotor. The armature current, by means of transformers and transmission circuits, is also coupled with the other generators connected to the power system. The result is that the rotors of all the generators are electromagnetically locked together.

Again refering to Figure 2.14, in order to transmit power from location *A* to location *B*, the phase angle of the voltage at *A* must be *advanced* or must *lead* the voltage at *B* by some angle. This is accomplished by increasing the mechanical power input to the generator at *A*. To increase the power input to the generator, the amount of steam input to the turbine driving the generator at *A* must be increased. There will be a momentary increase in speed of the generating unit at *A* until the new angle is reached or stabilized. Conversely, a reduction in the steam input to the turbine will result in a decreasing of the power angle and a lowering of the power transmitted from *A* to *B*.

To increase the reactive output of a generator, the magnitude of the output voltage of the unit must be increased. As seen in Equation 2.6, the magnitude of voltage generated is proportional to the magnetic flux in the machine. The magnetic flux is, in turn, proportional to the mmf in the unit. The mmf of the rotating field is proportional to the current in the field winding. Therefore, an increase in voltage is accomplished by increasing the field current of the generator. As indicated in the following section, increasing the magnitude of the output voltage of one or both of the generators of Figure 2.13 will also affect the permissible power that can be transmitted between the two locations.

2-2.3 Stability Concepts

The amount of real power that can be transmitted over a transmission line or system is shown in Figure 2.15.

For a given condition, the amount of power is the product of the voltage magnitudes at the two locations and the sine of their angular difference, divided by the reactance of the circuit or circuits between the two locations or voltage sources.

$$\text{Power} = \frac{E_A * E_B}{X} * \sin \delta \qquad (2.15)$$

In the above equation, δ is the angular difference between voltages E_A and E_B, and X is the reactance between them.

This relationship is best shown by the power-angle diagram of Figure 2.15. It is apparent that the maximum power flow will occur when the source voltages are 90° apart. In this case, the voltages are the generated voltages; therefore, the internal impedances of the generators must be included in the total impedance of the circuit or system. The peak of this power-angle curve is the power limit or *steady-state stability limit* of that system. If an attempt were made to increase the phase-angle difference between the two locations above 90°, the units would lose the elec-

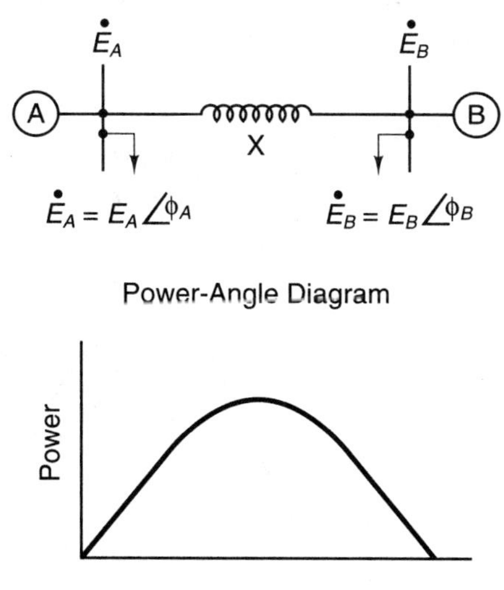

Figure 2.15 Power-angle relationships.

tromagnetic coupling between the generators. This loss of coupling is referred to as *losing synchronism* or *pulling out of step* and is further discussed in Section 2.4.1. The steady-state electrical power output of the generator must be supplied by the mechanical input from the turbine. If there is a mismatch between the turbine power and the generator output, the generator will either speed up or slow down. If the power level of the turbine is within the power transfer capability of the system, the change in speed may be only a momentary one until the conditions for the power-angle diagram are satisfied.

Figure 2.16 shows a simple system in which the power P_1 is being transmitted over two parallel lines from location *A* to location *B*. The upper curve on the power-angle diagram defines the power transfer capability of the system with the two parallel lines in service. The lower curve defines the power transfer capability with only one line in service. If a circuit breaker in one of the lines is opened, the power flow will instantaneously drop to the value P_2. Because the turbine output power (or generator input power), is still P_1, the imbalance between electrical power out of the generator and mechanical power into the generator at *A* will cause the generator to speed up until the angle at which power level P_1 intersects the lower curve is reached. In the meantime, the generator at *B* must absorb the power change. That will result in a momentary slowing down of that generator until generator *A* picks up the load. There may be power oscillations between the units until the system damping stabilizes the power transfer between the units.

If, in the above example, the power being transmitted from *A* to *B* had been above the maximum transfer capability of one line when the current in one line was interrupted, the units would have lost synchronism. That is, unit *A* would have increased in phase angle beyond the steady-state stability limit. In such a case, the two generating units will experience violent torque pulsations unless they are isolated immediately.

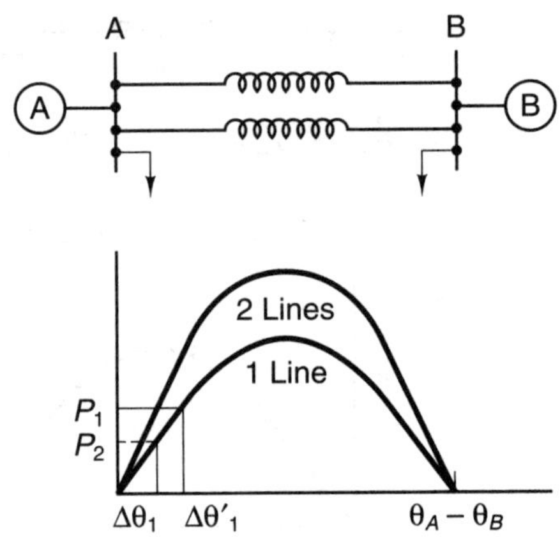

Figure 2.16 Power-angle relationship with two lines.

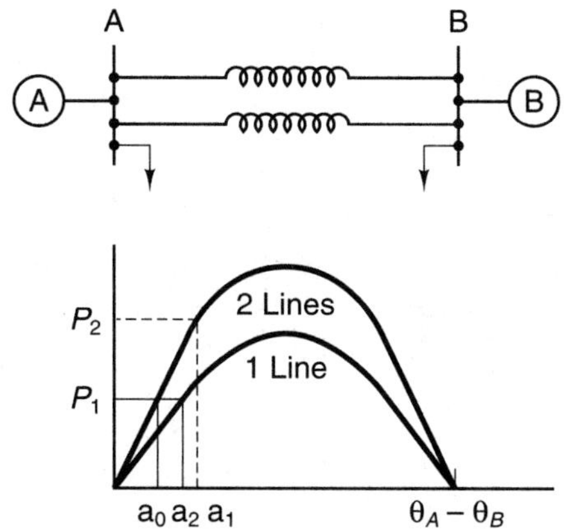

Figure 2.17 Fault and fault-clearing effects on power transfer.

Refer to Figure 2.17. Again, assume that power level P_1 is being transmitted across the two lines between *A* and *B*. Although most transmission line faults are single-phase-to-ground faults, for this example it will be assumed that a three-phase fault occurs on one of the lines near location *A*. No power can be transmitted during the fault because the voltage at *A* is zero. Until the circuit breakers at the terminals of the faulted line are opened to isolate the fault, the generator at *A*, and possibly at *B*, will speed up. When the fault occurred, generator *A* was at angle α_0. Assume that, during the fault, generator *A* has accelerated and is at angle α_1 when the fault is cleared.

The power transfer capability for one line is that corresponding to angle α_1 on the lower power-angle curve. It is further assumed that the turbine output at *A* cannot change significantly during the time of fault and fault clearing. Because the power out of *A* is now above the turbine input, that unit must slow down until the mechanical power input of *A* matches that power level on the power-angle curve. However, because of mechanical inertia, the unit will continue to increase in angle until the retarding torque of the load is effective in slowing the unit down. As long as the maximum power transfer capability of the one-line system is not exceeded at any time, the unit will ultimately stabilize at point α_2, usually after one or more oscillations about angle α_2.

It should be apparent that the faster the fault is cleared, the lower the amount of increase in angle that unit *A* will experience. Furthermore, because most transmission line faults are temporary, high-speed reclosing circuit breakers on the faulted line will result in the higher transfer capability of the two lines in parallel. As will be pointed out in Section 2.4.8, any switching of a transmission line, particularly near a generating station, may cause some mechanical shock to the generating unit. For this reason, some utilities do not employ high-speed reclosing on lines terminating at the generating station.

As stated earlier, most transmission line faults are single-phase-to-ground faults. In this case, the unfaulted phases of the line will permit power transfer during the fault, and the generating unit will not accelerate as much as it would with a three-phase fault. However, during the time that the circuit breaker is open, the effect on the unit is the same as described for the three-phase fault condition.

Because the power transfer capability of a line is inversely proportional to the reactance of the line, reducing the reactance of a line will increase its power or stability limit. Typically, a line with a single conductor per phase will have a reactance of approximately 0.8 ohm per mile per phase at 60 Hz. Lines with bundled conductors have a reactance of approximately 0.6 ohm per mile. By using bundled conductors, the power limit of a line can be increased by about one-third; that is, the ratio of the power limit of a bundled conductor line of the same length as one without bundled conductors is four-thirds.

In the western part of the United States, the very long transmission distances (and, therefore, lines) result in very high values of line inductance. In such cases, series capacitors are sometimes used to reduce the net reactance of the lines. The capacitive reactance cancels out some of the inductive reactance of the line, thus improving its power transfer capability. These series capacitor banks must be shorted or by-passed during a fault. The reduction in net reactance results in higher values of fault current on some portions of the line. Also, during the fault, the voltage across the capacitors may exceed the voltage rating of the capacitors. Usually, some fairly elaborate air-gap systems are used to by-pass the capacitors during faults, and compressed air is used to blow out the shorting arc of the gaps to reinsert the capacitors in the line after the fault is cleared. The application of series capacitors to enhance stability must be carefully studied to determine whether or not a condition known as *subsynchronous resonance* will occur. This will be discussed further in Section 2.4.9.

It is beyond the scope of this book to discuss the many aspects and types of system stability and stability enhancement measures that can be employed. It should be noted, however, that fast fault-clearing and high-speed reclosing are the most often used methods of ensuring stability. Modern protective relaying systems and power circuit breakers permit detecting and interrupting faults within a few cycles. Because most line faults are single-phase-to-ground faults, some users employ single-pole tripping and reclosing with independent pole-operated circuit breakers. Tripping only the faulted phase permits the flow of power through the unfaulted phases, increases the stability margin of the system, and minimizes the effects on the generator.

High-response excitation systems (described in Chapter 7) are also used to improve the transient stability, and, in a few cases, fast operation of the turbine intercept valves is employed to quickly reduce the mechanical input power to the generator. This last technique is called *fast valving* or *early valving*.

2.3 THE PRACTICAL SYNCHRONOUS AC GENERATOR–RATINGS AND PERFORMANCE

AC generators are rated in apparent power (*kilovolt-amperes,* or kVA) rather than in kilowatts. The generators will also have a voltage rating and a minimum power factor at which rated kVA can be supplied.

Typically, fossil steam turbine generator units in the United States are rated at 800 megawatts or less, although there are a few cross-compound units rated at up to 1300 megawatts. Because most large generators have a power factor rating of 0.90 or 0.95, this means that the MVA ratings will range from 900 to 1450. The largest nuclear units are approximately 1300 megawatts. The voltage ratings of the large generators are typically between 18 kV and 26 kV, but smaller units have lower voltage ratings. For a given kVA rating, the current in the generator windings is inversely proportional to the winding voltage. By raising the voltage of large generators to the highest practical values, the amount of current in the windings, and therefore the size of the conductors, can be reduced. The large generators are always connected to a unit transformer, and the transformer voltage rating can always be selected to accommodate the generator voltage. Medium-size units will typically be rated at 13.8 kV, while the very small units may be rated as low as 2.4 kV. A typical generator rating, therefore, might be 800,000 kVA, 21 kV, 0.9 PF, 60 Hz, 3600 r/min.

Because of practical considerations involved in the design and manufacture of generators, proper operation of these machines is best described by their manufacturers. Operation of these units within the limits identified by the manufacturer will ensure their reliable operation. These operating characteristics are represented by the manufacturers through a series of curves. These curves are described next.

2-3.1 The Generator Saturation Curve

The voltage generated in the armature winding is proportional to the strength of the magnetic field, which, in turn, is proportional to the current in the field winding. When the generator is delivering load, the current in the armature circuit also establishes a magnetic field, or mmf. A component of this armature mmf is in opposition to that of the rotor mmf. In addition to this demagnetizing effect of the armature current, there is an internal impedance voltage drop in the generator winding. At high flux densities the permeability of the magnetic circuit is reduced because of partial saturation of the steel.

The effects of armature current and saturation are shown in Figure 2.18, which is a typical excitation curve for a large steam turbine generator at rated speed. At rated load and PF, the field current required to maintain rated terminal voltage is much greater than that required for rated voltage at no load. It also should be noted that the field current required at a given load is a function of the load PF; for example, the current required in the rotor field winding for rated load

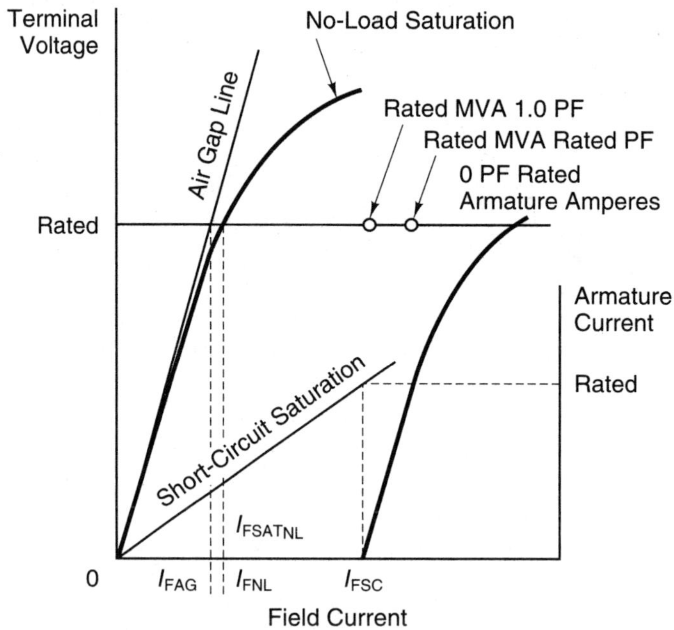

I_{FAG} = Generator field current required to produce rated voltage on the air gap line.
I_{FNL} = Generator field current required to produce rated voltage at no load.
I_{FSC} = Generator field current required to produce rated armature current with generator terminals short-circuited.

Figure 2.18 Typical saturation curve for a large steam turbine generator.

at rated PF (lagging) is more than that required for rated load at unity PF. Thus it is obvious that the generator must be supplied by a variable source of DC.

Because the generator field current requirements vary greatly from no load to full load, an *automatic voltage regulator* (AVR) is used to maintain a constant preset value of terminal voltage on the generator. The AVR will be discussed in more detail in Chapter 7.

2-3.2 The Generator Capability Curve

The application of excitation in the rotor field makes the synchronous generator a variable Mvar device. The MW and Mvar ability is defined by the synchronous generator capability curve. This graph is universally recognizable by power plant personnel, and it is usually mounted in the control room near the synchronous generator controls. In fact, many power plants have the synchronous generator capability curve programmed into the plant computer data acquisition system (DAS) and video display terminal. Figure 2.19 illustrates a typical synchronous generator capability curve supplied by a synchronous generator manufacturer. Note that the capability curves are typically labelled in MW and Mvars, while ANSI C50.10–1990 requires that the unit nameplate ratings be in kW and kvar's.

Synchronous generators can be rated for a range of lagging (overexcited) PFs. The most common PF ratings are 0.90 and 0.85 lagging. For a given MW rating, the lower the lagging PF rating, the greater the cost of a generator, because larger field windings and a higher power exciter are required. Lower lagging PF machines have greater Mvar ranges and may be needed for electrical systems in which the synchronous generator is electrically remote from the load center.

Note in Figure 2.19 that the region at the top of the capability curve to the left of the rated PF break point (0.85 PF in this case) is the *rotor field winding current*

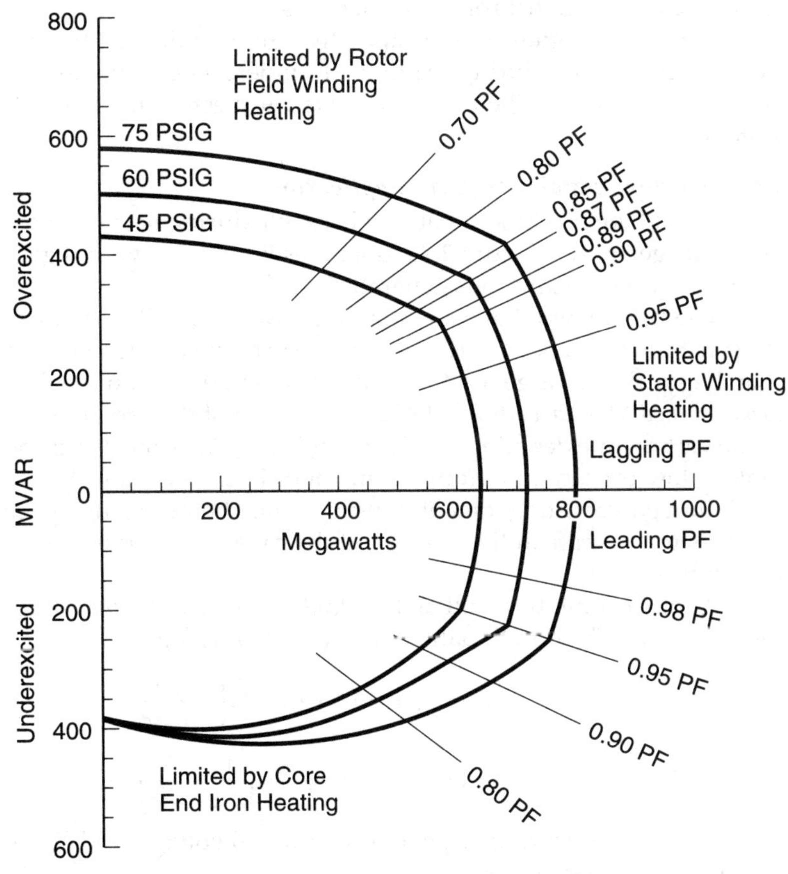

Figure 2.19 Typical manufacturer's synchronous generator capability curve.

limited region. At a specified cooling gas pressure (in this case, hydrogen), the synchronous generator capability can be approximated by an arc with its center at a value about equal to the *short circuit ratio* (SCR) in per unit on the negative Mvar axis. The radius can then be approximated by E_i/X_d, where E_i is the internal synchronous generator voltage or the rotor field excitation voltage with the synchronous generator terminal voltage at one per unit. X_d is the direct axis synchronous reactance of the generator in per unit on its own rating. Note that this region tends to expand for decreasing stator terminal voltages and to contract for increasing stator terminal voltages.

The region on the right-hand side of the capability curve is the *stator winding current limited region.* For a specified hydrogen cooling gas pressure, the synchronous generator capability can be approximated by an arc with its center at the origin. This capability is constant from 0.95 per unit stator terminal voltage to 1.05 per unit stator terminal voltage as specified by ANSI C50.13–1989, "Requirements for Cylindrical Rotor Synchronous Generators."

The lower region on the capability curve is limited by the stator end iron heating that occurs during leading PF underexcited operating conditions. This phenomenon is described in more detail in Section 6.1.1, which addresses the stator core.

Large machine capability curve application

Not all the machine capability is usable during normal operation of the synchronous generator. Figure 2.20 illustrates limitations with regard to the capability of the large synchronous generator.

The turbine will have a maximum power capability that usually will match the synchronous generator real power (MW) rating at rated lagging PF. This means that the synchronous generator cannot be operated continuously beyond the *gross demonstrated MW capability* (GDC). The portion of the capability curve to the right of the vertical line denoting GDC is not utilized. Also, most large boilers cannot operate below certain minimum steam conditions continuously, which results in a minimum power rating for a unitized system. Therefore, this synchronous generator does not operate in this portion of the capability curve except during start-up and shut-down.

Another limitation will be the *steady-state stability limit* in the leading PF region. This is defined in Equations 2.16 and 2.17 as follows:

$$\text{Center} = j0.5 * E_t^2 * \left(\frac{1}{X_e} - \frac{1}{X_d}\right) \tag{2.16}$$

$$\text{Radius} = 0.5 * E_t^2 * \left(\frac{1}{X_e} - \frac{1}{X_d}\right) \tag{2.17}$$

where E_t is the synchronous generator terminal voltage, and X_e equals the equivalent external system reactance.

Whereas an AVR can increase the stability range of the machine, the manual regulation characteristic is used on the capability curve to address those situations in which the AVR may be malfunctioning and must be taken out of service.

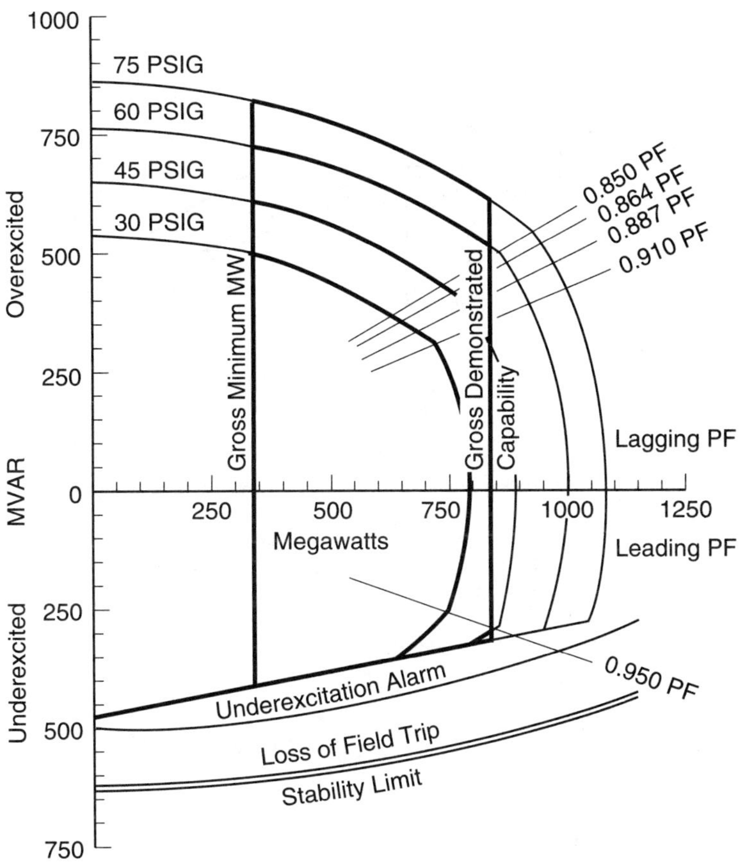

Figure 2.20 Typical large machine capability curve application.

The system conditions for which X_e is computed must also be evaluated. An electrically stiff system (one with relatively low external impedance) results in a more stable machine operating characteristic than one with a relatively large value of X_e. Accordingly, credible contingencies should be taken into account when computing X_e.

Once the steady-state system stability limit is plotted, a loss-of-field relay (discussed in Section 12.6.1) operating characteristic can be determined, which will protect the synchronous generator from operating at such a low excitation that it loses synchronism with the power system. Typically, an alarming underexcitation relay will be applied for backup protection. The voltage regulator will have either an *underexcitation reactive ampere limiter* (URAL) or a *minimum excitation limiter* (MEL) that will prevent the excitation from going below the specified level when the automatic voltage regulator is in service. Some original equipment manufacturers may have different names than these for their underexcitation limiting

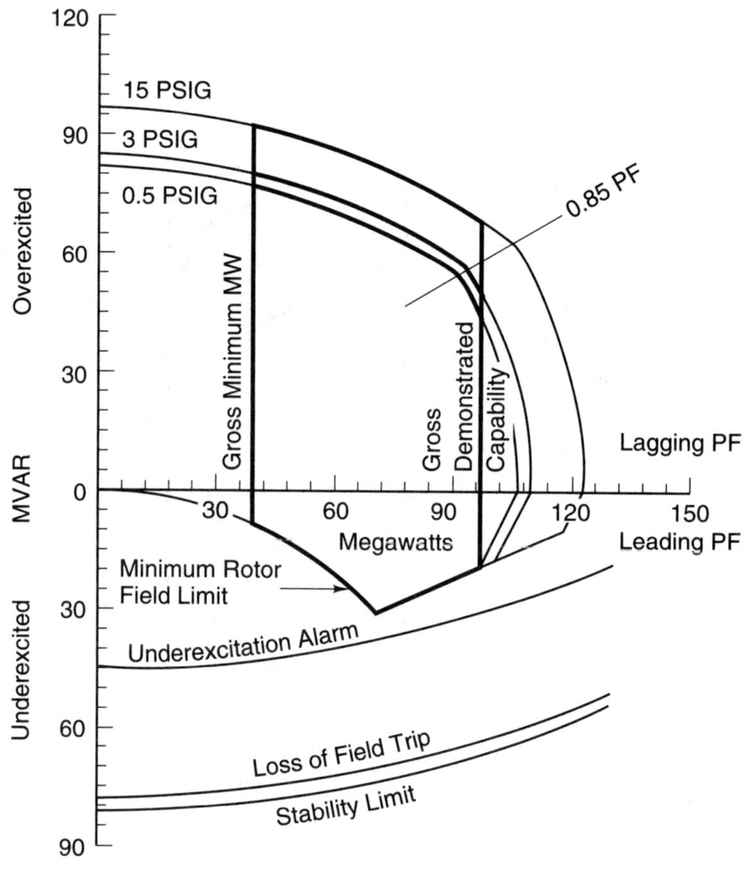

Figure 2.21 Typical older machine capability curve application.

devices. Figure 2.21 shows the URAL characteristic, a straight line ramp. The MEL has a curved characteristic. These are discussed in greater detail in Chapter 7.

There are other limiting operating conditions, but most of them do not need to be described in detail at this point. Briefly, one common practice is to operate a 75 lbf/in^2 hydrogen-cooled synchronous generator at 60 lbs/in^2 of hydrogen pressure because the windage losses and the hydrogen loss (often called *consumption*) are lower. The trade-offs are higher I^2R losses and less capability at lower lagging PF. Also, some utilities have an operating philosophy that prohibits running synchronous generators at a leading PF when the AVR is out of service.

Older machine capability curve application

Older machines often have some additional limits on the usable range of the capability curve. Figure 2.21 shows some of these additional limitations.

The machine shown in Figure 2.21 does not have an automatic limiter in the leading PF region. Accordingly, greater margin is required between the under-excitation alarm characteristic and the loss-of-field tripping characteristic than was necessary in the previous case. Also, this machine has a minimum excitation capability that limits the ability of the synchronous generator to take in Mvars at a low PF.

2-3.3 The Load-Excitation Curves and the Vee Curves

Load-excitation curves are plots of per-unit armature current or kVA (because they are the same at rated per-unit voltage) versus field current for constant power factors, with the generator operating at rated armature voltage. These curves come to a common point at zero armature current with the field current corresponding to open circuit operation at no load. It is common to plot also *vee curves* on the same plot, which are curves at constant power, for selected values of kilowatts. Also shown on the load-excitation curves are lines defining the reactive capability of the generator, perhaps at each of several selected hydrogen pressures. A vertical line corresponding to the field current at rating (or the capability at a lower hydrogen pressure) shows the farthest to the right on the curve that the generator should be operated. A horizontal line at the rated armature current (or the capability at a lower hydrogen pressure) shows the highest position on the plot that the generator should be operated. A third line or curve is provided to indicate the maximum armature current at which the generator is capable of operating, at low values of field current, in the leading PF range. Figure 2.22 displays a load-excitation and a vee curve.

2.4 ABNORMAL OPERATING CONDITIONS

There are a number of abnormal operating conditions in the power system that can affect the generating unit and jeopardize its integrity or reliability. Usually more often than one of these conditions occur at the same time. These operating conditions and their causes will be discussed here, but protection for them will be discussed in Chapter 12.

2-4.1 Excessive System Load

Excessive load on a system or portion thereof is the result of a mismatch between the available generation and the load. It is usually caused by the isolation of a portion of a system following a disturbance that opens the interconnections between areas of the systems. If the *island* (the newly isolated power subsystem) has enough *spinning reserve generation* (difference between the present output of generators currently on-line and their rated full-load outputs), it may recover after the combustion and steam systems have a chance to respond. The response of these systems is relatively slow compared to the speed with which electrical events occur. If the condition is sustained, it will usually be accompanied by reduced frequency and

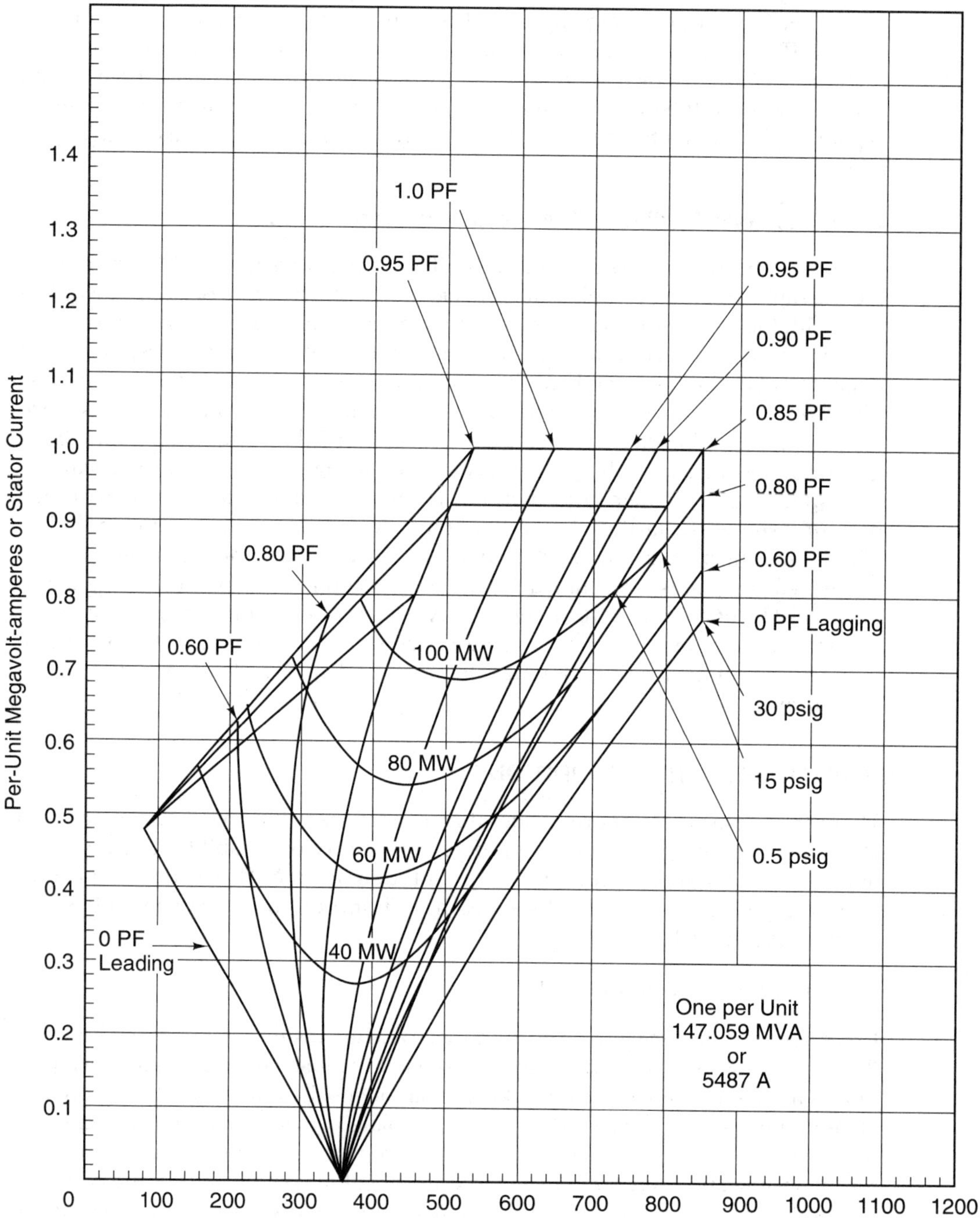

Figure 2.22 Generator load-excitation and vee curves.

depressed voltage, which can be damaging to some loads as well as to generating equipment.

A sustained overload on a generator can result in the potential for thermal distress of both the armature and field windings. If the voltage regulator is in the automatic operating mode, it will act to increase the field current in an attempt to maintain the preset generator terminal voltage.

The short-time thermal capabilities of both the armature and field windings are defined in ANSI C50.13–1989 and are given in Chapter 12. In lieu of automatic means of limiting the duration of field and armature currents, as discussed in Chapter 12, it is up to the operator to take appropriate action to minimize the hazard to the unit.

2-4.2 Depressed System Frequency

As in the case of excessive system load, depressed system frequency is usually the result of a mismatch of available generation and load. In general, operation slightly off from rated frequency is not a severe hazard to the generator if it is not accompanied by excessive armature or field current. If this condition is not accompanied by high electrical load, there is a potential for excessive flux density in the generator core if the voltage regulator maintains the preset voltage at too low a frequency, as shown by rewriting Equation 2.10 as 2.18 as follows:

$$B = \frac{V}{K * f} \tag{2.18}$$

Off-frequency operation, however, does represent a potential hazard to the turbine blades, particularly the long tuned blades of the low-pressure elements. These blades may suffer fatigue damage if such off-frequency operation is sustained. Turbine manufacturers provide limits for off-frequency operation, usually in terms of the accumulated time of operation within defined frequency bands.

Most utility companies and operating or reliability groups have policies regarding automatic load-shedding and operating procedures during depressed frequency conditions. The time limits for which a generator is operated at reduced frequency may be based not only on the turbine limitations, but also on the ability of electrically driven auxiliaries, such as boiler feed pumps, to maintain acceptable operation of the unit without jeopardizing equipment. As will be discussed in Chapter 12, protective measures should be applied to limit the field and armature currents of the generator and to limit the duration of operation of the unit at potentially damaging frequencies. When the time limit for operation at reduced frequencies has been reached, the generator must be separated from the system.

It may be assumed that if an "islanded" system has an area in which the load exceeds the generation, another portion may have an excess of generation. If that portion is also an island (that is, not connected to the larger grid), an overfrequency condition may exist. In that case, means must be employed to prevent too high a frequency.

2-4.3 Depressed System Voltage

Depressed system voltage can be the result of excessive system load, loss of a large generating unit, or malfunction of a generator AVR. The effect on the generator will depend on the reason for the condition. If the AVR is in service and not the cause of the depressed voltage, it will attempt to raise the voltage to the preset level. In this case, the generator field current and the armature current will be increased. The increased armature current will be primarily reactive; that is, the kvar output of the generator will be increased.

As stated in Section 2.4.1, the short-time limitations or capabilities of the field and armature windings are defined in ANSI C50.13–1989. Corrective action to limit these currents should be taken, preferably automatically. How this can be accomplished is discussed in Chapter 7.

2-4.4 Unbalanced Phase Currents

It is typically assumed that the three-phase currents on a transmission system are balanced; that is, they are equal in magnitude and displaced 120 electrical degrees apart. In general, however, this is not the case. There is usually some unbalance in the phase currents, and, during disturbances, there may be severe unbalances. The normal unbalance is usually due to untransposed transmission lines or unbalanced loads on the lower voltage distribution systems. In most cases, utility companies attempt to keep the connected loads on the distribution lines balanced as well as possible. If the lines were uniformly transposed throughout their entire lengths, the impedance of each phase would be identical. Transpositions complicate the physical construction of a line and also constitute a potential for phase-to-phase faults; therefore, most utility companies do not transpose the lines. As a result, the impedances of the three phases are not identical. Even though the generators may generate balanced voltages, and even if the loads are balanced, the unequal impedances between the generators and loads cause the currents to be somewhat unbalanced.

There are cases where large loads such as arc furnaces may cause momentary or continuous current unbalance. In these cases, special agreements between the utility and the users are developed.

A high impedance or open conductor in one phase will also cause unbalance. There have been incidents of high-resistance contacts or open contacts of one phase of a disconnect switch in a switchyard or substation. Of course, the most severe current unbalances occur during unbalanced faults, such as single-phase-to-ground or phase-to-phase faults.

The effect of unbalanced currents on the generator is discussed in detail in Chapters 5 and 12, but it should be noted that the negative-sequence components of unbalanced currents can cause heating of the rotor forging. The negative-sequence capabilities for large turbine generators are defined in ANSI C50.13–1989 and are given in Chapter 12.

In addition to the negative-sequence protection to be described in Chapter 12, some means for continuous monitoring of these currents should be included in the monitoring and diagnostic system. When the sustained negative-sequence current exceeds a normal level, the source of the unbalance should be determined and eliminated, if possible.

2-4.5 Faults on Transmission System

Because the generator is electrically connected to the transmission system by means of the step-up transformer, short circuits, or faults, on that system will affect the generator. As stated earlier, most of these faults occur between one phase and ground and are usually temporary ones that will be cleared by tripping and reclosing the line circuit breakers. Some utilities, however, do not employ high-speed reclosing at generating stations because of the potential mechanical hazard to the generator.

These faults are caused by lightning or foreign objects, such as tree limbs, kites, airplanes, cranes, and so on, contacting the conductors. The effect on the generator will depend on the type, severity, and duration of the fault. Faults usually result in high currents in the generator armature, mechanical shock to the unit, and possible loss of synchronism if they are not promptly cleared. As will be discussed in Chapter 12, the generator protection must include backup relays to remove the unit from the system if the transmission system relays do not isolate a faulted line from the generator.

2-4.6 Faults in Equipment Connected to Generator Terminals

The most severe effects on the generator are from those faults that occur in equipment connected to the generator terminals or between the high voltage terminals of the generator step-up transformer and the unit circuit breaker(s). Not only are they more severe because of the electrical proximity to the generator, but also, because of the absence of an interrupting device in the line, the fault current will continue to flow until the magnetic flux of the generator decays to a low value. Terminal faults are also severe because the inability of the generator to export real power to the system will result in an overspeed of its shaft system as the excess power is converted to rotational speed. This may result in serious damage to the unit.

These faults may be caused by failures in electrical equipment such as the *generator step-up* transformer (GSU) or a *unit auxiliaries transformer* (UAT), flashover of insulators in the switchyard (generally because of contamination of the porcelain insulator), or mechanical damage from cranes or other foreign objects. Unlike the transmission line faults, which are caused by lightning and are usually temporary, these faults are typically permanent. The connection between the generator terminals and the terminals of the GSU is usually *isolated phase bus* work. Each conductor of the bus is surrounded by a grounded enclosure. This limits the exposure of the bus to single-phase-to-ground faults. Because the generator neutral is grounded through a very high impedance, the ground fault current is limited to a few amperes.

Faults on the low voltage side of the UATs require special consideration, not because of the effect on the generator, but on the transformer. This subject is also discussed in Chapter 12, but some comments bear mentioning here. Because the ratings of these transformers are usually only a small percentage of the rating of the generator, and because the faults may not be cleared by the low side circuit breakers, the transformers may be subjected to severe thermal distress. In the plant design, the secondary circuits of these transformers should be considered as part of the unit and should have the same level of electrical and mechanical integrity as the generator main leads.

2-4.7 High System Voltage

High-voltage on a power system or a generator may be caused by the loss of a large load on the system, by islanding, during which there is an excess of generation for the load on the island, or during light load conditions when the capacitance of high voltage transmission lines is not offset by lagging PF loads.

If the high voltage is caused by a mismatch of load and generation following a disturbance, it will probably be accompanied by high system frequency. If the mismatch is not too great, the AVRs and turbine steam and combustion controls may be able to reduce the voltage and frequency to normal values. Unfortunately, the turbine or combustion controls may not be able to respond fast enough to prevent tripping of generators, and the island may end up with a deficiency of generation and possible ultimate system shutdown. The voltage regulator can usually reduce the voltage fairly rapidly, but if it attempts to reduce the field excitation too much, the MEL, URAL, or any equivalent device in the AVR may prevent further reduction of excitation, and the situation may not be completely corrected. In this case, corrective action by the operator may be required. At some locations on the power system, the voltage may be reduced by means of a *load tap changer* (LTC) on power transformers. However, this will not necessarily alleviate the condition on the generator. If the AVR is not in service, or if the MEL/URAL is not coordinated with the loss of excitation relay, that relay may trip the unit.

If the high voltage is caused by too much line-charging current for normal light-load conditions, then remedial measures must be taken in the system operation and/or configuration, or shunt reactors may have to be used to absorb the reactive power. Usually, the need for such measures is determined by computer studies before new generators or transmission lines are installed.

In general, high voltage within a reasonable range is not detrimental to the generators or transformers unless it is accompanied by high volts-per-hertz and corresponding high flux densities.

2-4.8 Transmission System Switching

As indicated in Section 2.3, the power angle diagram for a given system condition must be satisfied. When a transmission line is switched into or out of service, the power-angle diagram becomes instantly revised for the new system configuration.

The generating units cannot change speed or phase angle instantaneously. The result will be an instantaneous or step change in generator output power and a corresponding mechanical shock to the unit. A similar effect occurs when a generator is suddenly tripped while carrying load. Because the power transfer equation or power-angle diagram must be satisfied at all times, the other generators in the station must instantaneously pick up the load. If the generator that is tripped is in the same station as a much smaller unit, the shock to the smaller unit may be excessive because the instantaneous change in load (torque) imposed upon it is greater than the shaft system is capable of withstanding.

A generating unit consists of many rotating elements connected by shafts. For example, a typical unit may include a super-pressure turbine element, a high or intermediate-pressure turbine element, one or more low-pressure turbine elements, a generator, an exciter, and a pilot exciter. This is a complex rotating spring-mass system with numerous natural torsional frequencies. In addition, the turbine blade systems will have natural frequencies of vibrations.

For a single switching operation, the natural frequencies of the unit will be excited and after a while will be dampened by the mechanical losses in the system. If the shock is not too severe, there will be no fatigue damage to the shaft or blades. If there are multiple or sequential switching events whose effects are reinforcing, the stress to the unit may be amplified. If, for example, a switching event shocks the unit into oscillation, and a subsequent switching operation occurs such that the new shock to a shaft element is in phase with an oscillating stress, the stress on the shaft will be amplified.

In the 1970s a technical committee of the Institute of Electrical and Electronic Engineers (IEEE) was established to provide a guide for the acceptable step change in power to which a steam turbine generating unit should be subjected. The recommendations of this committee were based on a study of the design strengths of many generating units in service. The permissible step change in power was defined in terms of per unit of rated generator kVA. The committee recommendations apply to a single switching event. The results and conclusions of this committee can be found in IEEE [1980]. In this reference, a screening level of 0.5 per unit step change in power, based on the generator MVA rating, is suggested for a single switching event.

If analysis of switching events indicates that the shocks caused by switching are too severe for a generating unit or units, special arrangements for normal line switching may be required. In some cases, electric utilities have given consideration to modifying their automatic line reclosing practices. In some cases, equipment has been installed to monitor the torsional stresses on a generating unit.

2-4.9 Subsynchronous Resonance

As indicated in Section 2.3.3, series capacitor banks are sometimes used to increase the power transfer capability of a transmission line. In some cases, these lines connect a mine mouth power plant with a remote load center. A complex phenomenon known as *subsynchronous resonance* may occur in such cases. Although the capacitor bank cancels out some of the inductive reactance of the transmission line, it may also constitute a series resonance circuit with the generator impedance. If this is the case, and if the difference between the power system frequency and the natural electrical frequency of the generator-capacitor circuit is equal to one of the natural torsional frequencies of the generating unit, sustained oscillations will occur and will increase in magnitude. These sustained and amplified oscillations can destroy the shaft(s) of the generator, turbine, or exciter. Some shaft elements are more susceptible to damage than others.

If it is determined that a series capacitor installation is likely to cause subsynchronous resonance on a system, other means of stability improvement must be employed, or means must be taken to dampen the oscillations if they occur. Usually corrective filters or *dynamic stabilizers* must be installed at the generating site.

2-4.10 Loss of Synchronism

Section 2.3.3 discussed stability concepts and how a generating unit may lose synchronism with the other generators on a power system. Although system stability is one of the power system conditions most widely studied, there have been many incidents in which the generators have lost synchronism with one another.

When a generator pulls out of step, the unit will increase in speed because the average electrical power output of the generator does not equal the turbine input power. The electromagnetic coupling, or lock, described in Section 2.3.2 is lost. The magnetic poles of the rotor and stator are rotating at different speeds. This condition is referred to as *slipping of poles*. With each pole slippage, the generator rotor and stator experience violent torque pulsations. These torque pulsations cause severe stresses on the rotating parts and the armature winding. If the slip frequency approaches or reaches a natural torsional frequency, the shaft stresses are amplified, and severe shaft damage may occur. If, however, the unit is isolated from the power system during the first slip cycle, damage to the unit is unlikely.

2-4.11 Loss of Load

Sometimes a generator is isolated from the power system while fully loaded. This may be the result of operator error or of incorrect switching of transmission lines, either manually or by protective relays. Full load rejection causes a mechanical shock to the unit, but if the turbine and excitation control systems are in the automatic operating mode and do not malfunction, they should control the speed and voltage of the unit to prevent any damage.

If the AVR is not in service for some reason, excessively high magnetic flux densities and high generator voltage could occur. As discussed in Section 2.2, a generator at rated conditions requires much more field current than that required to generate rated voltage at no load. When the demagnetizing effect of the armature current is suddenly removed, the voltage will gradually increase to the no-load voltage corresponding to the field current. The magnetic flux would be correspondingly high and could damage the core of the generator and the cores of transformers connected to the generator terminals. It is for this and other reasons that the AVR should be in service at all times, unless there is a serious reason to take it out of service. In this case, it is imperative that volts/hertz relays, set to protect the generator and GSU, be in service.

SUMMARY

This chapter described the basic elements of power generation from the basic laws of physics, such as Faraday's Law and Lenz's Law. From these basic fundamentals, an elementary AC generator is described, which is thereafter extended to a three-phase machine.

The chapter also covered the basic aspects of power systems as they affect and are affected by the generator. The main topic discussed is the concept of stability of a power system and the maintenance of synchronization among the various units composing it.

Lastly, the chapter discussed the various abnormal power system conditions that can cause problems for the generator, such as system faults, transmission line switching, system overloads, and others. This discussion should provide background for the generator protection described in Chapter 12.

It is hoped that this chapter has given the reader a basic overview of what power generation is in its most fundamental terms without involving highly theoretical issues and the accompanying extensive mathematical equations.

REFERENCES

[IEEE, 1980] *IEEE Screening Guide for Planned Steady State Switching Operations to Minimize Harmful Effects on Steam Turbine Generators.* Committee Report. Transactions IEEE, Vol. PAS-9, July–August 1980, pp. 1519–1521.

CHAPTER 3

Major Generator Sensors and Monitoring Systems

Monitoring devices and systems for generators can be classified under the following five categories, as described in Hetrick [1984]:

1. Remote indication to present quantitative information to a plant operator on the status of the generator at the instant of observation.
2. Apparatus specifically intended to monitor a generator and provide an alarm when the parameter exceeds a certain threshold.
3. Protective equipment that monitors a parameter, which, when exceeded, will automatically initiate a unit shutdown. This equipment falls under protective relaying and is covered in Chapter 12.
4. Equipment that stores historical information that can be used to locate areas in the generator that were not functioning normally prior to unit removal from service.
5. Equipment similar in function to that of item 4, except that the information retained is used to diagnose the cause of the malfunction.

The objective of this chapter is to introduce some of the major special purpose sensors and monitoring systems that fall under classification items 1 and 2 in the preceding list. These systems have been expressly designed to detect critical abnormal activity or operating conditions in a generator while it is on-line. All systems described are either commercially available at the time of this writing or are expected to be by the time the book is published. The names and contact points for the manufacturers of the systems contained herein will be supplied in the summary section of this chapter expressly for the convenience of the reader. If an instrument is made by more than one manufacturer, this book does not endorse one manufacturer over another. Thus, the descriptions contained herein are intended to be as general in nature as is reasonably possible, with no explicit or implicit preference of one vendor over another.

It is not the purpose of this chapter to present and discuss all the different monitors available for use in a generator. Rather, it serves as an introduction to the types of monitors that will be discussed further in subsequent chapters. In this chapter we concentrate on the first two items of the preceding list.

3.1 THE RADIO FREQUENCY MONITOR

Various electrical malfunctions exist in a generator that may be characterized by electrical *arcing* in or around the generator area. Of particular interest is the breaking of stator coil strands. While not a common problem, it can be quite serious, and some generator designs are particularly susceptible to this problem. The coil strand cracking failure mode itself and its consequences are described in greater detail in Chapter 4, but a discussion of how its presence can result in the emission of radio frequency (RF) signals will be described here as a way to introduce the *Radio Frequency Monitor* (RFM).

An arc is a self-sustaining discharge of electrical energy through an ionized column of gas. Arcs are characterized by low voltage drops and high conductivity. The presence of an arc depends upon the voltage developed across it, the gap distance, and the conductivity of the liquid or gaseous medium (i.e., air, hydrogen, oil) immediately surrounding the gap. High voltage levels can cause the insulating medium in the gap to break down and establish the gaseous conducting plasma through which the arc current flows. It can take hundreds of volts to cause a breakdown of a gap if no arc had been previously established across it in the very recent past.

An AC arc will be extinguished twice each current cycle as the current wave approaches its natural zero crossing, called the *current zero* (when the AC current changes polarity). Whether the arc remains extinguished depends on the length of the gap, the voltage impressed across the gap immediately after the current zero (called the *transient recovery voltage,* or trv) and the amount of residual gases and particulates still present from the past arcing. These residuals, called the arc byproducts, represent conductive gases, which reduce the insulational integrity of the medium surrounding the gap.

As a copper strand in a generator bar cracks and eventually breaks, a small gap is introduced between the previously joined parts of the strand. The voltage across this gap is a function of the slightly altered magnetic flux being cut by each strand [Twerdochlib, 1989], as well as of the number of broken strands in a particular location.

The arcing may be continuous if the combinations of the three parameters (e.g., the size of the gap, the trv levels, and the presence of arc byproducts) are of such magnitudes that the arc is reignited after each current zero. More likely, however, the arcing will be intermittent, as the vibrations in the coil cause continuous making and breaking of the contact in such a way as to extinguish the arc temporarily during the contact opening and reestablish it when the gap distance decreases. This phenomenon is driven by 120-Hz end-turn vibrations and may be manifested as a short spike. As the arc collapses, the rate of current decay with respect to time (di/dt) can be quite high, thus producing a near-instantaneous release of the inductive energy stored in the coil. This energy will manifest itself as a very short (30 ns) voltage spike that has frequency components that fall in the radio frequency portion of the electromagnetic spectrum.

Alternatively, if the initial break results in a large gap at the outset, the result could be a single spike created when the initial separation took place, which is unable to reignite because of its large gap size. This is similar to turning off a light switch once, which creates an arc that is quickly extinguished, never to be reestablished until the switch is again turned on.

In these ways, arcs produce an radio frequency (RF) signal that is proportional to the intensity of the arcing. In a generator, this signal propagates throughout the electrical conductors in the generator. Therefore, the presence of arcing in a generator can be detected by the measurement of the amplitude of the RF signal in its conductors. The high inductance of the generator winding, however, serves to significantly attenuate the amplitude of the pulse as it travels from the gap to the generator neutral where it is typically measured. The RFM is designed to detect these RF signals emitted by the arcing within the generator. It can provide a general warning of electrical malfunctions in and around the generator that may be associated with an arcing condition. However, it cannot discriminate between the signals given off by various sources of RF. This is analogous to a smoke detector alarm, which identifies the presence of smoke, but can neither pinpoint the location of the source of the smoke nor determine the severity of the problem. The RFM thus represents a potential early warning of conditions that can lead to catastrophic failures in a generator. Figure 3.1 depicts a typical trace of RF activity [Twerdochlib, 1987a].

3-1.1 RFM Operating Principle

A *voltage coupler* (VC) or a current transformer (CT) is used to measure the amplitude of the voltage or current, respectively, of a 4- (or 3-) MHz frequency signal at the generator neutral leads, just above the neutral transformer. This location is chosen partly because of the low 60-Hz voltage levels seen in the neutral during normal generator operation. The voltage coupler is typically used when the neutral transformer is a dry-type distribution transformer. The current trans-

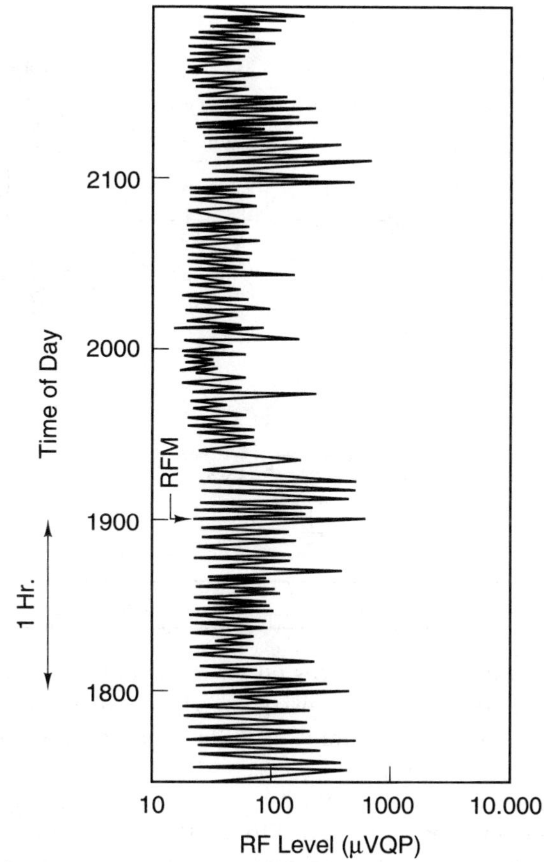

Figure 3.1 Typical trace of RF activity [Twerdochlib, 1978a].

former is used when an oil-filled transformer is used as the neutral transformer. The widely different internal RF reactances of the two different types of transformers is the primary reason for he different parameters monitored. Figure 3.2 shows how an RFM may be connected to the generator as a current coupling device [Twerdochlib, 1987a].

The RFM may be functionally represented as an RF preamplifier and fuse, a receiver board, an interface board, and a control board. The CT produces an equivalent voltage spike in response to a current spike in the neutral leads, using the following relationship:

$$V_{\text{spike}} = 5 \text{ ohms} * I_{\text{spike}} \tag{3.1}$$

where the 5 ohms represent the CT transfer function.

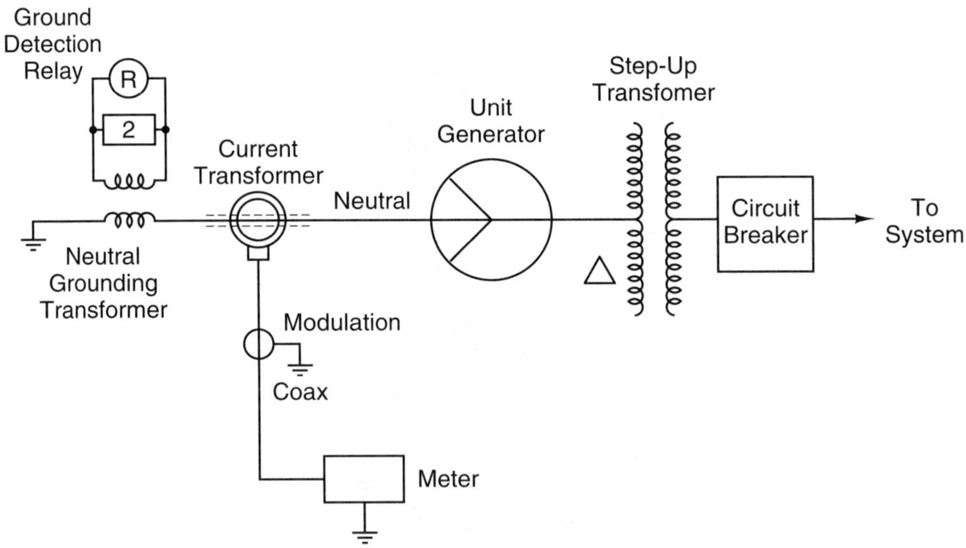

Figure 3.2 Basic radio frequency monitor installation in the generator (Twerdochlib, 1987a).

The preamplifier receives either a current or a voltage signal and provides amplification to the signal. It also serves to isolate the instrument when the neutral leads are exposed to machine-level 60-Hz voltage spikes during faults in the power system near the generator.

The receiver board provides the filtering required to bandpass the 4- (or 3-) MHz frequency that is of interest to the RFM. The two frequencies are used so it is possible to move away from a continuous wave RF signal frequency radio station harmonic, if such is present, and close to one of the two frequencies. The control board provides the monitoring capability of the RFM. It uses a microprocessor to monitor the amplitude of the 3- and 4-MHz signals, as well as to control all the functions of the RFM, including all diagnostic self-tests, alarms, and input-output functions. Finally, the interface board provides all the input and output terminations for the RFM.

3-1.2 Inherent Limitations of the RFM and Their Mitigation

Because of its high frequency, the RF signal propagation in the generator conductors attenuates rather quickly in the high inductances of the generator coils. Unfortunately, it does not attenuate quickly enough to avoid some unidentified readings. Other sources of RF signals exist in the generator, as well as in the neighboring power system (e.g., plant switchyard, secondary plant electrical systems) that may not represent dangerous conditions to the generator. These sources are often transmitted to the generator neutral through the neutral ground cable in the building. These signals can be picked up by the RFM and mistaken for alarm conditions unless the limitations of RF monitoring are mitigated.

Sources of RF signals potentially detected by the RFM can be conveniently divided into three groups [Twerdochlib, 1992]:

1. Signals that emanate from the generator conductors and whose presence and level the RFM is designed to monitor.
2. Signals that come from generator subsystems, such as the exciter brushes, shaft grounding brushes, rotor ground detection brushes, shaft seal rubs, core seals, and others. These signals generally do not represent serious malfunctions in the generator; therefore, it is not typically desirable to alarm when signals from these sources are detected to be high. Nevertheless, these signals may be indicative of abnormal conditions that may require attention at some point.
3. Signals received by the RFM from all sources external to the generator and its subsystems, such as motors, switches, transformers, arc welders in the plant, fluorescent lighting, and even nearby radio stations. These signals are strictly extraneous to the generator and do not represent any abnormal conditions in the generator or its subsystems. It should be noted, however, that some of the most destructive failures outside of the generator have been preceded by an RF alarm. Thus, the RF monitor has, in some cases, become an important monitoring device for problems that occur outside the generator.

Most RFM designs cannot distinguish among these types of RF signals. It is up to the users to do this based on their knowledge of the generator as well as of the phenomenon. The signals received from categories 2 and 3 above can usually be considered background noise. The signals of category 1, therefore, must be isolated from the noise in order to be identified. Unfortunately, the background noise is different for each generator and must be determined based on the operating experience of the unit in question. RF signals that exceed the background noise level should then be considered "real" signals that indicate a detectable malfunction. Thus, the performance level of the RFM is established by the signal-to-noise ratio of the signal emitted by the malfunction whose detection is desired. If the amplitude of the signal caused by a malfunction is small when compared to the ambient noise level, then the sensitivity of the RFM to that malfunction is compromised. Likewise, if the ambient noise is not very constant, it can then be difficult to determine when the "real" signal is being received. In order to make the RFM a useful instrument, these problems need to be mitigated. Some useful techniques follow.

The Find-and-Eliminate Process

Find-and-eliminate is the simplest and most successfully used procedure. If an unusual RF signal is received that is construed to be abnormal arcing from some source in or around the generator, all possible external sources (categories 2 and 3) of the RF signal should be inspected to determine whether they are producing an RF signal. If one of these is found to be a source, then the cause of the unusual signal has most likely been found and can be eliminated through repair or maintenance. If the unusual signal disappears after the maintenance work, then the problem has been solved. If it persists, however, then another potential source of arcing must be inspected in a similar fashion. When all potential sources have been inspected and none is found to be the source of the unusual RF signal, then through the process of elimination, it can be determined that the true source is inside the generator and should be treated with the seriousness it deserves.

In stationary exciter equipped generators, a primary source of extraneous arcing are the main field brushes. Contamination of the brushes can result in intermittent contact with the slip-ring surface, resulting in arcing. This can be easily eliminated by either cleaning the contacts or replacing the brushes.

The next most common source of extraneous RF signal is the set of shaft grounding brushes. Contamination of the brushes can result in intermittent contact with the slip-ring surface, resulting in arcing. Thus, if arcing still persists after the main field brushes have been eliminated as a source of RF signal, the shaft grounding brushes should be inspected next for arcing and cleaned if arcing is indeed found. This last step is important in order to maintain good grounding on the shaft and avoid arcing at the bearings, which can be destructive, as will be seen later.

Another potential source are the set of field ground detector brushes on brushless exciters (to be discussed in Chapter 7). Because these brushes are generally activated into contact once every 24 hours, and then only for a period of a few seconds, a correlation between plant records and the RF signal history will reveal the validity of the brushes as a source of the unusual signal.

Other potential sources of arcing that should be checked are the use of arc welders in the plant, the presence of arcing in the switchyard, or neighboring radio stations.

Load Variation Procedure

The arc voltage developed across a gap is proportional to the magnitude of the arc current. As the current is decreased, not only are the 120-Hz vibrations decreased, but the arc will also persist for a shorter time during each cycle, resulting in a lower integrated amplitude of the RF signal. Thus, if the arcing is a result of cracked strands or connection failure, then the RF signal would be expected to decrease in amplitude as the unit load is reduced.

Alternatively, some other sources of arcing are a result of voltage-induced arcing and are not dependent on the unit load. Some of these sources are the core seals and the hydrogen gland seals. Arcing from these sources will remain constant as the unit load is varied. Other sources outside the generator will also be insensitive to changes in generator loading. Thus, if the RF signal varies directly proportionally to unit load, then the source is likely to be an undesirable malfunction.

Another potential source of RF signals in the generator are the high frequency spikes produced by the thyristors and/or diodes in the excitation control system. These are effectively coupled to the generator shaft and often picked up by the RF monitor.

The 120-Hz Repetition Rule

Most strand breakage will occur at the end-turns (see Chapter 4). The vibratory forces in this region that govern the winding vibrations occur at 120 Hz. Thus, the make-break contact frequency will be 120 Hz, which can be detected by an oscilloscope connected to the RFM. Presence of a 120-Hz signal modulation typically indicates an undesirable malfunction as the source of the arcing.

The Turbine Generator Seals Procedure

The hydrogen seals in the generator will arc in the presence of a seal rub (see Chapter 9). Because seal rubs are often (although not always) caused by low-temperature seal oil, an indication of the plausibility of seal rubs being a source can be evaluated by monitoring the seal oil temperature. If the seal oil temperature is raised, the RF activity should diminish, if this is the problem.

3-1.3 Serious Problems Do Not Go Away

Identification of the RF signal as being produced by a serious problem within the armature conductors can often be misleading as to what corrective action to take. This is because the RFM is designed to monitor only the RF signal amplitude at the neutral of the generator. This amplitude represents a single variable and may by itself be insufficient to be able to relate the RFM signal to a diagnosis. For example, a rapid fall in RF signal amplitude from alarm level to below alarm level does not necessarily indicate that the abnormal conditions have improved. In fact, it can mean just the contrary.

As noted before, the presence of arcing in a cracked conductor strand is a function of many parameters, such as the size of the gap, the voltage developed across the gap, the conductivity of the cooling gas immediately surrounding the crack, and the armature current level. The arcing may be present in a continuous form, intermittently, or only during the time of initial cracking and separation. A cessation of the arcing could mean that the gap has increased, thus placing additional mechanical stress on the other unbroken conductors. This indicates a worsening of the situation, not an improvement.

To maximize the useful information provided by the RFM, its historical data should be carefully examined at all times to look for trends. Seemingly isolated and intermittent spikes of high RF activity could be a result of individual breaks in the conductor, which can only accumulate. It must be remembered that once a conductor cracks or a connection becomes loose, its condition will only get worse.

Improvements to the RF monitor described previously have been made that have resulted in a digital version of the monitor. See Twerdochlib [1993] for details.

3-1.4 Alternative Designs of the RFM

An alternative to the RFM described previously is also available, which can mitigate the disadvantages described automatically, by monitoring various locations in and around the generator for RF signals. The relative strength of the signals coming from the various sensors will indicate the locations where the arcing is most likely taking place. Preferred coupling locations are the surge capacitors in the generator phase leads, the generator neutral, and the exciter or generator bearings. Other locations include the high-voltage bushings of the generator set-up (GSU) and unit auxiliary transformer (AUT).

3.2 THE GENERATOR CORE/CONDITION MONITOR

Excessive heat buildup in the insulated conductors in the generator field or armature can result in thermal decomposition of the winding's groundwall insulation. Such buildup can be caused by strand breakage, connection failure, or cooling system deficiencies, among other things. Thermal decomposition of the insulation of the individual core laminations can also result in core hotspots when organic insulation is being used. Decomposition will take place in the range of 340° to 500° C, depending on the insulating material. Inorganic types of core lamination insulation, however, which are used by some manufacturers, decompose only at very high temperatures and are not likely to decompose under the worst-case hotspot conditions in a generator core. Nevertheless, even for those generators, decomposition of paints in or around the core may be detectable.

Tagging compounds, or paints, whose decomposition onset temperature can be as low as 200° C, are commonly used to detect hotspots at an early stage, before any danger is posed to the insulation. Furthermore, by varying the chemical nature of the paints, different parts of the generator internals can be "painted" with different compounds to detect the location of the hotspot.

This thermal decomposition of winding or lamination insulation causes the formation of byproducts in the form of small particulates, often less than one micron in diameter, which are captured by the hydrogen stream on the outer surface of the conductor or lamination and blown throughout the generator.

The primary objective of the *generator core/condition monitor* (GCCM) is to identify such excessive buildups of heat in the winding/core insulation before they result in catastrophic failures of the insulation, and thus of the generator itself. The GCCM accomplishes this by detecting the thermal decomposition particulates in the hydrogen stream. The following section will describe how this is generally done.

3-2.1 GCCM Operating Principle

The GCCM operates by monitoring the hydrogen gas for the presence of particulates. A small sample of the gas is diverted continuously to the GCCM, which is typically located outside of the generator. The flow is typically driven by the main generator blower differential pressure.

The concentration of particulate matter is measured by its effect on a small flow of electrons through the hydrogen gas. The electrons are generated by a weak radioactive current source (Thorium 232) and are driven by a 10-volt potential difference [Twerdochlib, 1987a]. When decomposition particulates are present in the hydrogen stream, the electrons will stick to the particulates. The increased mass of the charged particulate means that it is not influenced as much by the driving potential difference as are the free electrons, and is thus removed from the electron current. The GCCM measures this current flow and produces an analog signal directly proportional to the electron current level (typically picoamperes). A threshold point can be set to alarm when the current flow decreases to a preset level. Figure 3.3 describes the schematic diagram of one particular design of a GCCM.

Generally, two alarm points are available, a warning alarm and a shutdown alarm. However, the real value of the GCCM as a diagnostic tool comes from closely analyzing its level over a period of time, rather than from using it as an instantaneous warning of high particulate levels at any one time, because the nature of the decomposition may be quite different in many cases. Typically, once the heat buildup begins, particulate matter will be continuously generated in increasing levels because of the worsening of the situation. However, situations where the particulate generation is either intermittent or takes place only in a single, and rather short, occurrence have also been seen. This is because the particulate generation is dependent not only on the temperature and the size of the hotspot, but also on the location and the rate at which the heat buildup takes place. For example, particulate matter may be produced for a short time until thermal decomposition of the insulation is complete. Once this happens, the source of particulate matter is expended, and no further emission will take place, returning the GCCM signal back to normal. Yet, the heat buildup continues because of the malfunction still exists. Another abnormal case is when the heat buildup rate, and the subsequent emission of particulates, is so slow that the concentration of particulates never increases significantly. The danger involved with this type of occurrence is that an operator may mistakenly associate a cleared alarm with the absence of internal problems. Some monitors allow the extraction of a gas sample volume for use in off-line chemical testing. This is used to verify the presence of particulates more accurately.

3-2.2 Inherent Limitations of the GCCM and Their Mitigation

The primary objective of the GCCM is to indirectly identify a potentially catastrophic heat buildup in either the windings or the core laminations by monitoring the thermal decomposition byproducts of the hotspot insulation. As with any instrumentation system, however, the GCCM has some limitations. A number of these can be mitigated effectively through auxiliary sensors and secondary tests performed by an operator, while others are inherent in the phenomenon on which

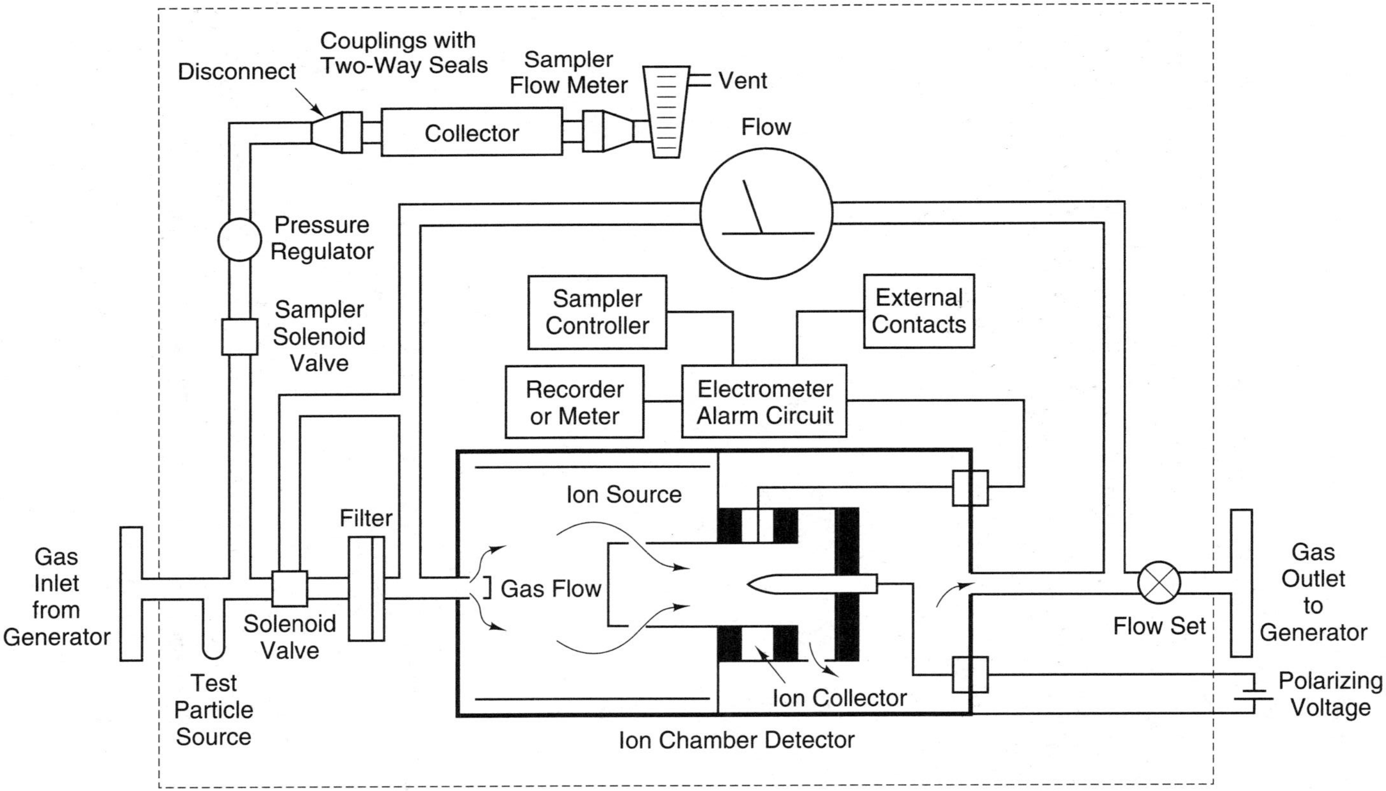

Figure 3.3 Schematic diagram of a generator core condition monitor (Twerdochlib, 1987a).

the concept of the GCCM is based and are, therefore, not as easily mitigated. These aspects will be discussed in the following sections.

Undetectable Decompositions

As mentioned previously, some decompositions are not detectable by monitoring electron current through the hydrogen stream that carries decomposition byproducts. There are a number of reasons for this.

- The rate of heat buildup can be so slow that it does not cause a significant concentration of particulates to be emitted at any one time. The particulates that are generated could settle out and adhere to the generator internals before reaching a critical concentration, thus causing no decrease in electron current. Such a phenomenon will cause the problem to be transparent to the GCCM; there is no practical way to mitigate this problem.
- The properties of the insulation material may be such that decomposition does not occur at the temperatures typically faced, even under the worst conditions. This is true of the inorganic insulation material used for core lamination by some manufacturers. While this is a significant improvement in the integrity of the insulation system, it decreases the possibility of detecting the heat buildup by the GCCM. This, however, can be easily addressed through the use of tagging compounds that decompose at lower temperatures.

Relationship Between Alarm Level and Malfunction

The GCCM signal may not always reach alarm level, even in the presence of a potentially dangerous hotspot. Figure 3.4 shows a GCCM trace obtained from an actual operating generator [Twerdochlib, 1987a]. Note that the two incidents of decreased electron current were registered by the GCCM, which indicated the presence of a hotspot in the generator (one at 15:30 hours and the other at approximately 17:30 hours). These were intermittent, and neither reached alarm level. Yet, they represented an advance warning of a heat buildup, which became clearly evident at 19:00 hours when the GCCM finally went into alarm. This problem can be mitigated by automated monitoring of the GCCM signal to look for patterns that indicate low level (and early!) signs of existing problems. It should be noted, however, that momentary changes in hydrogen flow to the GCCM can also be the cause of these fluctuations in GCCM level.

Another aspect of this issue is that, as an indirect sensor of the cause of the heat buildup, there is a weak relationship between the GCCM reading and the cause or location of the heat buildup. This has been mitigated by the use of various tagging compounds that have different identifying "signatures" when analyzed off-line in a laboratory. By applying these somewhat different compounds in different locations throughout the generator, the location of the problem, and therefore its probable cause, can be identified by determining that particular tagging compound was applied. However, this process typically requires a 36- to 48-hour delay for the off-site analysis.

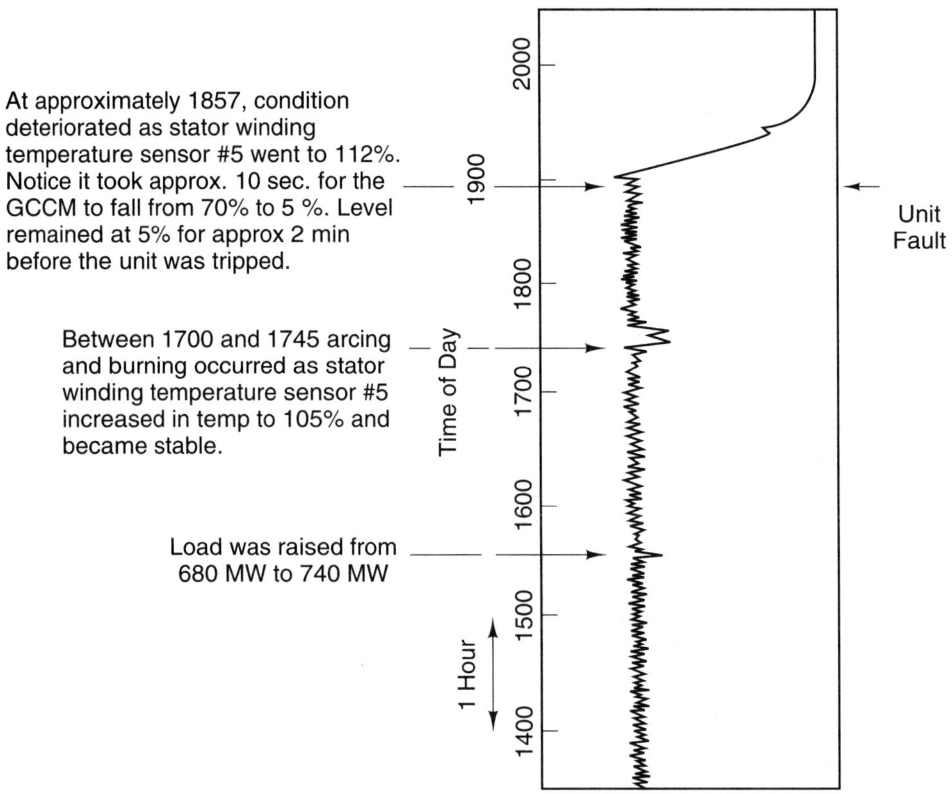

Figure 3.4 Strip-chart trace of generator core condition monitor output prior to a generator trip (Twerdochlib, 1987a).

False Alarms

The electron current measured by the GCCM is directly dependent on the hydrogen gas flow rate. If the flow is decreased because of blockage, clogging, or incorrect installation after an outage, the amount of current will decrease and possibly cause a false alarm. This problem can be mitigated by two separate means. The first is prevention of clogging or blockage through the improved design of components in the flow path. Correct installation and periodic maintenance will help minimize this problem.

The second means is to monitor the flow rate of the hydrogen gas through the ion chamber and alarm separately when the flow is reduced for some reason. The GCCM may carry out an internal test to check the status of the hydrogen flow when an alarm takes place.

Another source of false alarms is submicron particles of oil that may be formed when the oil comes in contact with hot surfaces within the generator. This problem can be mitigated by heating the gas before it enters the ion chamber, or within the chamber itself, but this reduces the sensitivity of the measurement. The heat will vaporize the oil particles without affecting the decomposition byproducts.

Lastly, water buildup in the hydrogen line to the GCCM through condensation will lead to spilling of this condensate into the instrument, which will also give a false alarm.

3.3 The Armature End-Turn Vibration Monitoring System

As generators age, the end-winding support structure tends to relax. As the relaxation process begins, the compressive preloads that were built into the support structures are relieved, resulting in reduced resistance to the 120-Hz vibratory forces naturally found in the stator end-turn region (see Chapter 4). As the vibration amplitudes increase, the stator coil end-turns are progressively more fatigued from a mechanical standpoint. The increased fatigue may move the natural frequencies of the end region closer to a value near the 120 Hz, thereby worsening the situation through the resonance amplification mode. Figure 3.5 shows the vibratory amplitude of the stator coil end turns of two generators of identical design. Generator A has a nearly steady vibratory amplitude of approximately 4 mils, while the corresponding amplitude for B varies between 8 and 22 mils. The difference is that A has no component in the end-turn region with a natural frequency of 120 Hz, and B does.

One utility has found that when replacing older thermoplastic insulation systems with modern thermosetting systems, in almost all cases and regardless of size, problems are encountered with end-winding resonance unless specific precautions are taken. Their rewind specifications require dynamic testing of all completed

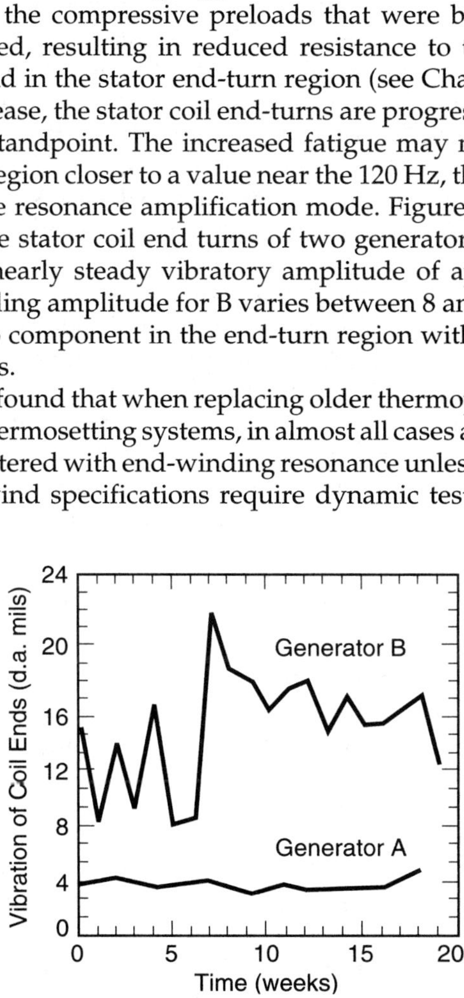

Figure 3.5 End-turn vibration amplitude comparison of two generators of identical design (Twerdochlib, 1987b).

rewinds, and they specify a frequency range that must be avoided. Thus, this may not be solely an aging problem, but may also arise as windings are upgraded.

One final result of this vibratory problem is the fatigue cracking of the copper strands, which leads to increased heating of the coil (bar) and, ultimately, its failure (see Chapter 4). One utility reports that these vibratory forces were the probable cause of a failure of a connection to the parallel circuit (a hollow copper pipe type of connection).

Protection against this type of problem generally consists of three instruments. Each of these will detect the problem at a different stage in its progression. The first is the temperature monitoring system to be discussed in Chapter 4. It serves to indicate the presence of heat buildup because of the conductor breakage by detecting evidence of abnormal heating in the conductor coolant medium (hydrogen or water). It looks for a failure that has already taken place, albeit at relatively low levels. The RFM described earlier in this chapter is the second one, and it detects the immediate indications of strand breakage. This occurs at a point chronologically prior to the heat buildup that causes the ultimate failure, thus providing significantly earlier indication than the temperature monitor. Nevertheless, the failure has already taken place.

The third instrument, the *end-turn vibration monitoring* system, is the topic of this section. It measures the vibratory amplitude at various points in the end-turn region of the stator winding. This represents not a malfunction, per se, but, rather, the presence of conditions that can lead to a malfunction. In this way, the end-turn vibrations monitor can provide a true predictive indication of potential failures, rather than merely react to an incipient fault at an early stage in its development. If the warnings of the end-turn vibration monitor are heeded in a timely manner, breaking of conductors may be altogether avoided.

3-3.1 End-Turn Vibration Monitor Operating Principle

The vibration sensor is sensitive to vibratory motion in one direction and designed to function at one frequency. Because the dominant driving frequency in the end-turn region of the stator winding of 60-Hz generators is 120 Hz, this is the designated frequency that the end-turn vibration monitor is designed to detect. For 50-Hz machines, the frequency monitored is 100 Hz. The monitor operates by mounting a small grid of evenly spaced slots on a reed tuned to resonate at a frequency far enough from 120 Hz to achieve a reasonable level of resonance amplification, typically 132.5 Hz. The reed is cantilevered within the sensor housing, which is, in turn, mounted directly on the stator bar ends. The amplitude of the reed vibration will be proportional to the amplitude of the bar end-turn vibration. This amplitude is captured by directing a continuous beam of light through the grid at a right angle to the grid's vibratory motion. As the grid vibrates, some slots (depending on the vibration amplitude) will pass through the continuous beam, causing pulses of light to shine on an optical cable on the other side of the reed. The pulses are passed through an optical diode and counted electronically in the instrument. The number of pulses counted in a given period of time is proportional to the amplitude of vibration. Figure 3.6 describes the construction of the sensor.

One significant advantage to using fiber optics is that the sensor and all cabling to and from it are nonconducting. This eliminates the need to insulate them from the high voltages inherent within the generator.

Typically, 12 points are monitored in a generator, six on the turbine end and six on the exciter end. Top phase coils are normally selected for the measuring

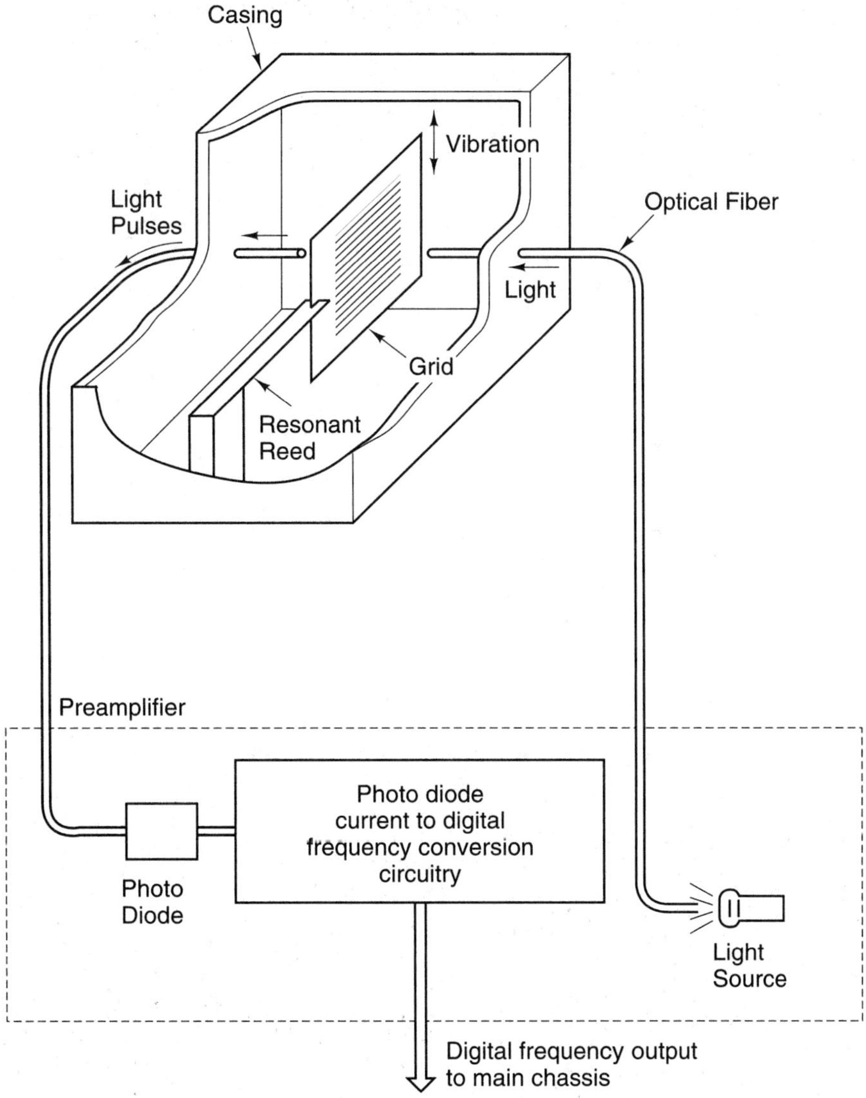

Figure 3.6 End-turn vibration monitor (Twerdochlib, 1987b).

points. The locations chosen are typically the top coil that registers the highest amplitude vibration on each parallel. Figure 3.7 depicts this configuration. The bar in which the highest amplitude in each parallel is found in determined through off-line dynamic impact testing performed prior to the installation of the system. More recently, experience has shown that it is not critical that the worst case locations in each circuit be determined, but rather, that the sensors be evenly spaced around the stator end region.

3-3.2 Inherent Limitations of the End-Turn Vibration Monitor and Their Mitigation

Very few limitations apply to the end-turn vibration monitor. It is a rather simple concept that is easy to implement. One of its advantages is that the electrical isolation of the optical sensor allows it to be placed directly on the vibrating surface. This provides very direct and safe access to the information.

Nevertheless, one limitation of the end-turn vibration monitor is the need for off-line dynamic impact tests to determine the location of the highest amplitude vibration. However, this needs to be done only once for each frame design.

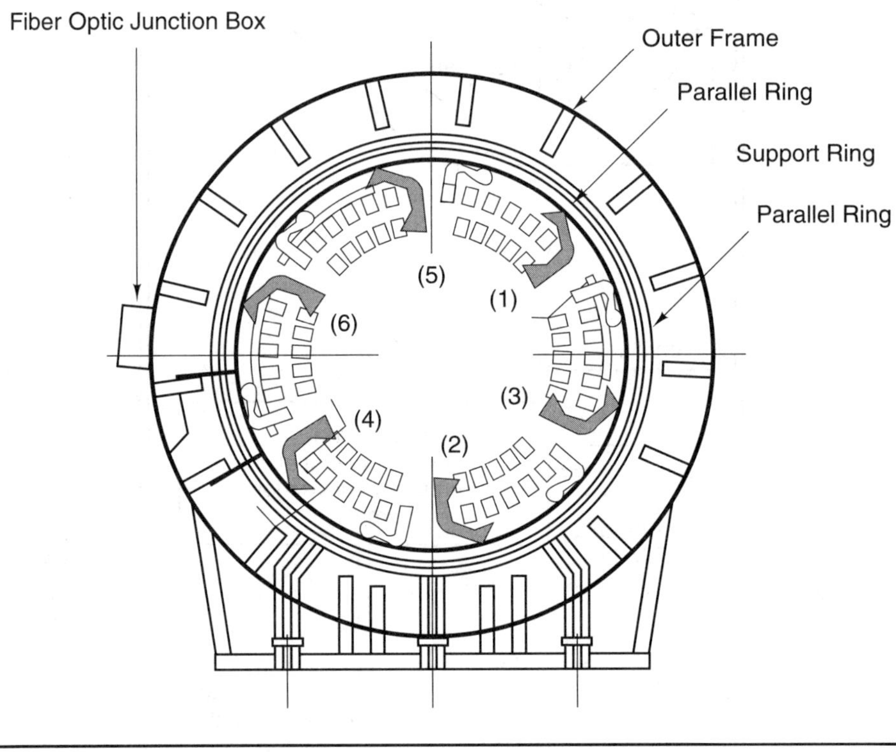

Figure 3.7 Typical installation locations of the end-turn vibration monitor (Twerdochlib, 1987b).

Although the tests are not complicated, they require special know-how and equipment. Furthermore, the elastic properties of the end-turn region are somewhat different at running temperature than they are at ambient temperatures, such as when the impact tests are done. Thus, there is no guarantee that the worst location selected through the impact testing will be the worst location during regular operation. This makes its installation comparatively difficult to carry out in existing machines. Nevertheless, as mentioned previously, in some cases the tests are considered unnecessary in light of recent experience. The reader should consult the manufacturer of the generator as well as of the end-turn vibration monitor to determine whether the tests are necessary.

Another limitation is that the monitor does not provide phase information. This, however, is not an inherent limitation of the fiber-optic measuring device, but only of the electronic processing of the raw sensor data. The sensor signal does, in fact, contain the phase information, although there is a $\pm 180^\circ$ ambiguity.

Lastly, hydrogen leakage though the penetrations of the fiber optic cable could result.

3.4 THE FIELD INTERTURN SHORT-CIRCUIT MONITOR/ GENERATOR FLUX PROBE

As will be seen in Chapter 5, short circuits between turns in the rotor field winding of the generator can cause abnormal operating conditions, which may limit unit loading. There are many generator rotors in service with known shorted turns that continue to run without seriously affecting overall generator operation. Uneventful operation with shorted turns present in the rotor will depend on their severity, their location, and the capability of the generator excitation system. Since shorted turns can result in abnormal operating conditions and loading restrictions, the ability to monitor and identify their presence can assist in operations, trouble analysis, and outage planning.

These abnormal conditions include the following:

- Rotor unbalance varies with field current changes (thermal sensitivity). Unsymmetrically located shorted turns can cause temperature gradients resulting in a rotor bow and, consequently, high vibrations.
- If the number of shorted turns is high enough, the excitation system can be overloaded trying to supply enough excitation to maintain the required generator terminal conditions.
- High field currents result in higher operating temperatures. It should be noted that under the conditions of shorted field turns, the volts-ampere field temperature monitor (to be described in Chapter 5) may erroneously indicate a lower operating temperature. This is because the temperature calculation algorithm assumes constant field winding resistance, a value that in reality has decreased because of the shorted turn.

Forced outages due to isolated shorted turns are not common because, as additional shorted turns develop in the winding, the performance of the generator deteriorates such that corrective action is typically planned and carried out before a catastrophic fault takes place. This deterioration is in the form of excessive bearing vibrations and/or overloading of the excitation system. However, in the past, because definitive shorted-turn detection systems were not available, operating personnel would continue operating units with shorted rotor field turns for a long time without taking cost-effective corrective action.

3-4.1 Field Interturn Short-Circuit Monitor Operating Principle

Shorted-turn detection technology exists that can identify the coils with shorted turns and the number of shorted turns in the affected coils. The test data can be obtained while the generator is at normal load conditions. No special off-line set-up is required. If a baseline set of test data is obtained before a critical shorted-turn condition develops, the sensitivity of the detection system can be significantly enhanced, but this is not absolutely necessary.

One type of shorted-turns detector makes use of a simple, permanently mounted air-gap search coil positioned on a stator wedge. The coil is sensitive to changes in radial flux density magnitudes generated by the rotor winding as it passes by the search coil. Analysis of the search coil waveform can detect these critical shorted-turns conditions. Knowing this, rational corrective action can be taken.

Figure 3.8 shows two different air-gap probe designs. Which one is used depends on the air-gap length of the generator to be monitored. The probe must be installed during a major inspection when the rotor has been removed from the machine. The probe tip is a full 2 inches from the rotor surface, thus avoiding any interference with the retaining rings during assembly and disassembly. The probe is wired to the instrument terminal gland, which is mounted on the outer diameter (OD) of the generator frame. The wire routing is similar to that of the stator bar resistance temperature detector (RTD) wiring.

Figure 3.9 shows the cross section of a rotor and defines the stationary search coil location. The tests can be performed under two different conditions: short circuit and load.

Running the test under short-circuit conditions, when the terminals of the step-up transformer are shorted and approximately 30% of full stator current is generated, will yield the most sensitive results. Under these conditions the armature reaction almost exactly balances the main field flux, leaving only the slot leakage flux as the measurable parameter, resulting in voltage spikes of the same height. Furthermore, by varying the speed of the rotor, flying short circuits (shorted turns that disappear at different speeds) can be detected. The disadvantages is, however, that the unit must be off-line.

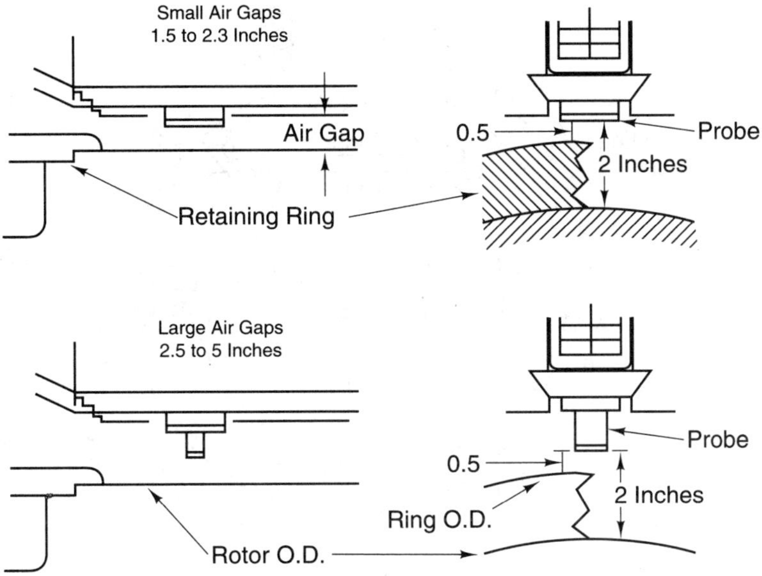

Figure 3.8 Probe design and casing gland. (Courtesy of Generatortech, Inc.)

Performing the test under load provides almost the same sensitivity if the load is varied between zero and full load. This is done because the region of greatest sensitivity is where the actual air-gap flux goes through zero; this point can be varied by varying the load and the PF. Additionally, the fact that the generator does not need to be off-line has been one reason for the utilities' preference for performing the tests under load.

Figure 3.10 shows a waveform from a search coil taken under load conditions. The nomenclature of the rotor and the associated waveform are identified. Note that the voltage spikes are directly associated with the field slots. The search coil, which is sensitive to radial flux variations, provides voltage reversals at the center-line of the slots (labelled) and the centerline of the teeth. These voltage spike magnitudes are a function of the active turns in the slots (slot leakage flux) plus a distortion factor because of the air-gap flux density waveform.

Figure 3.11 illustrates test data from another generator, taken while it was at no load and open-circuited. Note the air-gap flux density waveform (60 Hz) passes through zero at the quadrature axis (90° from the direct, or pole, axis). The flux density waveform can be obtained by integrating the search coil waveform. In the following displays, the flux density waveform has been inverted to minimize overriding the pertinent search coil data.

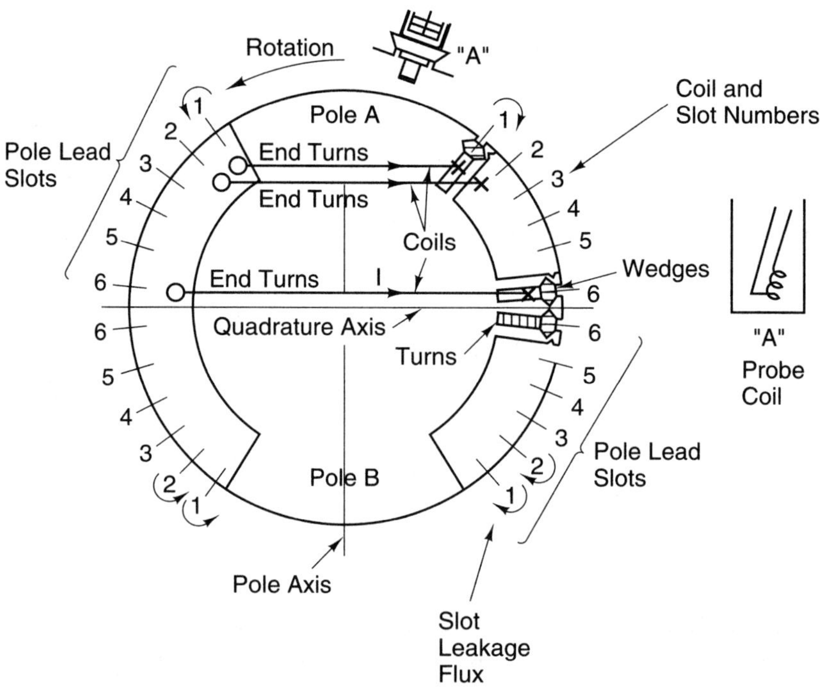

Figure 3.9 Field nomenclature and stationary search coil A. (Courtesy of Generatortech, Inc.)

Text and analysis of waveforms when the generator is under load show that the distortion factors mentioned above are minimal at the zero-crossing of the air-gap flux density. It has been further shown that the zero-crossing traverses from the quadrature axis towards the #1 coil slots leading the poles as load is increased. This phenomenon is analogous to power-angle shifts with load. Thus, by taking roughly five sets of test data from no load to full load at unity PF, excellent shorted-turn detection sensitivity can be obtained at all of the coil lead slot locations. Figures 3.12 and 3.13 show test data from the same unit as Figure 3.11. There are shorted-turn conditions in coil #6 on one pole and coil #4 on the other pole. These shorted turns were defined best at the loads where the zero-crossing of the air-gap flux density was aligned with these slots.

If the data are recorded using a 12-bit y-axis resolution digital oscilloscope or an analog-to-digital computer interface board, it is possible to reduce the data in a very descriptive qualitative and quantitative manner. Data from one pole can be inverted and aligned with similar data from the other pole. Figures 3.14 and 3.15 are examples of computer analysis displays of the base data in Figure 3.10—low intermediate load. Two turn shorts out of six turns in coil #6 were defined at this load.

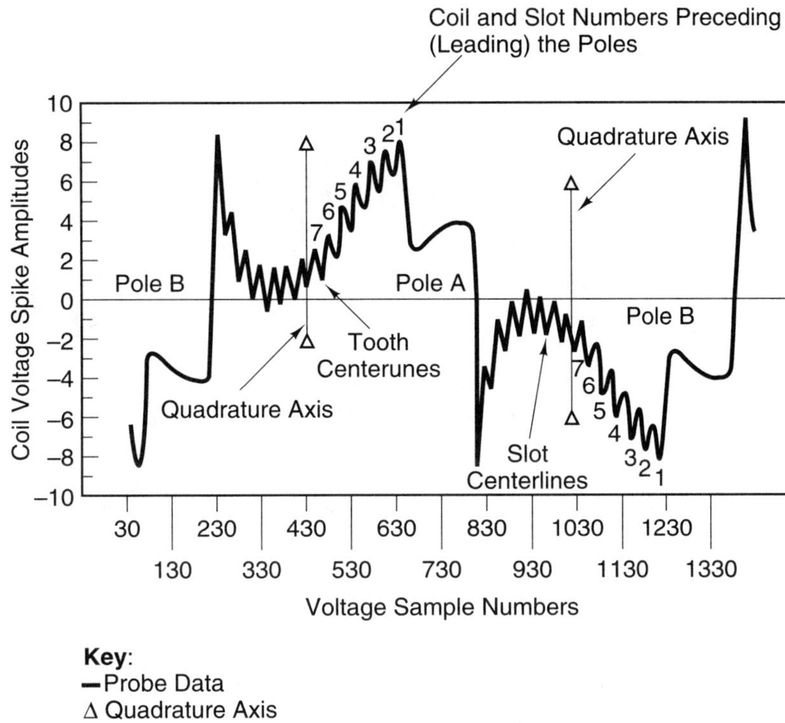

Figure 3.10 Sample load data: No shorted turns found. (Courtesy of Generatortech, Inc.)

Computer reductions of the digital data can also provide descriptive quantitative signal analysis. The voltage reversal y-axis values can be found and used to measure the voltage spike heights of each coil. Figures 3.16a and 3.16b are the coil voltage spike height signature analysis of the same generator at 25% and 75% loads, respectively. Note that the definitive sensitivity for detecting shorted turns is at the flux density zero-crossings. A computer analysis program provides all of these displays for a single load point in less than four minutes after the data are collected.

3-4.2 Inherent Limitations of the Field Interturn Short-Circuit Monitor and Their Mitigation

Engineering judgment must be used to identify the specific number of turns shorted in an affected coil when load test analyses are performed. The coil voltage spike height has two components: slot leakage flux and a distortion factor because of the flux density waveform. If a 10% shorted-turn condition develops in a coil, the voltage spike height decreases only 6%–8%. It is likely that a resistive turn-to-turn short circuit will become 100% effective; that is, the resistance will ultimately

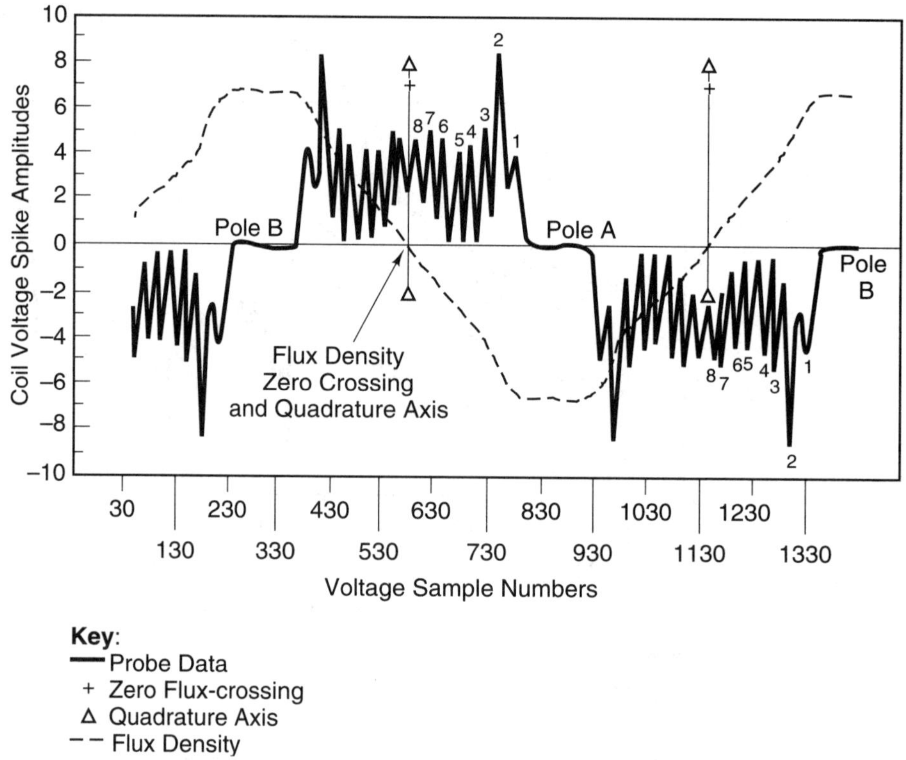

Figure 3.11 Open-circuit load data. (Courtesy of Generatortech, Inc.)

become very small in magnitude. Therefore, it is possible to estimate the number of short-circuited turns by selecting the next higher integer turn short-circuit percentage. The sensitivity of the test can be improved by having a baseline set of data when the winding is not experiencing an operating problem (i.e., a healthy winding). For example, shorted turns in identical coils of both poles may not be detected by using coil-to-coil data evaluations. However, comparing the data to a healthy winding baseline provides an opportunity for evaluation.

Experience has shown that the sensitivity is sufficient to detect critical shorted turn conditions, such as in the 15%–30% range. Such conditions generally produce operating problems whose corrections must be addressed by the unit owner. Once again, these are general numbers. The generator manufacturer should be contacted for more specific numbers.

Another drawback is that significant expertise is required to interpret the output of this monitor. It does not result in a simple alarm that indicates when there are significant shorts in the field. While this section provides some indication of how to interpret these results, the reader should seek other professional expertise when attempting to interpret the readings.

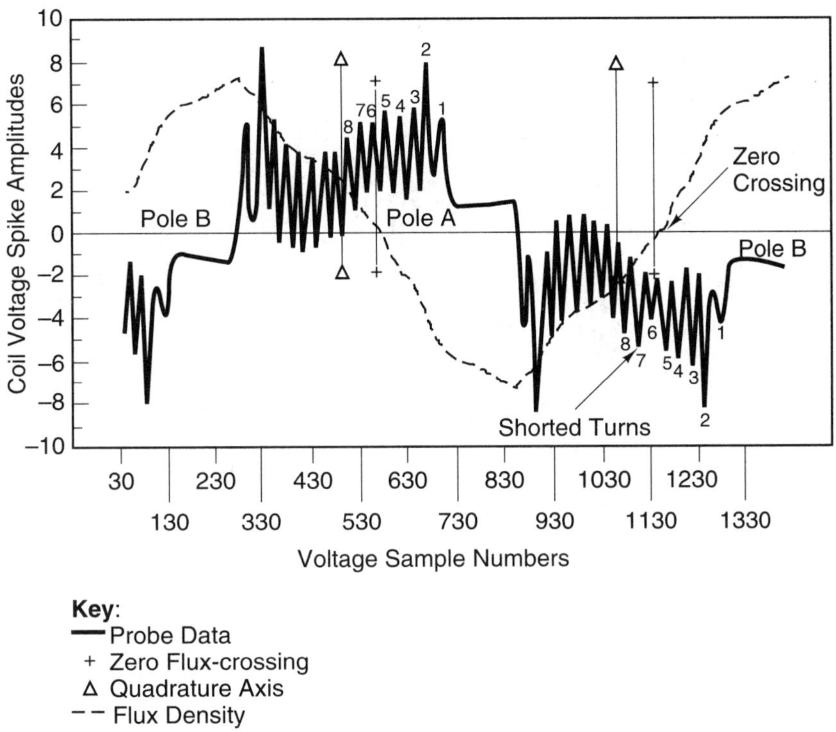

Figure 3.12 Low intermediate load, shorted turns, coil #6. (Courtesy of Generatortech, Inc.)

3-4.3 Other Types of Field Shorted-Turns Detectors

The system described above is only one type of device that is used to detect shorted turns in the rotor field. The objective of this section is to briefly describe other such devices.

One alternative is a search coil embedded in the slot above the top coil, underneath the wedge. The coil is placed far enough away from the end of the core to eliminate end leakage effects. This has the advantage that the coil is not subject to damage when the rotor is removed. One such device is shown in Figure 3.17.

The concept of this device is similar to the one featured in Sections 3.4.1 and 3.4.2. It responds to the rate of change of the flux produced by the rotor field turns as they go by. By feeding the voltage produced by this flux to an oscilloscope, the peaks produced by the flux change can be compared across each rotor slot, thus detecting any shorted turns. The waveform under no-load conditions will be relatively uniform except for coils having shorted turns. Under load conditions, the waveform will be distorted somewhat, but the peaks can still be measured and compared. The normal procedure suggested by the manufacturer is to record waveforms at no-load as well as at various load points when the search coil is first installed to use as points of comparison for future tests.

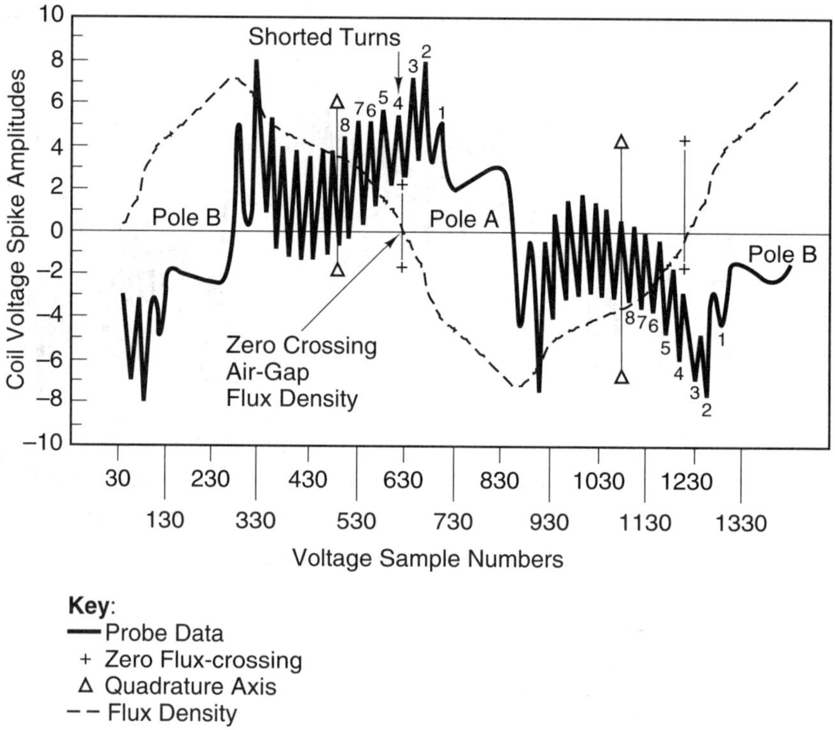

Figure 3.13 High intermediate load, shorted turns, coil #4. (Courtesy of Generatortech, Inc.)

Another manufacturer supplies a long thin wand type of flux probe, which is pushed in and out of a tube on the side of the generator.

3.5 THE PARTIAL DISCHARGE MONITOR

A *partial discharge* is a phenomenon in which the insulation capability of a gaseous substance breaks down locally because of high intensity electric fields. When a partial discharge takes place in air, the phenomenon is referred to as *corona discharge.*

In an electric generator, this gas breakdown phenomenon normally takes place in gas-filled pockets that are surrounded by solid insulation boundaries in high-voltage stator windings. Complete breakdown is prevented by the presence of these boundaries of solid insulation that has approximately 100 times the insulation strength of the gas. Partial discharges can be characterized by sparks within this gas-filled pocket of insulation. The spark generates a fast current pulse that travels through the stator winding. The larger this pulse, the higher the current and voltage pulses that reach the terminals of the winding. The detected pulse may be considered to be in direct proportion to the size of the discharge spark. Pulse de-

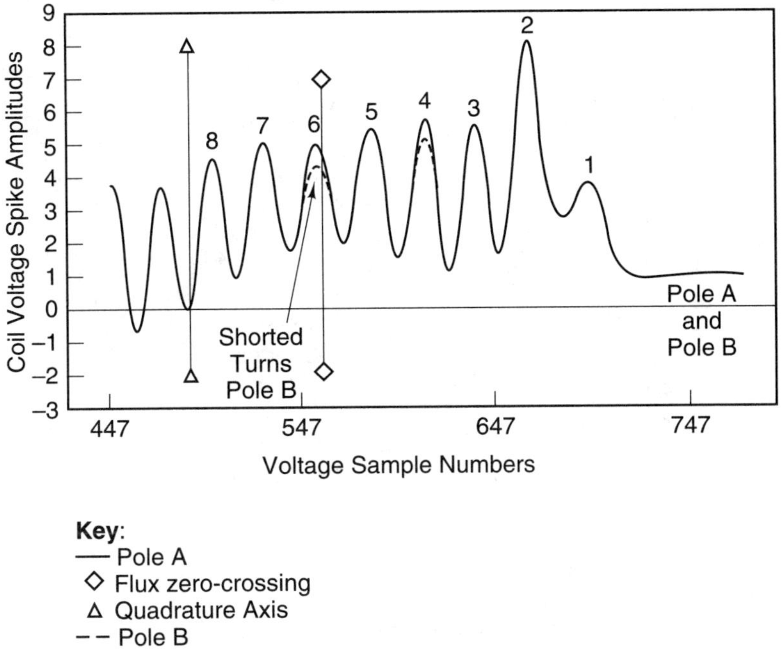

Figure 3.14 Lead slots data comparison. (Courtesy of Generatortech, Inc.)

tection can be done by monitoring the voltage and the currents at the high frequencies involved.

Partial discharge is characterized by a threshold voltage under which partial discharges cannot exist. This voltage corresponds to the electric field gradient required to ionize the molecules of the gaseous insulation surrounding the electrical conductor. This electric field gradient can be expressed more generally as a voltage level. The implication here is that only the high-voltage windings of high-voltage generators (5 kV or more) need examined for partial discharges. The type of gas present and the shape of the electric field will also have an influence on the partial discharge onset voltage.

Partial discharges will typically take place within voids in the groundwall insulation, which can be caused by poor impregnation of the bonding varnish or resin or by various thermal and mechanical aging mechanisms. Therefore, the incidence of partial discharges in a generator can be indicative of the existence of these voids, which, in turn, implies deteriorating insulation. Thus, partial discharges represent a measurement of the health of the insulation of the generator's high voltage windings.

More significantly, however, the physical action of the discharges themselves will accelerate the further deterioration of the insulation. "There exist various partial discharge emission mechanisms in a generator. These take place in various

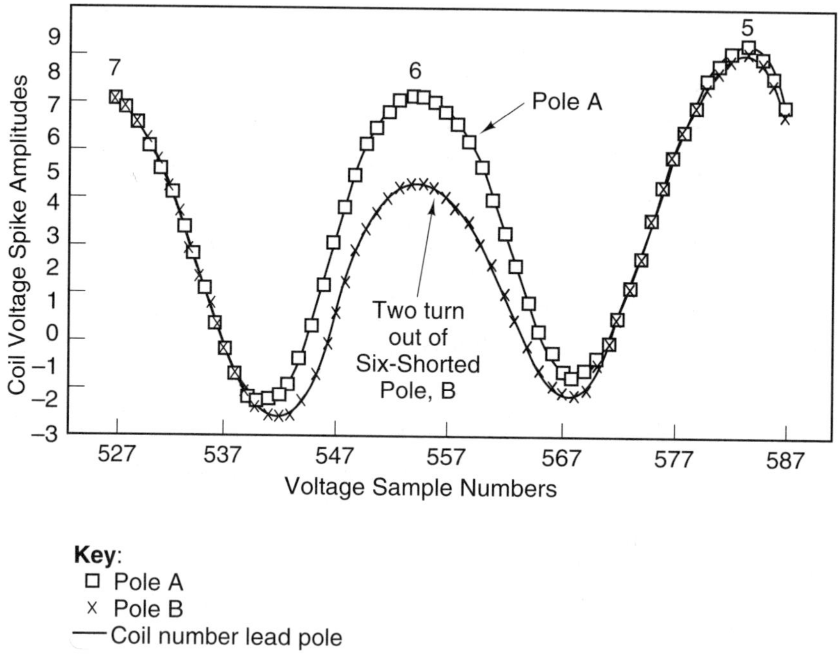

Figure 3.15 Zoomed data of coil #6. Two turns out of six are shorted. (Courtesy of Generatortech, Inc.)

locations within the generator winding, and their effects on the winding insulation is different." Discussion of these phenomena is beyond the scope of this chapter, but the reader is referred to Chapter 3 of Volume 16 of the *Power Plant Electrical Series* [Culbert, 1989] for an in-depth discussion of this subject.

Testing for partial discharges has traditionally been done off-line. This test requires subjecting each of the stator winding phases individually to its normal line-to-ground voltages. The phases, ungrounded and isolated from one another, are tested one at a time. The test consists of slowly increasing the voltage until partial discharge pulses are first noticed. This is called the *discharge inception voltage* (DIV). When the test voltage reaches machine line-to-ground voltage, the highest pulse magnitude is recorded from the screen. As the voltage is decreased, the voltage at which pulses disappear is also noted. This is called the *discharge extinction voltage* (DEV). The health of the machine is measured in terms of the DIV and the DEV. Usually, the lower the DIV and DEV, the greater the deterioration present in the winding insulation.

A similar on-line test can be performed. However, with the on-line test, the generator provides its own discharge excitation voltage. It can be done either open-circuited or connected to the power system. Although it is a more realistic test than the off-line version, the neutral is not adequately tested in the on-line case because of the low stress placed on it.

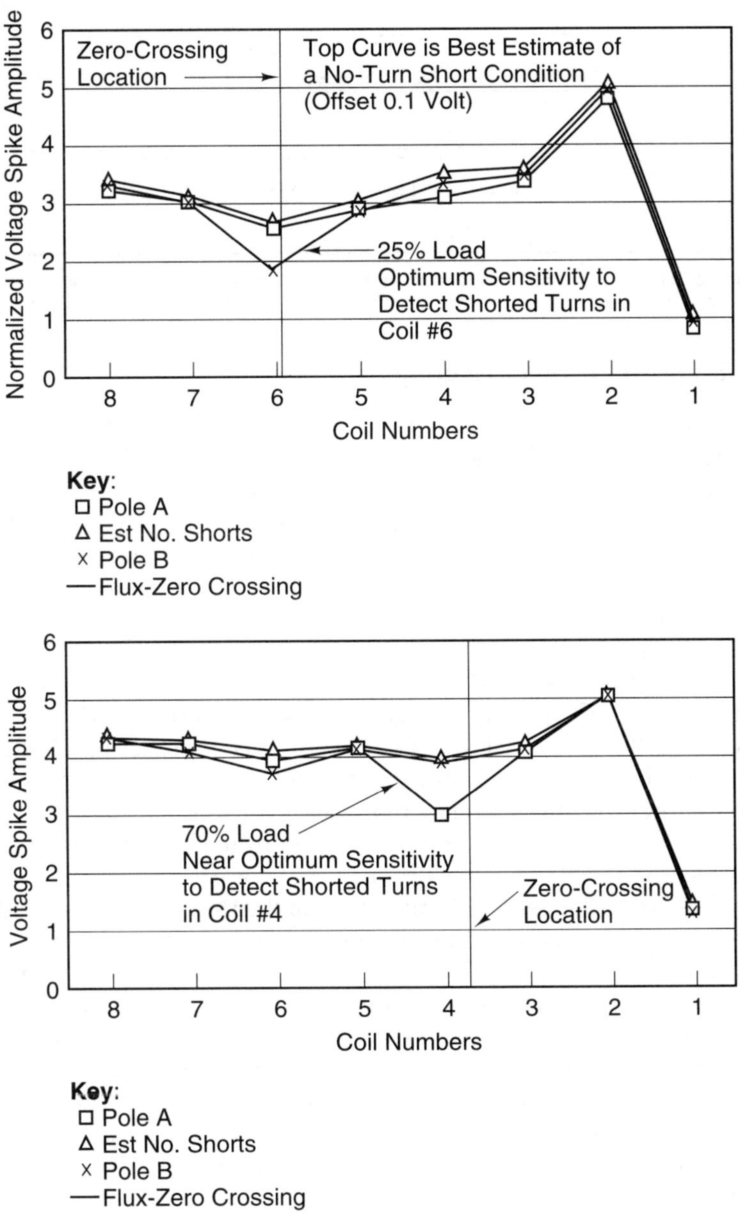

Figure 3.16 Quantitative signature analysis graph at 25% and 75% loads.

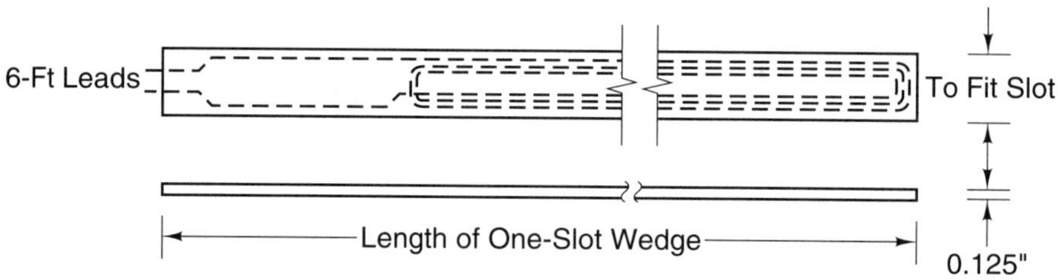

Figure 3.17 Search coil shorted-turn monitor. (Courtesy of Siemens Power Corporation)

3-5.1 PDA Tests for Hydrogenerators

The idea of on-line partial discharge monitoring by plant personnel grew out of a need by operators of air-cooled generators to monitor their machines more conveniently and with less expense than allowed by the conventional on-line tests, which required experts to conduct. Because air will ionize at a lower voltage than high-pressure hydrogen, air-cooled generators are considerably more susceptible to partial discharge damage than hydrogen-cooled units. For this reason, a *partial discharge analyzer* (PDA) was developed in the mid-seventies for use in hydroelectric generators.

The PDA uses high-voltage blocking capacitors to sense the high-frequency signals in the generator's circuit. These are permanently installed in the stator winding. The principal feature of the PDA is its ability to reject the electrical interference that typically dominates the high-frequency signal emitted by the partial discharge. The interference is reduced by forming a bridge-like circuit from two capacitative couplers that are installed on different parallels in each phase. Figure 3.18 shows how the blocking capacitor can be connected. The PDA sensing element makes use of the very high frequency of the partial discharge pulse and the time it takes for the electrical pulse to travel around the generator's circuit ring bus to eliminate common-mode electrical noise from the measurements [Culbert, 1989]. PDA tests are not continuous, but are carried out approximately twice per year. It is relatively easy to distinguish healthy windings from deteriorating ones, as the latter show a doubling of discharge activity approximately every six months [Lyles, 1993]. Unlike the off-line and on-line tests described above, PDA tests do not require an individual highly skilled in partial discharge testing to carry out the test. PDA couplers have been installed in more than 1000 hydrogenerators around the world.

3-5.2 Partial Discharge Tests for Turbine Generators and Large Motors

While the PDA's noise control features are suitable for hydrogenerators, they are clearly inadequate for high-noise environments. Thus, the significantly higher

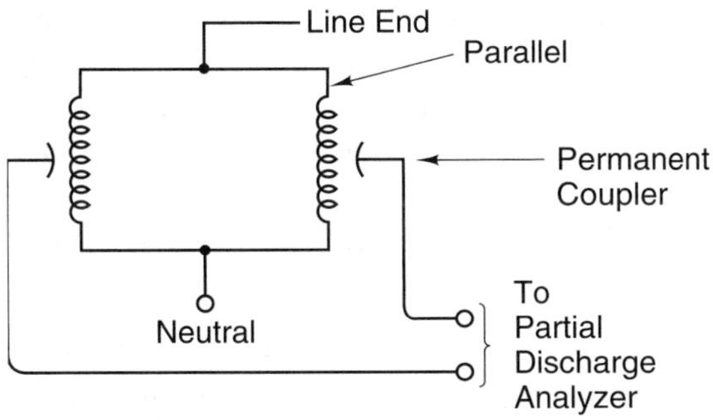

Figure 3.18 Permanent coupling capacitor installation for PDA tests (Culbert, 1989).

noise levels found in turbine generators have precluded the use of this instrument for partial discharge testing of the latter. A sensor called the *Stator Slot Coupler* (SSC) can serve to distinguish between noise and partial discharge signals in turbine generators. Nearly 85 generators are currently in service with these couplers installed, spread over 15 different utilities.

The SSC is essentially a directional electromagnetic coupler [Sedding, 1991]. It consists of a ground plane and a sense line with coaxial output cables at each end. Its dual-port nature permits the determination of the direction of the pulse. The SSC, in effect, acts like an antenna, picking up the high-frequency electromagnetic signals in and around the generator. Figure 3.19 shows a simplified schematic of the SSC [Sedding, 1991].

Partial discharge pulses have subnanosecond rise times and pulse widths of 1–3 ns. Thus, a bandwidth of approximately 350 MHz is required in order to be able to characterize the partial discharge pulses. Such high-frequency electrical signals will attenuate very quickly in the high-inductance windings of the generator.

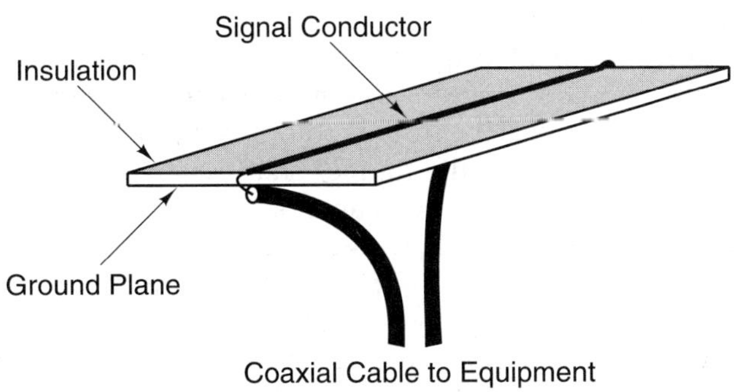

Figure 3.19 Simplified schematic of the SSC [Sedding, 1991].

Therefore, the location of the SSC must be very close electrically to the source of the partial discharges. Because only the pulses originating close to the SSC location will have their high-frequency component intact, the stator winding tends to act like a natural filter of external noise, such as that from the grounding bushes.

3-5.3 Inherent Limitations of the PDM and Their Mitigation

Since October 1989, the SSC sensor has been under prototype testing the SSC sensor has been under prototype testing in six generators in the range of 300–500 MW. Although it has been quite effective in discriminating between partial discharge pulses and noise [Sedding, 1991], all of the details about its limitations are not completely known at this time.

The pulse signals from the SSCs are measured with an instrument called the *turbine generator analyzer* (TGA). The TGA discriminates stator winding partial discharge pulses from electrical noises pulses on the basis of pulse width. Partial discharges have a width of 8 ns or less, whereas the width of a noise pulse is greater than 8 ns. As with the PDA, the TGA determines the number, magnitude, and phase position of the stator winding partial discharge pulses. Deteriorated stator insulation because of thermally-induced insulation delamination and loose stator bars in the slot has been measured to have a discharge activity 30 times higher than new windings [Campbell, 1993].

3.6 THE ROTOR SHAFT TORSIONAL VIBRATION MONITOR

Electrical disturbances in the power system, such as line faults or transmission line switching operations, can radically alter the electromagnetic torques seen by the turbine generator shafts during these episodes. More specifically, certain types of disturbances are capable of exciting torsional oscillations in the shaft system that can result in a reduction of the useful life (called loss-of-life) of the turbine-generator shaft. Shaft loss-of-life is a cumulative process, such that a series of relatively minor individual incidents can amount to a significant loss-of-life, and resultant shaft failure.

During torsional torque episodes, the turbine generator shaft system acts like a lumped torsional spring-mass system. The masses represent the large rotating inertial elements, such as the turbine rotors or the generator forging, whereas the torsional springs represent the thinner shafts connecting the rotating elements. The specific inertia and spring constants of the masses and springs, respectively, depend on the design of the various elements.

Mode shapes reflect the relative deflection of the rotor elements for oscillations at each natural frequency. Thus they describe the manner in which the rotor system will tend to respond at each natural frequency. A time-dependent description of the electromagnetic transient that excites the natural frequencies in the shafts is also necessary in order to calculate the loss-of-life.

The shaft section with the greatest amount of torque is the one with the most relative deflection between its ends. Because transient torques are induced on the generator shaft, the most readily excited modes are those that have large generator deflections. Figure 3.20 shows the mode shapes for the four-element T-G set shown therein.

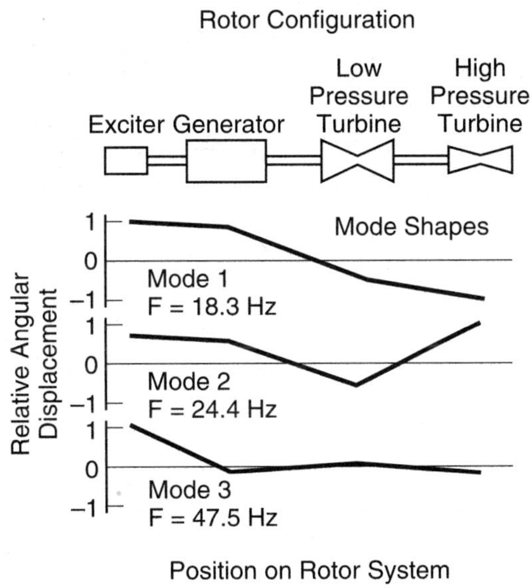

Figure 3.20 Mode shapes for a four-element T-G set (Hurley, 1979).

The definition of fatigue in materials science states that an increment of damage occurs when a complete stress-strain hysteresis loop is completed. Whereas counting such loops is rather easy for simple oscillatory waves, the fact is that shaft torsional torques exist in very complex waveforms for which no easy counting system exists [Hurley, 1979]. Figure 3.21 shows the traces of the torsional oscillations for the T-G set shown in Figure 3.20, after a line closing operation.

The loss-of-life due to torsional oscillations on the shaft for each episode can be estimated by monitoring the torsional oscillations and counting the number of cycles at a certain torque magnitude. A method called the *range pair method,* which is a practical implementation of the *rainflow method* commonly used by metallurgical engineers for metal fatigue studies, can be used to count the cycles of torsional oscillatory torque induced on the rotor system. Figure 3.22 shows a representation of a short duration of shaft torque oscillations. It also depicts how the range pair counting method can be used to count the torque oscillations. The following procedure is used:

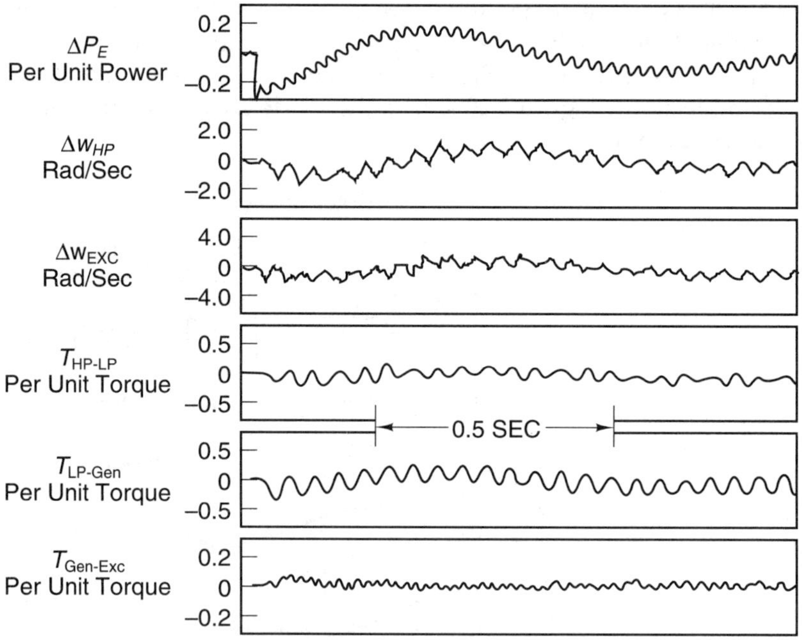

Figure 3.21 Traces of the torsional oscillations for the system in Figure 3.20. (Courtesy of Westinghouse Electric Corp.)

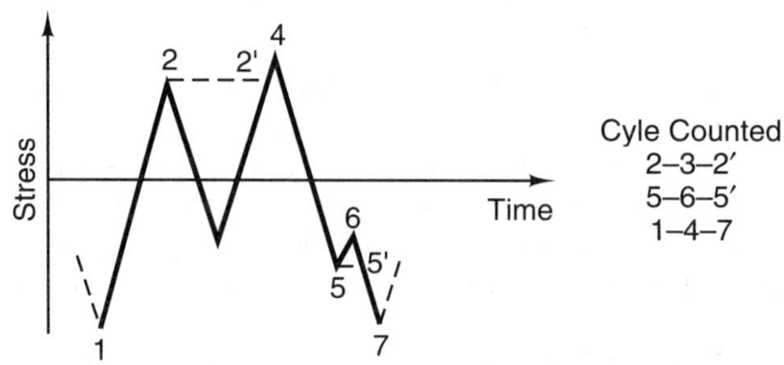

Figure 3.22 Torsional torque oscillations and the range pair counting method.

1. Measure and store the positive and negative peak values obtained from the signal. These include the relative maximums and minimums detected, as well as the overall peaks.
2. If the positive peak equals the absolute value of a negative peak, then match them as a pair and deallocate the storage for those two points.

3. If there are no signals in storage, then disregard all new inputs until there is a reversal.

The range pair method counts small reversals as momentary interruptions of large reversals. This is significant because of a large reversal can cause more damage than two smaller reversals that add up to the larger one.

By comparing this count to the shaft's *stress-life curve,* the accumulated loss-of-life for one incident can be calculated. The stress-life curve (also called the S-N curve) relates surface shear stress at the weak point along the generator shaft to the number of oscillation cycles estimated to cause fatigue failure. The y-axis contains the surface shear stress (S), and the x-axis the number of cycles (N). Both axes have logarithmic scales. The upper part of the curve can be described as the low-cycle fatigue limit, beginning with the single-cycle fatigue level. This is the stress level required to fail the shaft at the weak points after only one cycle. The low-cycle fatigue portion of the curve continues until the 1000-cycle point, where it changes slope and enters the high-cycle fatigue portion. This continues until the 1,000,000th cycle. Below the torque level corresponding to the 1,000,000th cycle point, the shaft will not fail, regardless of how many oscillations take place. This point is called the high-cycle fatigue limit. Figure 3.23 depicts an S-N curve for a particular T-G set. The loss-of-life is typically calculated in terms of percent loss of its total life after one incident, such as shown in Figure 3.21.

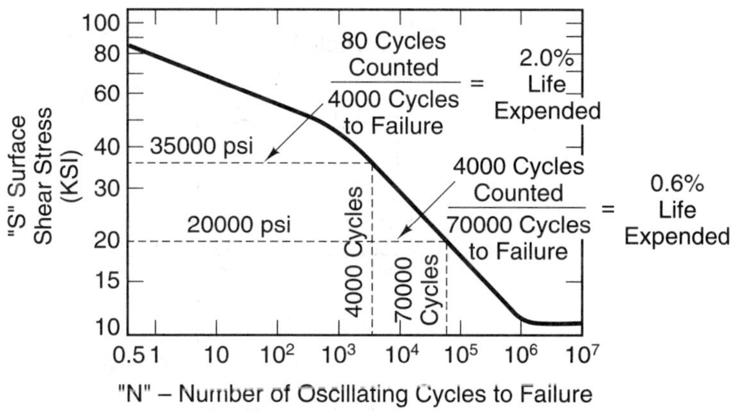

Figure 3.23 Typical stress-life curve for calculating loss-of-life in a generator shaft. (Courtesy of Westinghouse Electric Corp.)

3-6.1 Torsional Vibration Monitor Operating Principle

The direct measurement of torsional oscillations is a difficult task at best. Although strain gauges placed on the rotor would provide such a direct indication, maintaining, and extracting their signals from a rotating shaft introduces many difficulties.

The alternative used by one torsional monitoring system manufacturer is to model the T-G set by means of a lumped-mass model as described above and excite it with a measured signal of the unit MW and instantaneous shaft speed. A tooth wheel with a magnetic pickup and an FM decoder is used to measure the instantaneous shaft speed. One of these tooth-wheel/FM-decoder combinations is used at each end of the rotor system. Figure 3.24 shows a diagram of a tooth-wheel/FM-decoder system.

Figure 3.25 shows a simple schematic diagram of the torsional monitoring system. It shows the T-G analog model as well as the sensor inputs to the model.

A torsional monitor may alarm under two separate conditions:

- An instantaneous alarm will indicate when the torque levels reach very high magnitude levels. The level will correspond to a level on the upper portions of the S-N curve.
- A constant alarm will warn of a trend of growing oscillatory magnitude that could signal the onset of subsynchronous resonance (SSR). These unstable oscillations occur when the T-G shaft system resonates with a transmission line because of the latter's high level of series capacitance and have been known to very quickly cause fatigue failure of the shaft system. Other growing oscillations could result from an unstable interaction with a DC transmission line.

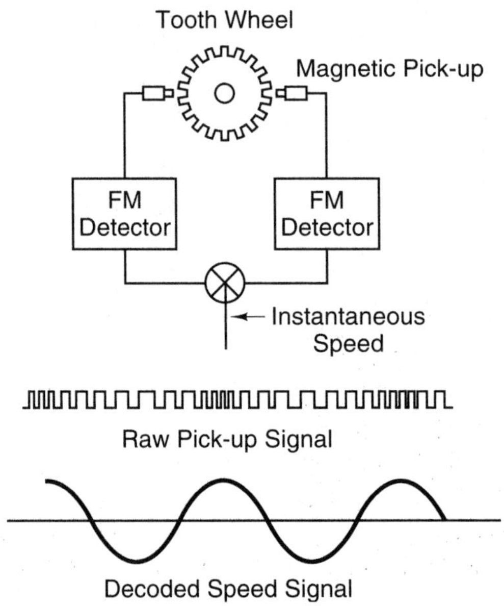

Figure 3.24 Instantaneous shaft speed detection system (Hurley, 1979).

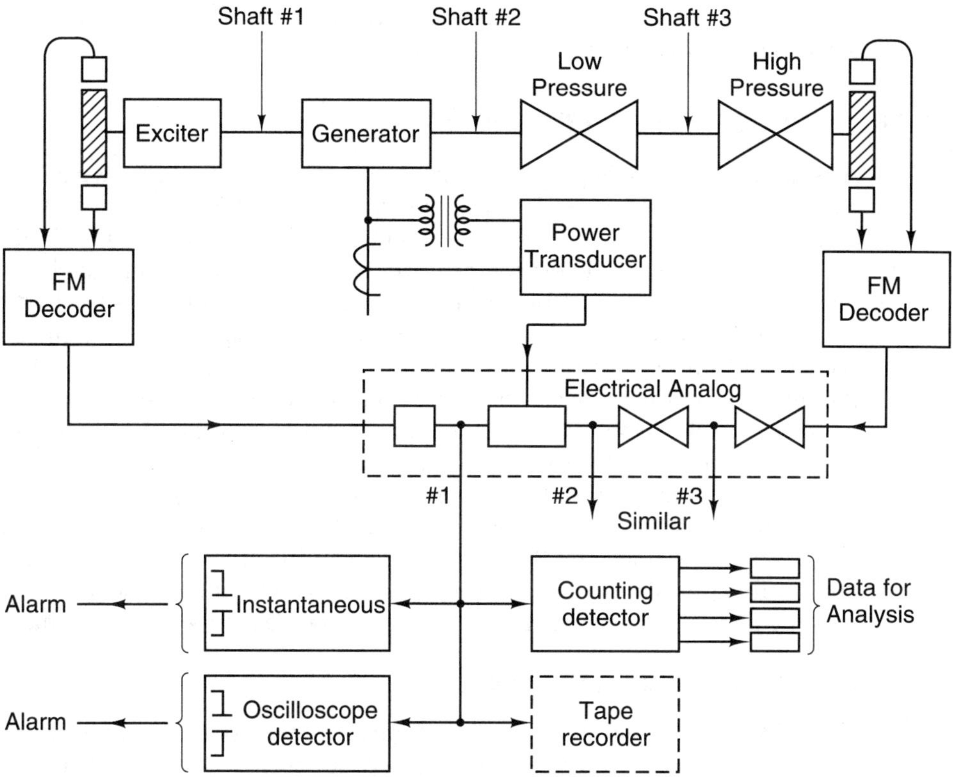

Figure 3.25 Torsional monitor block diagram (Hurley, 1979).

3-6.2 Inherent Limitations of the Torsional Vibration Monitor and Their Mitigation

By its nature, the torsional monitor is an indirect monitoring system because it does not measure the torsional oscillations directly. Rather, it simulates the effect of a measured transient disturbance in the power system that is applied onto the rotor shaft system. In order to provide accurate results, the torsional monitor must have the proper values for the rotational inertia and spring constant for the spring-mass model. Otherwise, the results could be skewed and show overly conservative loss-of-life calculations because the difference in the natural frequency of the model and of the actual T-G set. Computer simulations have shown that an error of 5% in the rotational inertia and/or spring constant between the model and the T-G set is tolerable [Hurley, 1979].

To mitigate this potential problem, tests can be performed in the field during the system installation to more accurately reflect the T-G set's specifications. This is done by performing line closing tests and measuring the response of the torsional monitor model. Using the field verification techniques, the model parameters can be set to be better than 0.5% accurate.

3.7 THE HYDROGEN GAS PURITY METER

As shall be seen in Chapter 8, maintaining a pure supply of hydrogen in the machine is very important to the safe, as well as efficient, operation of a turbine generator. Briefly, the safety issue is that pure hydrogen will not ignite in the presence of a spark. The efficiency issue is that pure hydrogen gas has very low density, thus minimizing the windage losses of the machine. As the hydrogen becomes less pure as it mixes with water vapor, oil vapor, or even air, its density rises significantly, as do the windage losses. Hydrogen purity should be held at the level recommended by the manufacturer, or higher, in order to minimize windage losses. This will probably result in maintaining hydrogen pressure at or above the range of 96%–98%.

As its name implies, the *hydrogen purity meter* (HPM) monitors the purity of the hydrogen gas inside the generator frame during operating conditions. There are various means to measure the purity: One is through conductivity measurements, and another popular one is through density measurement. The next section will discuss the density measurement technique and will describe its operating principle.

3-7.1 HPM Operating Principle

The HPM uses measurements of the hydrogen density to calculate its purity. The method utilizes the differential pressure produced by a small, constant speed fan called the *purity blower*. This differential pressure is proportional to the gas density, which, as mentioned above, is proportional to the gas purity. It is assumed that hydrogen gas is mixed with air during normal operation of the generator, and that an air/CO_2 mixture is present when the generator is being purged.

The density indicator is calibrated to read 100% air at atmospheric pressure. Densities obtained at higher pressures are compensated to the reading that would be obtained at the atmospheric calibration pressure. Thus, the uncompensated density reading is the reading of the differential pressure across the fan. The percent of uncompensated density, therefore, is the density measured with the blower differential pressure divided by the density of air at atmospheric pressure.

This value is then compensated by the following formula:

$$\%\text{Density} = \%\text{Density}_{\text{unc}} * \frac{P_{At}}{P_{Ab}} \tag{3.2}$$

where

$\%\text{Density}_{\text{unc}}$ = uncompensated density in %
P_{At} = Atmospheric pressure, and
P_{Ab} = Absolute pressure

The density of air and a mixture of another gas can now be calculated as follows:

$$\%\text{Density}_{\text{m}} = \frac{[(MW_G * X) + (100 - X) * MW_A]}{MW_A} \tag{3.3}$$

where

Density_m = Density of air/gas mixture
MW_G = Molecular weight of the gas
= 2.0159 for hydrogen gas
= 44.01 for CO_2,
MW_A = Molecular weight of the air
= 28.966
X = % gas in the gas-air mixture

Now, solving for X, the purity of hydrogen can be calculated by inserting the molecular weight of hydrogen during operating conditions and of CO_2 when purging. These formulas are as follows:

$$X_{H_2} = 1.0748 * (100 - \%\text{Density}_m) \tag{3.4}$$

where $\%\text{Density}_m \leq 100$

$$X_{CO_2} = 1.9254 * (\%\text{Density}_m - 100) \tag{3.5}$$

where $\%\text{Density}_m \geq 100$

It should be noted at this time that there are other methods for determining hydrogen purity that are not discussed here, including one that measures the thermal conductivity of the gas and compares it to that of pure hydrogen. However, these methods are not discussed in this book.

3-7.2 Inherent Limitations of an HPM and Its Mitigation

The HPM is a simple device that uses very exact and proven formulas from physics and chemistry. As such, it has no inherent limitations, per se. The only possible problem would be that the constant speed purity blower could somehow provide an inaccurate reading by running at an abnormal speed. This, however, is not very likely to happen.

An alternative to using the purity blower would be to measure the differential pressure of the main generator blower, which always rotates at the shaft speed. This could be used as a backup to the purity blower, as well as when the latter fails. However, there could be blades missing in that blower, which could affect the differential pressure across it. As a matter of fact, the differential pressure across the main blower will be used in Chapter 5 to detect that very problem.

3.8 THE HYDROGEN DEW POINT MONITOR

Determining the amount of moisture in the hydrogen gas in an important monitoring function. For reasons that will become more clear in Chapter 8, it is important to minimize the moisture content of the hydrogen gas. The physical qualities of the gases will be briefly discussed here so as to better explain the problem and its solution.

The ability of any gas to absorb water vapor is a direct function of its temperature. Therefore, at higher temperatures, hydrogen can absorb a greater amount of moisture than at lower temperatures. *Saturation* occurs when the gas has absorbed all the moisture that it is capable of doing at its particular temperature. Any further absorption of moisture will result in condensation of the water vapor as droplets. This is commonly seen in early morning fog. The *relative humidity* is a measure of how much of the vapor-holding capacity of the gas at the particular temperature has been filled. For example, at a relative humidity of 85%, the gas is holding about 85% of the amount of water vapor that it can hold without saturating at that particular temperature. If the temperature rises but the amount of moisture remains constant, then the relative humidity will decrease because the gas is now able to hold more water vapor. The constant amount of moisture now occupies a relatively smaller part of the overall moisture-holding capacity of the gas.

The hydrogen gas in the generator will very rarely be saturated because of the high temperatures involved during normal operation, even at less than full-load levels. The danger from excessive moisture in a generator arises from the fact that even though the general environment may be relatively hot, the temperatures are unevenly distributed. Specifically, the temperature of the gas at the outlet of the hydrogen coolers will naturally be significantly cooler than at its inlet. These localized lower temperatures cause a decreased moisture-absorbing capacity in the gas around the area of decreased temperatures and thus possible condensation in some sensitive elements of the generator internals. This is analogous to the wet glass of ice water. The reduced surface temperature on the outside of the glass, in effect, reduces the moisture-holding capacity of the air immediately surrounding it to below the saturation point. As a result, the moisture in the air condenses around the glass.

The relative humidity, however, does not serve well in determining whether this phenomenon is about to occur. Percent of moisture-holding capacity does not intuitively translate to an easy assessment of the probability of this happening. For this reason, the absolute humidity measurement, expressed as the *dew point temperature,* serves our purposes much better. The temperature of the gas is necessary in order to transform relative humidity to dew point temperature.

As the name implies, the dew point temperature is that temperature at which the gas will saturate with the amount of moisture present in it. Unlike the relative humidity, it is not a function of the ambient temperature of the gas. As such, it is much easier to compare the cold gas temperature (at the hydrogen cooler outlet) to the dew point temperature in order to determine whether there is any possibility of condensation taking place.

There are various means of determining the humidity of gases. One of the oldest and most reliable, if not the most accurate, is the *wet/dry bulb* method. The temperature readings of two thermometers are taken. One has a small cloth covering the bulb that is saturated with room temperature water. The other one measures the ambient temperature directly. By using these two temperatures and a look-up table, the relative humidity can be calculated. Such a method, however, is not practical for generator applications. The following section will describe other methods that are more applicable to generators.

3-8.1 Dew Point Monitor Operating Principle

This section title is a misnomer because there will be two methods presented here. Nevertheless, we shall concentrate on one more than the other.

The chilled mirror method uses a rather simple concept, which says that the reflectivity of the mirror is a function of the amount of condensation that has settled on its surface. The temperature of the mirror can be changed to determine the dew point. The problem with this method is that its reliability can be compromised if the gas stream is contaminated with dirt, dust, or oil vapor.

Another method is the use of thin film aluminum oxide to detect the moisture content [Panametrics, 1991]. This sensor consists of an aluminum strip, anodized by a special process to produce a porous oxide layer. A thin coating of gold metal is evaporated at the other end of the porous oxide layer. This, in effect, forms an aluminum oxide capacitor, with the aluminum surface forming one electrode and the gold coating forming the other electrode. The porous aluminum oxide layer acts as a dielectric. The electrical impedance of the porous material is proportionally related to the amount of water vapor that forms on it. Thus, the absolute humidity is calculated by relating it to the impedance between the electrodes.

3-8.2 Inherent Limitation of the Aluminum Oxide Moisture Sensor and Its Mitigation

The performance of the aluminum oxide sensor can also be compromised by the presence of conductive or corrosive particles in the stream of gas. Furthermore, compromise of its operation is likely if condensation of water is a possibility around the sensor. In such cases, sample conditioning is necessary in order to maintain the integrity of the reading.

Dew point measurement, as a rule, is a most difficult measurement to make, and it is often prone to error because of sensor failure or water collection. Thus, a dew point monitor demands an inordinate amount of regular maintenance and calibration.

3.9 GENERAL POINT TEMPERATURE SENSORS

While it may seem to the reader that point temperature sensors are quite common and not appropriate for discussion in this section, they are used in many of the generator subsystems and should be discussed in a chapter that is common to all of these subsystems. The topic of this section will, therefore, be thermocouples and resistance temperature detectors (RTDs).

RTDs are basically electrical wires whose resistance varies in accordance with their temperature, a well-known concept in electricity. Thus, if the characteristics of the wire are known and its resistance can be accurately measured, then the temperature of the RTD can be calculated.

RTDs are generally more accurate than thermocouples and exhibit greater linearity. They are typically used to measure a small area, rather than a point. One disadvantage is that bridges are generally required to accurately determine their resistance, which can be rather expensive. Furthermore, it is not common to find dual elements in RTDs.

Thermocouples are basically two wires of different material joined at the end. The difference in materials will generate a small voltage proportional to the temperature at the joint. By measuring the voltage accurately, the temperature at the joint can be calculated. Their advantage over RTDs is that simple millivoltmeters can be used to measure their output. These are generally less expensive than the bridges used to measure the resistance in RTDs. Thermocouples are true point measurements, and they can come in dual elements. However, they are typically not as accurate as RTDs.

There exist various materials for the thermocouple wires. Copper-constantan has been a popular type of thermocouple. It has relatively good output in terms of $mv/^\circ$ change. One drawback for this type of thermocouple is that it has a low temperature range (up to approximately 600° F) because of the low melting point of copper. They have in the past been used to measure bearing temperatures, which under most circumstances are within this range.

Iron-constantan is another type of thermocouple that is commonly used. It has higher temperature limits than copper-constantan, but its output is not as good. Furthermore, it can rust when used in humid environments.

Chromel-constantan addresses both of these issues. It has good output, it is linear in nature, and it can withstand temperatures in the range of 1300°–1400° F. It seems to be the thermocouple of choice for many applications.

One last type is a platinum-platinum/rhodium thermocouple. Its main advantage is its capability of operating in high temperature environments (up to 2500° F), but at the cost of poor output.

3.10 DISTRIBUTED FIBER OPTIC TEMPERATURE SENSOR

With very few exceptions, temperature measuring techniques in generator monitoring are of the point type; that is, a thermocouple or RTD is placed at different points to measure the temperature only at that point. It is hoped that the point selected is the most likely hotspot when a problem might develop. However, in some parts of the generator, the most likely hotspot location is difficult to predict because of the great length of the potentially overheating element. This is particularly true for hotspots in the stator windings, as well as in the core slots. Any problems that arise in a location sufficiently distant from the point sensor may not be detected early enough to prevent serious damage.

The *distributed fiber optic temperature sensor* was developed to address this situation by measuring temperature along a generator component such as a core slot or a stator winding coil, which are relatively long and narrow, and detect the location of a hotspot anywhere in its length. As in the fiber optic end-turn vibration monitor, one significant advantage of using fiber optics is that they are nonconducting and thus do not require insulation from the high voltages inherent in the generator.

3-10.1 Operating Principle of the Distributed Fiber Optic Temperature Sensor

There are a number of approaches to distributed fiber optic temperature monitoring. The most practical one, however, is referred to as *optical time domain reflectometry*, or OTDR. The OTDR technique is used to monitor backscattered light from the segments of fiber along the entire length of the fiber.

In this approach, the elapsed time of the returned pulse of light can be used to locate the cause of any changes in the backscattered light. The location of the cause of the change can be described as [Boiarski, 1991]

$$L = \frac{c * \tau}{2 * n} \tag{3.6}$$

where

L = the location of the cause of the change
τ = the time required for the light pulse to propagate back and forth
c = speed of light in a vacuum (3×10^8 m/s)
n = the refractive index of the fiber being used

Thus, by monitoring the change in the local backscattered light characteristics using the above equation, temperature can be measured along the entire length of the fiber. The changes are a difference in local loss or backscattering characteristics that are caused by changes in temperature at a particular location. The amount of change is proportional to the temperature value. A proprietary ultraviolet light curable cladding material is used that produces the change in the local backscattered light intensity with a change in local fiber temperature [Boiarski, 1991].

3.10.2 Applications of the Distributed Fiber Optic Temperature Sensor

This monitoring system has a spatial resolution of as much as 2 cm, with a temperature accuracy of ±2° C. The sensor can be as long as 20 meters. As mentioned previously, this system can be applied to the core slots and the stator windings. However, other applications exist in power transformers, power switchgear, and bus connectors, among others.

SUMMARY

This chapter presented the reader with ten special purpose monitors which can be used in generator monitoring. These systems tend to be more complex in nature than other sensors that will be discussed directly under the "Sensors and Monitors" section of each chapter.

The use of some of these monitors is not always recommended for all machines at all utilities, as they have quite specific purposes and can require special expertise to make good use of their information. This is particularly true of the radio frequency monitor, the generator core condition monitor, the fiber optic vibration monitor, the field interturn short circuit monitor, the partial discharge monitor, and especially, the torsional vibration monitor. Proper application of these monitors is very important if they are to be effectively used in diagnostics. It is important, therefore, that the reader become familiar with the operating principle of the various monitors discussed in this chapter, as well as with their limitations.

Information for the monitors was obtained from the following sources, which should be given recognition for their assistance in the preparation of this chapter. It must be reiterated that in most cases, these sources are not the only suppliers of this equipment. Where possible, the other manufacturers will be listed in this summary section for the convenience of the reader.

The information for the radio frequency monitor was provided by Westinghouse Electric Corp., in Orlando, Florida. They are a supplier of the RFM and originally developed the concept with EPRI funding. Additional information was provided by Siemens Power Corp., West Allis, Wisconsin, which is also a supplier of an RFM. Significant work has also been done on the RFM by American Electric Power Co. and Ontario Hydro.

Information for the generator core/condition monitor was provided by Westinghouse Electric Corp. of Orlando, Florida. They are a supplier of the GCCM. General Electric Co. of Schenectady, New York and Siemens Power Corp. of West Allis, Wisconsin, are also suppliers of a GCCM. Environment One has also done a significant amount of work on the generator core/condition monitor and is a supplier of it.

The information for the fiber optic vibration monitor was supplied by Westinghouse Electric Corp. of Orlando, Florida. They are a supplier of the end-turn vibration monitor. Siemens Power Corp. of West Allis, Wisconsin, is also a supplier of a similar type of monitor.

The information from the field interturn short circuit monitor was provided by Generatortech, Inc., of 31 Sutherland Dr., Scotia, New York. This company is a supplier of the field interturn short circuit monitor. A similar system is also supplied by General Electric Co., of Schenectady, New York, and by Siemens Power Corp. The information on the torsional vibration monitor was provided by Westinghouse Electric Corp., Asheville, North Carolina. Another supplier of the torsional vibration monitor is General Electric Co., Schenectady, New York, who developed a version of it under EPRI funding.

The information for the partial discharge monitor was supplied by IRIS Power Engineering, Inc., of 6380 Tomken Rd., Unit 1, Mississauga, Ontario, L5T 1Y4, Canada. Additional material was obtained from the EPRI *Plant Electrical Reference Series,* Volume 16 [Culbert, 1989]. Another manufacturer of partial discharge detection equipment is FES International Ltd., 60 Ironside Crescent, Unit 9, Scarborough, Ontario M1X 1G4, Canada.

The information for the hydrogen purity monitor was obtained from Westinghouse Electric Corp., Orlando, Florida.

The information for the dew point monitor was provided by Panametrics, Inc., Boston, Massachusetts.

The information for the distributed fiber optic temperature sensor was supplied by Battelle Laboratories, 505 King Avenue, Columbus, Ohio 43201-2693. This project was developed with partial funding from the Electric Power Research Institute, Palo Alto, California.

REFERENCES AND FURTHER READINGS

[Albright, 1971] ALBRIGHT, D. R., "Interturn Short-circuit Detector for Turbine Generator Windings," *IEEE Transactions on Power Apparatus and Systems,* PAS, 1971, pp. 478.

[Boiarski, 1991] BOIARSKI, A. A., and N. E. NILSSON, "New Fiber Sensors Take Power Plant's Temperature," *Photonics Spectra,* September 1991, pp. 92–94.

[Campbell, 1993] CAMPBELL, S. R., et al., "Practical On-Line Partial Discharge Tests for Turbine Generators and Motors," IEEE Summer Power Meeting, Paper Number 93SM 353-3EC, July 1993.

[Culbert, 1989] CULBERT, I. M., H. DHIRANI, and G. C. STONE, *Handbook to Assess the Insulation Condition of Large Rotating Machines,* Volume 16 of the *Power Plant Electrical Reference Series,* Electric Power Research Institute, EPRI Report EL5036, 1989.

[Hetrick, 1984] HETRICK, J. A., N. E. NILSSON, and T. E. HITCHCOCK, "Overview of Generator Monitoring Equipment, Operating Expertise and Performance Evaluation," *Proceedings of the Workshop in Generator Monitoring and Surveillance,* Electric Power Research Institute, EPRI EL-3564-SR, August 1984.

[Hurley, 1979] HURLEY, J. D., and W. H. SOUTH, "Torsional Monitoring Equipment for Turbine Generator Units," *Proceedings of the American Power Conference,* Chicago, Ill., 1979.

[Lyles, 1993] LYLES, J. F., T. E. GOODEVE, and G. C. STONE, "Using Diagnostic Technology for Identifying Generator Winding Maintenance Needs," *Hydro Review,* June 1993, pp. 58–67.

[Panametrics, 1991] "Panametrics Aluminum Oxide Moisture Sensor for Gases and Liquids," Descriptive Bulletin, Panametrics, Inc., Waltham, Mass., 1990.

[Sedding, 1991] SEDDING, H. G., S. R. CAMPBELL, G. C. STONE, and G. S. KLEMPNER, "A New Sensor for Detecting Partial Discharges in Operating Turbine Generators," presented at the 1991 IEEE Power Engineering Society Winter Meeting, New York. Scheduled for publication in the *Transactions of the IEEE,* 1991.

[Sheets, 1987] SHEETS, A. W. J., *Electric Generators,* Volume 1, *Power Plant Electrical Reference Series,* Electric Power Research Institute, 1987.

[Twerdochlib, 1993] TWERDOCHLIB, M., "Improved Radio Frequency Monitoring for Turbine Generators," EPRI Report TR 102352, June 1993.

[Twerdochlib, 1992] TWERDOCHLIB, M., unpublished technical report, 1992.

[Twerdochlib, 1989] TWERDOCHLIB, M., and C. C. REPPLIER, "On-line Monitoring as a Means of Turbine Generator Life Extension," Joint Power Generation Conference, Dallas, Tex., October 22–28, 1989.

[Twerdochlib, 1987a] TWERDOCHLIB, M., and C. C. REPPLIER, "Diagnostic Capabilities of On-line Generator Monitors," CMIC/CNEEE/Westinghouse 3rd Joint Technical Seminar, Quinto, People's Republic of China, 1987.

[Twerdochlib, 1987b] TWERDOCHLIB, M., R. L. OSBORNE, R. C. MILLER, C. C. REPPLIER, T. A. LEMAK, D. R. MORRISON, and D. J. BOOMGAARD, "Two Recent Developments in Monitors for Large Turbine Generators," IEEE Power Engineering Society Summer Meeting, July 12–17, 1987, San Francisco, Calif.

CHAPTER 4

The Generator Stator Winding

The generator converts the mechanical power derived from a steam, gas, or hydraulic turbine into electrical power. It consists of a *stator* and a *rotor,* all encased in a sealed *frame*. As discussed in Chapter 2, magnetic flux produced by a rotating magnetic field passes across the air gap and through the stator *core* to generate three-phase voltages in the stationary *armature winding,* also called the *stator winding*. The rotating magnetic field is created by passing current through the rotating *field winding,* also called the *rotor field winding*.

This chapter will discuss the stator winding. The first section below includes a description. The rotor and the core will be covered in subsequent chapters.

4.1 DESCRIPTION OF THE STATOR WINDING

The stator (or armature) winding is one of the most important components of the generator because the power is generated and conducted to the power system through this winding. Its function is to

- continuously carry the full-load current of the generator
- briefly carry significant overcurrents due to system faults
- operate at output voltage

The armature winding of any turbine generator is classified and described in terms of its type of cooling. Small and medium-size generators are *conventionally cooled;* that is, the heat generated within the armature conductors is conducted through the ground insulation and removed by a coolant gas flowing over the outside of the insulation. The armature windings of large turbine generators are *conductor-cooled,* meaning that the heat is removed by the passage of coolant (hydrogen or water) within the ground insulation, in close or direct contact with the conductors (strands) where the heat is generated.

The armature winding of a generator is made up of individual coils connected in series. In large turbine generators, these coils are commonly made as half-coils (also called *bars*) to ease the process of manufacture. For the purposes of this book, the labels "bars" and "half-coils" will be synonymous unless otherwise indicated. Two half-coils are assembled individually into the generator and connected to form a complete coil, which is roughly a turn around the core. Coils are, in turn, serially connected to form the complete winding. These half-coils are made up of copper strands that are individually insulated. Half-coils are laid two-deep in slots in the core, axially along the length of the machine. Each half-coil is composed of an inner part, called the *bare bar,* which contains the insulated copper strands bonded to form an otherwise uninsulated conductor. A bare bar, therefore, is a current-carrying conductor without ground insulation. The ground-wall insulation is then applied in tape form to the outside of the bare bar to complete the armature bar, or half-coil.

Two half-coils are connected in series and are each placed in slots that are *short-pitched* (less than 180 electrical degrees apart from each other in the core) to form one or more *turns,* arranged so as to link the magnetic flux produced by the rotating field. The number of slots in a core normally ranges from 30–96, and each slot carries two half-coils, one in the top and one in the bottom. The lower numbers of slots within this range represent two-pole generators; the upper portion applies to four-pole machines. Figure 4.1 shows a top and a bottom coil placed within a slot in the core.

Half-coils (bars) are connected such that a selected number of them in series (the number depends on the size of the generator and the number of poles) compose a *circuit* within a phase. For large machines, two or more of these circuits are connected in parallel to make up one phase. Therefore, there are typically six circuits in an armature winding, as shown in Figure 4.2.

An individual half-coil is composed of copper *strands* stacked in a bundle. The strands, which are individually insulated, are normally transposed in position so that over its length, every strand occupies each heightwise position in a bar, in its slot section. This type of transposition is called a *Roebel* transposition. All strands in each half-coil may be connected at the two ends of the half-coil. The purpose of any of these types of transpositions is to even out, among the strands of any one armature bar, the voltages that are produced by the magnetic fields in the winding, and thus reduce heat losses by reducing circulating currents. Figure 4.3 depicts Roebel transpositions in a bar.

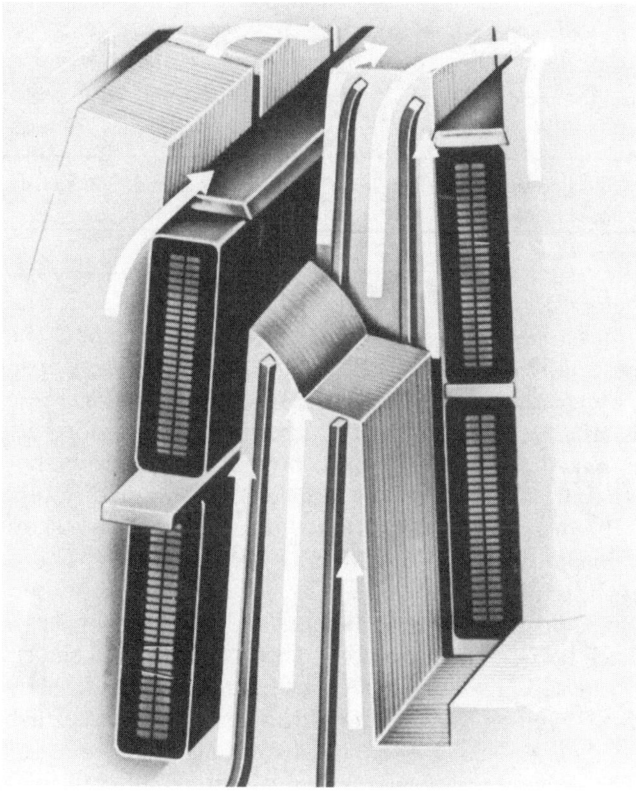

Figure 4.1 Top and bottom coil placement in core slots. (Used by permission from Westinghouse Power Generation, Orlando, Florida.)

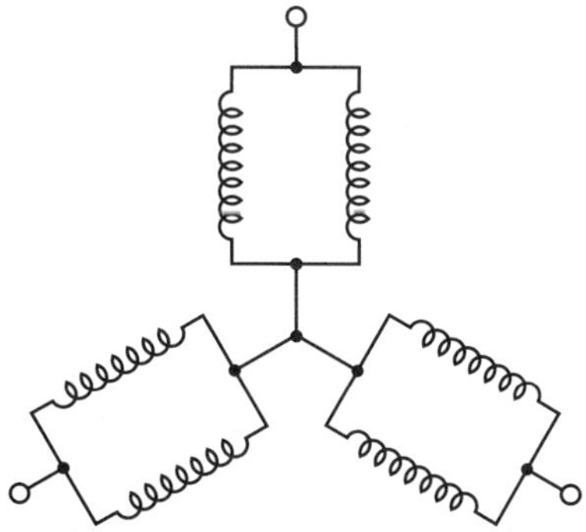

Figure 4.2 Circuit diagram of generator armature winding.

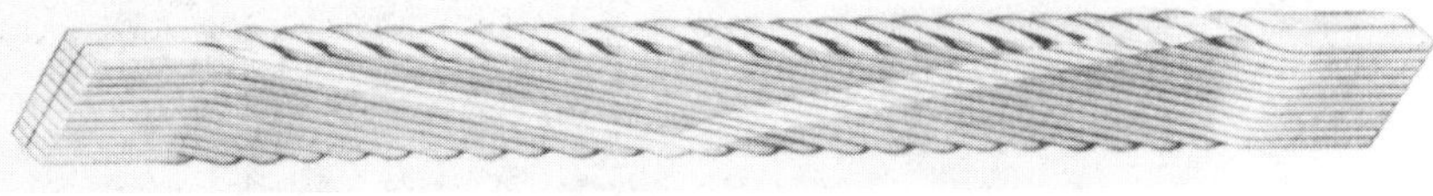

Figure 4.3 Roebel transposition in an armature bar. (Used by permission from Westinghouse Power Generation, Orlando, Florida.)

Alternatively, another type of transposition, informally referred to here as *group transposition,* forms groups of strands in each bar. The strands in each group are connected at the ends of each bar. The connection of successive half-coils in a circuit are made between the corresponding groups. The strands of the different groups are thus not connected to those of other groups at the end of the bar, but rather, make up an independent connection of strands that continues throughout the entire circuit. This further reduces the voltage difference between the strands due to the magnetic fluxes in the end regions and, consequently, also reduces the circulating currents and resulting losses and heating. This type of transposition allows the circuit formed by each group of strands in a bar to occupy the various heightwise positions within the bars of a circuit.

Each half-coil, composed of a bundle of transposed strands, is then wrapped in groundwall insulation. Figure 4.4 shows the Roebel transpositions in the strands, as well as the additional transposition accomplished by the connections at the ends of the bars, for a pair of half-coils that are to be connected serially.

The series connections of the individual half-coils in a circuit take place in the *end region* of the generator. In the core slot, each half-coil is built as a rectangular, flat, straight bar. However, as it exits the slot at the end region, the bar follows a complicated curve, such that all of the bars nest nicely together to form a conical arrangement. This is done to simplify the connections. Its conical shape and the weaving of the half-coils and the support structure make the end region look like a basket (see Figure 4.5), thus the commonly used name *end-basket*. The connections of the half-coils at the end region are also heavily insulated.

The first and last half-coils of each circuit in the same phase are commonly connected to two different *connection rings* that carry the current of each circuit in each phase and connect these half-coils to the *bushings* through the armature leads. There are normally two bushings per phase. The function of a bushing is to connect that particular phase of the armature winding to the external system. The bushing is connected to the connection ring though flexible armature leads or connectors on the internal side, and also through flexible connectors to the station *isolated phase bus duct* on the external side. The other bushing connects the other internal connection ring at the other end of the circuit to the generator neutral. All bushings are located in the *leadbox,* which is usually underneath the generator frame. Cooling for the bushings can be one of three types: (1) water-cooled for the larger ratings (2) gas-cooled for the middle ratings or (3) conventionally or exter-

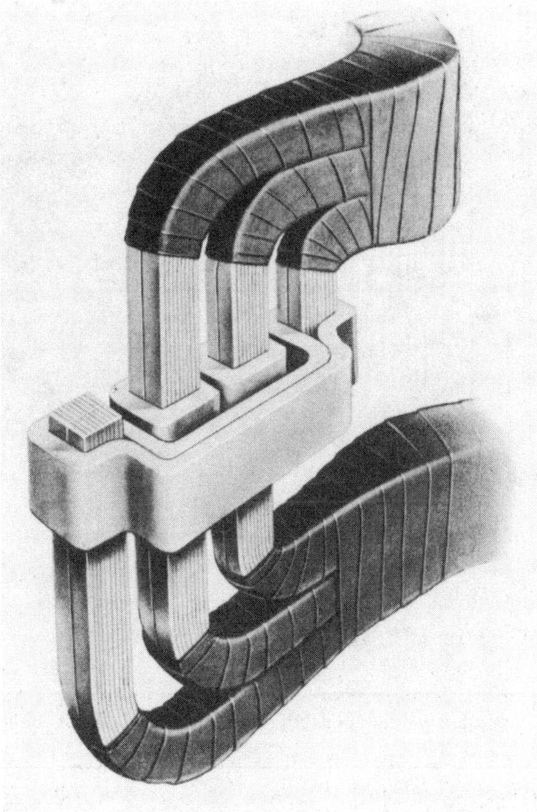

Figure 4.4 Group transposition of two serially connected bars. (Used by permission from Westinghouse Power Generation, Orlando, Florida.)

nally cooled for the lower ratings. Figure 4.6 shows a close-up of the generator leads and the bushings for a typical large generator. Figure 4.7 depicts a cut-away view of a two-pole generator.

The magnetic fields in the slot region and in the end region cause vibratory forces on the armature winding throughout the life of the generator. Also, on heavy-current transients, such as a sudden short circuit or synchronizing out of phase, very high momentary forces can be felt in both of these regions. The forces are higher in the slots, but the geometry lends itself to an effective support system. In the end region, where the geometry and voltages impose special requirements, bracing against these forces is a demanding design challenge.

Additionally, the large size of the armature current poses a severe heat removal requirement on the design of the windings of large machines and of the generator itself. The relatively high voltages involved pose an equally challenging

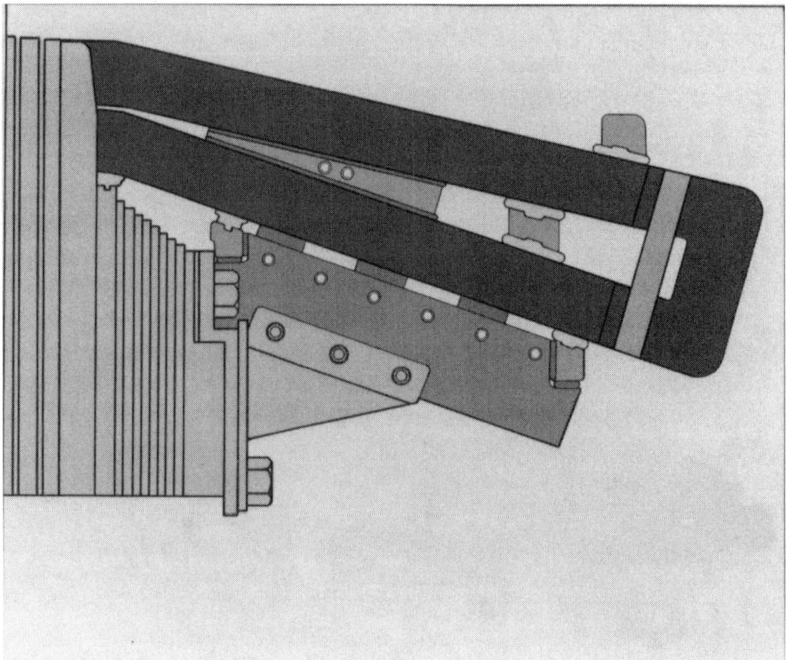

Figure 4.5 Armature winding end-turn region. (Used by permission from Westinghouse Power Generation, Orlando, Florida.)

insulation problem, which conflicts with the heat removal problem, because the more insulation used, the more difficult it becomes to remove the generated heat. Conductor cooling was introduced to aid in the heat removal for the large turbine generators. In fact, it was conductor cooling that made it possible to build generators of large ratings.

Thus, it should be no surprise that a large portion of serious problems in the stator winding is due to heat removal problems and to mechanical support failures. These are discussed in greater detail below.

4-1.1 Heat Removal

Coolant gases or liquids are used to extract the heat produced by the I^2R losses in the copper strands, as well core and rotational windage losses. In older and smaller generators, the heat is transferred by conduction through the copper and the groundwall insulation to the stator teeth and there removed through convection by a flow of cooling gas over the bars. In earlier days, the cooling medium was typically air.

But air is a relatively dense gas that creates significant windage losses at the surface of the rotor. While its electrical insulation capability is good, it is not con-

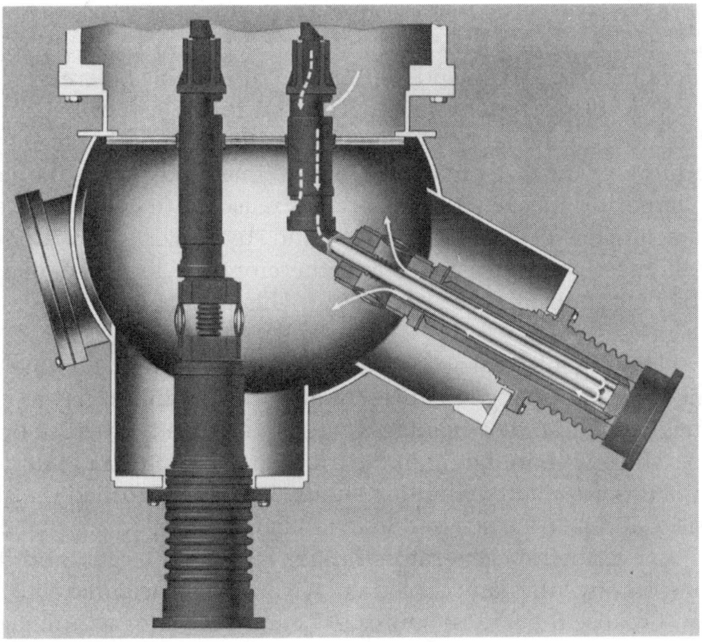

Figure 4.6 Generator leads and bushings. (Used by permission from Westinghouse Power Generation, Orlando, Florida.)

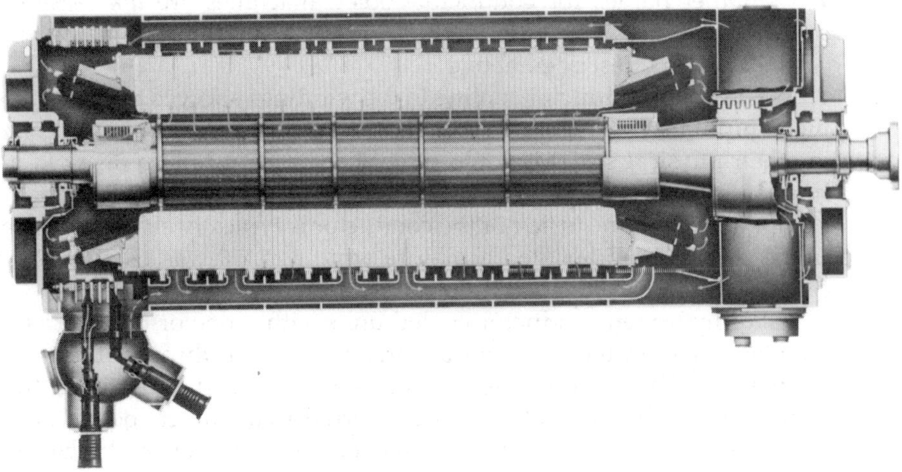

Figure 4.7 Cut-away view of two-pole generator. (Used by permission from Westinghouse Power Generation, Orlando, Florida.)

sidered to be excellent. Most importantly, however, it does not have a particularly high specific heat coefficient and is thus not an ideal cooling medium. As generators became larger and stator current densities increased, there clearly arose a need to use a better coolant. This led to the use of hydrogen gas, which was introduced in the late 1930s and the 1940s. It soon became the standard cooling medium for the large machines of the time. Today, it is the undisputed standard for large turbine generators for two reasons: first, it has the best heat transfer characteristics of any gas; and second, hydrogen's low atomic weight makes it the lightest gas in existence and therefore the lowest density of any stable gas (10% that of air at a purity of 98%). This results in the lowest windage losses of any gases.

Hydrogen's principal disadvantages are its high flammability when a significant amount of oxygen is present and the fact that it has to be sealed within the generator. These disadvantages were overcome by the use of high purity hydrogen, which will not ignite in the presence of a spark, and of a sealing system that maintains a seal at the rotating shafts. The latter is called the *gland seal oil auxiliary system,* and it will discussed in detail in Chapter 9.

Nevertheless, generators ratings still grew larger, and the use of hydrogen alone in conventionally cooled units was not sufficient to satisfactorily remove the heat. Because the largest source of heat is the stator winding, the next innovation was to cool the armature winding by passing the coolant in close or direct contact with the conductors where the heat is generated; that is, instead of relying solely on the groundwall insulation as the primary heat-flow path to the heat sink (the cooling gas for the entire machine), the coolant was forced to flow within the half-coil itself to directly cool the copper. Conductor cooling was introduced in the 1950s. Other names for conductor-cooled machines are *inner-cooled* and *direct-cooled* generators, as opposed to the *indirectly cooled* (another name for *conventionally cooled*) units described above.

The use of conductor cooling in the windings allowed manufacturers to build high output machines in the 1960s, 1970s, and 1980s without inordinately increasing their physical size, as would otherwise be expected, in order to maintain the heat losses and consequent temperatures at a tolerable level. The cooling medium pumped through the inside of the conductor is typically either hydrogen or water. In the earliest liquid-cooled armature windings, oil was used as the coolant because of its electrical insulating properties.

For hydrogen conductor-cooled units, either one or two stacks of *vent tubes* are placed within the bar itself to carry the coolant through the entire length of the bar. The heat is conducted to the vent tubes and is removed by the stream of cooling gas. These units are also referred to as direct-cooled, gas-cooled machines. Figure 4.8 shows a cross section of hydrogen conductor-cooled bars. The hydrogen is pumped by a shaft-mounted blower through each bar and is cooled by a set of coolers mounted inside the generator frame. There is no need

Figure 4.8 Cross section of hydrogen conductor-cooled bars. (Used by permission from Westinghouse Power Generation, Orlando, Florida.)

to keep the hydrogen isolated from anything else, because the entire generator is cooled with hydrogen. Therefore, the gas that flowed through the vent tubes of each bar will mix with the ambient gas at their outlet. The auxiliary system, which maintains the hydrogen at the proper pressure, temperature, purity, and humidity is the *hydrogen auxiliary system,* and it will be the subject of discussion of Chapter 8.

The superior attributes of water as a cooling medium, however, soon made it attractive for use as a coolant in conductor-cooled generators. Demineralized water is pumped through hollow strands interspersed throughout the bar cross section. Many older machines, however, have armature bars all of whose strands are hollow and carry water. With either type of conductor cooling, hydrogen is still used to cool the rest of the generator components that generate heat (e.g., the core, the rotor winding, and rotor surface windage). Figure 4.9 depicts a cross section of top and bottom water-cooled bars in core slots.

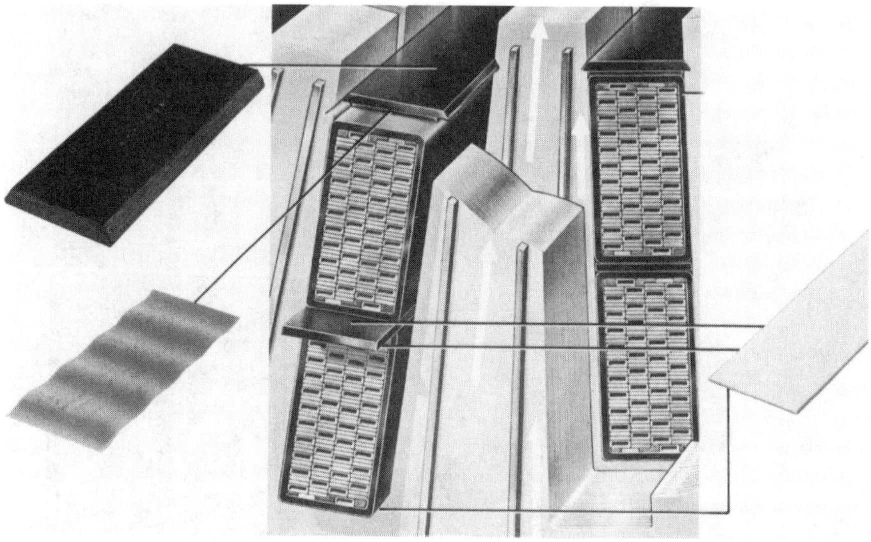

Figure 4.9 Cross section of water conductor-cooled bar. (Used by permission from Westinghouse Power Generation, Orlando, Florida.)

Water-cooled bars generally have water boxes and nipples on each end to carry the water into and out of the bar. Insulating hoses are used to conduct the water from the inlet water manifold to the inlet nipple, and from the outlet nipple to the outlet water manifold. Figure 4.10 shows a cross section of the ends of the water-cooled coils.

In water-cooled generators, water is pumped through either one or two bars before it exits to an external closed system for cooling, purifying, and re-pumping. The pressure of the water in the bars is generally lower than that of the hydrogen gas. This will cause the hydrogen to leak into the water system if any small ruptures occur, thus minimizing the likelihood of significant water leaks inside the frame. The external water supply system is called the *stator coil cooling water auxiliary system* (SCCW); it will be discussed in Chapter 10.

Most generator manufacturers will design a water-cooled unit to operate at a significantly reduced rating without any water flow. However, most will recommend that the water be removed from inside the bars and that they be vacuum dried if the unit is to operate under these conditions for an extended period of time. Because *no-water-flow* ratings vary among manufacturers, as well as their definition of extended operation, the reader is advised to consult with the manufacturer.

Some generators are designed with water-cooled connection rings, as well as water-cooled bushings. This is mainly true for generators with the largest ratings. Others have hydrogen inner-cooled bushings.

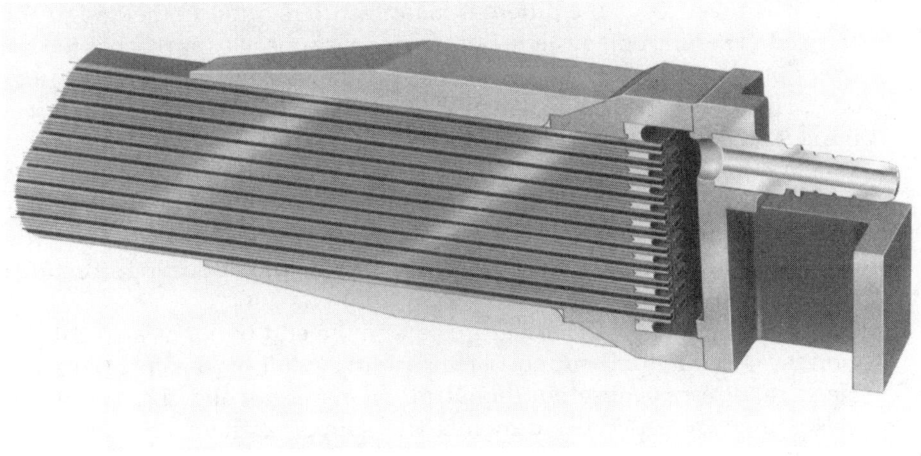

Figure 4.10 Cross section of the ends of the water-cooled stator bars. (Used by permission from Westinghouse Power Generation, Orlando, Florida.)

4-1.2 Winding Support Structure

Another critical design feature, as far as the reliable operation of a generator is concerned, is the mechanical support system for the bars. Mechanical support of the armature winding is critically important in the slot region and in the end region. The survival of the generator is dependent on the success of the winding supports.

The purpose of the armature winding support system is to

- Hold the winding in place
- Withstand normal vibratory forces
- Withstand forces during such high-current transients as sudden short circuit and out-of-phase synchronization
- Accommodate thermal motion resulting from load cycles

The winding support system, therefore, has to 1) prevent vibration damage to the winding; 2) prevent damage from sudden forces; and 3) prevent wearing of winding restraints, thus loosening of the winding from either the transient forces or thermal motion.

It is important that the support system remain tightly coupled throughout the life of the generator. Loosening of the support system will cause the natural vibrations of the generator to mechanically break the conductors or mechanically degrade the groundwall insulation. The mechanical forces themselves, however, will act to loosen the support system.

The practices of the different manufacturers differ in the design and materials used in the support system. But the purpose of each approach is the same. Relative vibration must be prevented because the survival of the generator depends on it. The conductor-cooled windings require the most effective support system. In the slot section, heightwise support is composed of *wedges* and *conforming material* and, in some cases, springs. Sidewise support consists of *packing* and, for large machines, *ripple* springs. These fill the heightwise and sidewise space in the slot and provide friction restraint and spring follow-up. Figure 4.11 shows a cross section of two bars, a top and a bottom one, in a slot with the supporting structure elements identified.

The support for the armature bars in the end region is also critical. This region lacks the natural support structure provided by the core slots themselves. Electrical stresses cause additional problems there, and the complex geometry makes the task especially difficult. The support system is composed of lateral blocks and ties, an outer bracing structure supported from the core, interlayer bracing and fillers, radial ties, conforming materials, antiabrasion materials, and bonding. Figure 4.12 shows an example of the support system for the end region of an armature winding.

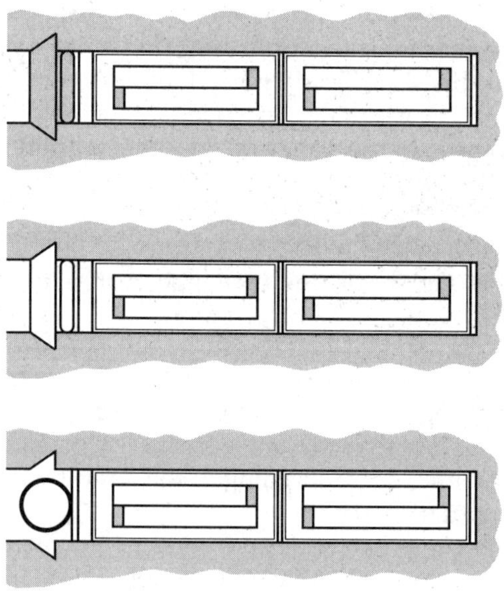

Figure 4.11 Cross section of a slot containing a top and a bottom bar. (Courtesy of General Electric Co.)

4.2 SENSORS AND MONITORS

The stator winding is a critical component whose failure accounts for the largest amount of generator unavailability and forced outage rate. Therefore, it is typically the most extensively instrumented component of the generator system. This does not mean that all generators need to be equally well instrumented, however. There are older, smaller generators that do not require the extensive sensing equipment employed by newer, larger machines.

There is a marked difference in the monitoring techniques employed for conductor-cooled versus conventionally cooled machines. Each of these will be described separately in the discussion of each particular sensor. Suffice it to say here that there is a need for more extensive monitoring of conductor-cooled armature windings. In particular, every cooling circuit within the armature bars may be monitored because each operates independently.

4-2.1 Slot RTDs for Indirectly Cooled Stator Windings

ANSI C50.10–1990 [ANSI, 1990] requires the use of at least six slot resistance temperature detectors (RTDs) in all medium and large machines with indirectly cooled stator windings. The standard recommends that resistance-type detectors be used, but allows the use of thermocouples. The location is intended to be "suitably distributed around the circumference, located between the coil sides and in positions along the length of the slots having normally the highest temperatures." The standard furthermore recommends that the detector be in contact with the upper and lower coils if possible, but at least with the upper coil.

These are the most important sensing elements for conventionally cooled stator windings. The RTDs are placed in an insulated liner in the slot, between the top and bottom bars, and they record the temperature of the groundwall insulation between the bars. Some manufacturers place one RTD in each slot, while others provide only one RTD per phase circuit. Whereas the slot RTDs can be effective in determining a number of bar overheating conditions, they suffer from the inherent lack of discrimination between problems in the top or bottom bars. In many cases, however, that is not a significant issue.

Additionally, slot RTDs are placed at only one point throughout the entire length of the bar. If the hotspot is localized and it doesn't coincide with the location of the RTD, it may never be detected. In an effort to address this problem, EPRI has sponsored the development of a continuous temperature monitor using fiber optics. With this monitor, the hotspot will be detected regardless of where it is. This system is described in Section 2.10.

Also, slot RTDs indicate a temperature that is substantially lower than the hottest copper temperature at the same axial location. The actual difference depends on the thickness of the groundwall insulation, the width of the slot, and the load that the machine is carrying. Therefore, the slot RTDs provide an indirect indication of what is happening within the ground insulation of the adjacent armature bars.

4-2.2. Outlet Gas Temperatures for Gas Conductor-Cooled Stator Windings

In gas conductor-cooled generators, the hydrogen is forced through the vent tubes within the bar itself, collecting the heat from within the conductor and consequently becoming hotter. The measurement of the temperature of the hydrogen gas as it exits the bar can be used to determine the amount of heat being generated within the coil.

It is common, although not universal, for the strands of gas conductor-cooled bars to be transposed using not only Roebel system in the slot sections, but also group transpositions. If one group carries an excessive current relative to the other groups in a circuit, overheating would occur in that group in every bar in the circuit.

Ideally, there would be one exit-gas sensor for every coil in the machine. However, as noted above, the construction of these bars may be such that each particular group of strands is kept electrically isolated from all other groups throughout the entire length of the phase circuit, that is, the groups are short-circuited only at the ends of the circuit itself, rather than at the end of each individual bar. This means that for some problems, such as breakage of a connection between bars (thus open-circuiting an entire group of strands throughout the entire circuit), the unbalanced heating condition of the conductors within the groups will affect the hot gas temperature of every bar in the circuit. In this case, the remaining healthy groups of strands would carry a higher-than-normal current density and thus cause the effluent gas to be at a higher temperature than that of the other circuits of all of the phases. Therefore, only one hot gas temperature sensor would be sufficient to detect such problems for each phase circuit. For this reason, most gas conductor-cooled generators have six hot gas sensors, one for each circuit. Some generator designs will have two parallel vent tube stacks per coil and are therefore fitted with one sensor for each tube stack in a circuit (total of 12).

The gas is monitored by directing the exhaust gas from one of the vent tubes in each stack of the monitored bar through a vent hose to a housing that contains the sensor. The gas is forced to flow over the sensor and exits at the other side of the housing to the generator's internal hydrogen environment.

The exhaust gas temperature is quite close to that of the cooper in the bars feeding the exit-gas sensor and provides a useful indication of the hottest temperature in the portion of the armature winding that it monitors. When compared to the slot RTDs, the cooling gas temperature is closer to that of the copper because the gas flows within the groundwall. RTDs are typically used for this purpose.

4-2.3 Slot RTDs for Gas Conductor-Cooled Stator Windings

While the ANSI C50.10–1990 standard [ANSI, 1990] does not require the use of slot RTDs in generators with conductor-cooled stator windings, many manufacturers choose to install them anyway. They have been helpful in finding certain problems, even in conductor-cooled armature windings. They basically perform the same function and are installed in the same fashion as those for machines with conventionally cooled windings.

4-2.4 Outlet Water Temperatures for Water-Cooled Windings

This concept of instrumentation is similar to that of gas conductor-cooled stators in that the temperature of the outlet water is measured as a way to detect any problems that may cause additional heat to be generated within the bar itself. It will also detect the effects of a water flow reduction because the water would experience a higher rise in temperature under such a condition.

The temperature of the water from every outlet hose is commonly monitored by a temperature detector, such as an RTD or thermocouple. These outlet water temperature monitors are usually located at the end of the hose that carries the water from the coils to the outlet water manifold. Some manufacturers will install one for each top coil as well as one for each bottom coil. Other manufacturers will install only one to measure the temperature of the effluent water coming from the top and bottom coils as it mixes together. In the latter case, the outlet water temperature as measured in the effluent hose is close to the average of the temperatures of the two bars from which the water flows.

It is useful to know that the temperature of the water in an armature bar is within a degree of the temperature of the copper strand through which it is flowing. Therefore, in a bar with mixed hollow and solid strands, the water temperature is very close to that of the hottest copper in the bar.

4-2.5 Slot RTDs for Water Inner-Cooled Stator Windings

"Embedded" slot RTDs are also typically supplied for water-cooled generators. In addition to being useful in identifying trouble in certain conditions, they can be used as in a conventionally cooled machine, when it is necessary to operate without water flow at reduced output.

4-2.6 Inlet and Outlet Bulk Coolant Temperatures in Stator Winding

There are a number of other sensors that may be used, depending on the type of stator winding cooling system. Many times, it is more significant to know the temperature rise of the coolant as it goes through the bar than it is the temperature itself. For this reason, the bulk cold (inlet) coolant temperature is almost always measured. By taking the difference between the coolant inlet temperature and the outlet temperature for each bar (or pair of bars) that is monitored, the temperature rise for that bar can be calculated. The bulk coolant inlet reading can also be used to indicate the effectiveness of the external cooling system for the coolant itself. A high reading of the coolant inlet temperature may indicate problems in the cooler and, if severe enough, may affect the system's cooling ability, resulting in possible conductor damage.

Also useful is the outlet bulk temperature of the water coolant. It can be used in detecting gross changes in the heat being generated, such as in overloads, or when the overall flow of coolant is reduced significantly. An abnormally high bulk outlet water temperature may indicate the presence of a problem with the coolers, the pump, or plugging of the filters in the external water treatment system. It also serves as a check for the average of all outlet coolant temperatures.

For water-cooled machines, the bulk coolant temperatures are typically taken at the inlet and outlet water manifolds. A thermocouple is generally used for each. For hydrogen conductor-cooled machines, the situation is somewhat more complicated. Each cooler section, and there are either two or four of these, will be monitored for the cold gas temperature. This represents the temperature of the hydrogen as it exits each of the cooler sections and is basically equal to the temperature of the gas as it enters the coils. For a machine with several coolers, an a verage temperature may be used for temperature rise calculations, even though the inlet to each cooler is measured. RTDs are generally used for this task, and there are typically either two or four of them, depending on the number of cooler sections. The temperature difference between the cold gas temperatures of the cooler sections must be maintained within a narrow band (2° C) in order to ensure consistent cooling of the generator internals.

The hot gas temperature is located at the entrance to the cooler sections. This temperature at the inlet to the cooler(s) commonly represents the average gas temperature from the entire machine, and not just from the armature winding.

4-2.7 Hydrogen-to-Water Leak Detector for Water-Cooled Stator Windings

As mentioned above, the hydrogen pressure in a water-cooled generator is commonly designed to be higher than the water pressure. Any cracks in the hollow, water-carrying conductors or other small leaks will tend to result in a hydrogen leakage into the water system, rather than a water leak inside the generator frame. The latter could lead to a serious failure. The types of failure that have been experienced include failure of the armature groundwall insulation because of permeation by water when the insulation is not water resistant. Polyester insulation, as used in some of the early thermoset insulation systems, is especially sensitive to this problem.

Monitoring the hydrogen leakage into the water system can serve to detect such cracks in the hollow conductors and other small leaks. There are various ways to detect this hydrogen leakage; these are covered in Chapter 10. For the purposes of this chapter, suffice to say that a sensor exists that can provide information as to the presence of a hydrogen leak into the water through cracks in the hollow strands.

4-2.8 Coolant Flow Measurement

The rate of coolant flow within the conductors is significant because the temperature of the copper will depend on these factors. For gas conductor-cooled stators, the flow through the armature bars is normally constant because it depends on the shaft-driven blower. However, there may be some blockage within the bars or in the gas circuit, which could impede the gas flow. There have also been occasions whereby a significant bypass to the gas circuit has reduced the flow through the useful circuits. This may be due to an omitted or malfunctioning baffle or removed access covers. In addition, on rare occasions, a blower may lose a blade, which will reduce its pumping ability and thus reduce the hydrogen flow. Nevertheless, flow monitors are rarely employed in gas conductor-cooled machines.

For water-cooled generators, the flow is generated through external pumps and directed through valves, piping, filters, and coolers into the generator and its armature winding. All of these factors combine to potentially make the water flow rate deviate from the intended value. Thus, the flow rate of water is generally measured in water-coded generators.

Some manufacturers employ direct measurement of the water flow, whereas others will measure the differential pressure of the water across the entire stator winding as an indirect measurement of the flow. However, this is not as accurate a measurement of flow because the resistance to flow has to be known, and that information is typically not available.

4-2.9 Temperature Monitoring Techniques

Monitoring all the temperatures in the stator winding is not an easy way to detect faults in the windings, because the temperatures themselves are subject to many variables that are not significant to determining the health of the unit. For example, the coolant outlet temperatures will rise and fall in direct proportion to armature currents and thus to the output MVA of the unit. But they will also vary with coolant flow and with the accuracy of the instrumentation. Therefore, it is important to develop algorithms that will discriminate between abnormal rises in the temperatures and normal ones.

The most commonly used algorithm is to compute the difference between the lowest and highest bar coolant outlet reading (called the *high-low differential*). When this difference reaches 8° C, an alarm should be recorded. However, this technique by itself can mask problems.

For example, assume the sensor that measures the outlet temperature of a faulty coil is calibrated such that it normally reads low in comparison with the other sensors—not the lowest, but barely above the lowest. If a problem arises within the bar that is monitored by this sensor, the temperature of the coil has to rise almost the full 8° C to even register a high-low differential alarm. As a result, it is possible that a full failure will occur before the full rise is registered. Additionally, the technique does not work well when the generator is operating at partial loads. In such cases, the temperatures may be lower, and the difference between the highest and lowest may be proportionally smaller. Yet, the problem may still be present. Lastly, false alarms can sometimes occur if the temperature of one sensor decreases instead of rising. Although this does not often indicate serious problems, it is misleading because the operators will typically look for a temperature rise, not a decline.

Nevertheless, the high-low difference algorithm has been successful on many occasions in detecting serious problems within stator bars.

Other algorithms have been developed to overcome these inadequacies. A popular one is to establish a historical record of the temperature, or the temperature rise, at several load levels, including selected loads and PFs. Thus, a profile for outlet coolant temperature reading relative to load level can be developed, and the actual temperature at any point in time can be matched against this profile. Any deviations within a previously determined error band from this profile can indicate a problem. The advantages of this technique are that the generator does not have to be at full load in order to detect problems and that one reading will not become buried among the others, such as may be the case in the high-low difference method. A disadvantage is that a profile has to be developed experimentally, and the stator bars must be healthy when the profile is being calculated. Another disadvantage is that when a sensing instrument is replaced or recalibrated, the profile for that particular bar may become invalid.

A second, and very similar, alternative is to calculate the deviation from the average of all readings. If the deviation for one particular coil reading is significant, that indicates a problem with that coil or its instrumentation. This technique makes the reading less sensitive to other parameters, such as load or coolant flow. Correction factors for the different sensing instruments may be necessary. They are generally easier to obtain than the aforementioned temperature/load profile.

One disadvantage of this method is that for gas conductor-cooled generators, the different hydrogen cooler sections will cool a particular part of the windings almost exclusively. If the temperatures of the cold gas for all the cooler sections are not balanced, then one side could have significantly higher outlet gas temperatures, which could result in a false alarm. Thus, is this technique is used, the coolers must be kept balanced within the guidelines determined by the manufacturer.

4.3 FAULT MONITORING AND DIAGNOSTICS

As mentioned before, there are many different abnormal conditions in the stator winding that can cause it to fail, often with serious consequences. This section discusses the various failure modes of a stator winding and how they can be detected. It also suggests courses of action that may be taken in order to minimize the damage. The suggested courses of action are general in nature, however, and may not be applicable to all generator designs. The reader is referred to the manufacturer for specific recommendations directed to the actual generator in question.

As with all diagnostic situations, the validity of sensors should not be unquestionably assumed. Uncalibrated or malfunctioning sensors can be the cause of a misdiagnosis and its attendant costs. The reader is advised to ensure, to the greatest degree possible, that all readings are accurate before any diagnosis is made.

4-3.1 Problem Area #1: Armature Winding

This section deals with problems that affect the armature winding as a whole, as opposed to those that affect individual bars. Problems with individual bars is the topic of discussion in the next section. The following topics are covered here:

- Generator overload
- General cooling system deficiency
- Current unbalance
- Damaged or broken connection ring

Generator Overload

When the generator is being operated above rated current levels, the result will typically include overheating of the entire stator winding. This means that operation is most likely outside the unit's capability curve. Because the heat losses in a conductor increase in proportion to the square of the current through it (I^2R), the high current levels produce more heat than the cooling system is designed to remove. Consequently, the temperature of the windings will rise accordingly. Thus, relatively small overcurrents can produce significantly higher amounts of excess heating. This is discussed in Diagnostic Box 4.1.

■ DIAGNOSTIC BOX 4.1 ■

PROBLEM: GENERATOR OVERLOAD

SENSOR INDICATION:

- **The electrical parameter meters,** such as the MW, MVA and phase currents, will detect the overload much sooner than will the temperature sensors. Nevertheless, it is the high temperature that does the damage, in most cases, not the current itself. The most direct way to detect the overload is to examine the operating point as it relates to the capability curve for the unit and for the hydrogen pressure at which it is operating.
- **The outlet coolant temperature indications** will be higher than normal for all sensors (assuming conductor-cooled stator). The algorithms described in Section 4.2.9 are designed to detect individual deviant bars and, as such, would not be sensitive to this problem. For water-cooled generators, the best sensor to use for overall heating of the armature winding is the bulk outlet coolant temperature sensor. The reading should be confirmed by the average of the sensors for the temperature of the outlet water from all of the armature bars. Because of possible fluctuations in the water flow rate the temperature rise should *not* be used exclusively. Note that water-cooled windings are normally kept at a temperature appreciably below the boiling point of water and, therefore, well below the temperature capability of the insulation. Thus, for these machines, temperatures because of overloads are typically not a problem, only the

consequence of boiling water, whereby the presence of steam can block the passage of sufficient water to keep the copper temperature at a reasonable level. A runaway condition can quickly develop under these circumstances. Because of the operating temperature margin, this should not happen unless extreme overcurrent or starved water supply should cause temperatures to reach the boiling point.

Because the flow of hydrogen gas is typically constant in a gas conductor-cooled unit, the average temperature rise of the gas across the coils should be used instead of the average of the available outlet gas temperatures from the armature bars. That will give a better indication of the true amount of heat being generated by the current.

Unlike for the water-cooled unit, however, the bulk outlet gas temperature should not be used in the same comparison as the absolute outlet gas temperatures. Because it includes gas that is cooling other parts of the generator, the bulk outlet gas temperature will typically be somewhat lower than the coil outlet gas temperatures.

For either type of generator, the important question is whether *all* of the temperature indications are high.

■ **The slot RTDs** will also register high absolute temperatures. This is the only way to detect the problem from a heating standpoint for conventionally cooled machines. They may also serve to confirm the problem for conductor-cooled generators. Once again, all the temperatures should be similarly high.

CORRECTIVE ACTION: If only one of the above indications is received, then the validity of the sensor should be verified. If the sensor is found to be valid, reduce load to normal level.

Cooling System Deficiency

Deficiency of the cooling system can also result in a general overheating of the entire stator winding. The excessively high temperatures discussed in Diagnostic Box 4.1 could be caused by the heat removal system not working properly, rather than excessive generation of heat as a result of an overload. This can be confirmed by normal readings of MW, MVA, or phase currents. The most likely characterization is insufficient coolant flow to properly remove the heat normally present.

The problem is especially likely in water-cooled machines, where the coolant is pumped externally through a system of piping, valves, coolers, filters, and tanks. Because of the typically constant flow of cooling gas driven by the shaft-mounted blower, this is less applicable to hydrogen conductor-cooled or conventionally cooled machines. However, the temperature indications could also be used to detect missing blades in the blower or other causes of hydrogen flow deficiencies, such as a bypass or short circuit of the cooling flow paths. These problems are covered in a later chapter, as are the various causes of impaired flow of water to the generator.

■ DIAGNOSTIC BOX 4.2 ■

PROBLEM: GENERAL COOLING SYSTEM DEFICIENCY

SENSOR INDICATION:

- **Temperature measurements** should be as in Diagnostic Box 4.1.
- **Electrical measurements, such as MW, MVA, and phase currents** as described in Diagnostic Box 4.1 should be within limits of the capability curve.
- **The water flow rate measurement** (if water-cooled) through the generator will read low for an analog sensor. The reading for an alarm switch would depend on the alarm setting.
- **The water differential pressure drop measurement** (if water-cooled) across the stator would indicate low for an analog sensor. The reading for an alarm point switch would depend on the alarm setting. Some generators have two alarm points, "low" and "very low."
- **The hydrogen gas pressure reading** may be low if the problem is due to low gas pressure.
- **The hydrogen gas purity reading** may be low if the problem is due to impure hydrogen. If the purity of the gas is so low that it significantly impairs its heat removal capability, this could indicate a more serious problem, such as explosiveness. This problem is covered in Chapter 8.

CORRECTIVE ACTION: If only one of the above indications is received, then the validity of the sensor should be verified. If the temperatures are above the manufacturer's recommended maximum, then the load should be reduced until the temperatures are brought within limits. This will help to ensure safe operation of the unit. Alternatively, the unit may be brought off line to correct the problem. Contact the manufacturer for specific recommendations.

Current Unbalance in the Power System

Current unbalance is generally not a serious problem for the stator winding, unless the unbalance is so severe that the temperature in the overloaded phase exceeds the temperature limit for the windings. Current unbalance is a much more serious problem for the rotor. This problem is covered from that standpoint in Chapter 5. There may be cases, however, where the current magnitude in one of the stator phases is sufficient to cause overheating of the coils in that phase, and this is the subject of this section.

Current unbalance causes negative-sequence currents to flow in the stator. Unless a negative-sequence relay is used, the unbalance may be difficult to detect. An algorithm to calculate the unbalance, implemented in the plant computer as an on-line function capable of alarm, is also capable of detecting an unbalance. This algorithm is described in Chapter 5. The current unbalance problem as it may affect the stator is described in Diagnostic Box 4.3.

■ **DIAGNOSTIC BOX 4.3** ■

PROBLEM: CURRENT UNBALANCE IN THE POWER SYSTEM

SENSOR INDICATION:

■ **The phase currents** will be significantly different, but the amount of unbalance can be deceiving. This can be calculated by hand, but it is best to implement this algorithm in the plant or other monitoring computer to continually calculate these values. See Chapter 5 for the description of this algorithm.

■ **The bulk coolant outlet temperatures** may not indicate anything out of the ordinary because the other two phases will be carrying less current, thus running cooler. Because the coolants from the various phases mix, the average temperature measured by this sensor would probably indicate normal temperatures.

■ **The bar outlet coolant temperatures**, whether gas or water conductor-cooled, will serve to indicate a current unbalance problem. In this case, all the readings in the same phase will be close to each other, but the average values for the different phases will deviate from each other significantly. The winding temperature algorithms discussed in Section 4.2.9 may indicate a problem because one of the phases will be running hotter for the load level, whereas another one will be running cooler than normal. Additionally, if the deviation from the average is used, that may also exceed the normal values. The high-low differential limit may also be exceeded.

■ **The slot RTDs** will show a similar pattern to that described for the coolant outlet temperatures.

CORRECTIVE ACTION: If only one of the above indications is received, then the validity of the sensor should be verified. The current unbalance must be corrected, or damage may be incurred on the overloaded phase (as well as on the rotor). If the unbalance cannot be corrected, then the load must be decreased until temperatures are brought within the normal range. See the corresponding sections in Chapters 5 and 12 for the impact of current unbalance on the generator rotor. Contact the manufacturer for more specific recommendations.

Damaged or Broken Phase Ring or Phase Ring Connection

The point of vulnerability addressed here may be a flexible connector between a connection ring and the lead to a bushing, or it may be a connection to an armature bar. These are more likely than a severe failure of a connection ring itself, although the latter would have the same effect.

If a connector, becomes damaged to the point that it severely limits or cuts off the current flow from its circuit to the connection ring, much of the phase current will be shunted to the other circuits in the same phase. The stator bars and the connection rings in the other circuits that pick up the extra current are generally not designed to carry that high of a current density and will severely overheat. For a one-circuit-per-phase armature winding, cutting off the current to an entire phase leads to very high negative-sequence currents, which were covered in the last section, as well as serious problems in the power system. When this has happened in the past, the arcing resulting from the continuation of the current as the connection breaks has evaporated major lengths of the connection ring and/or lead to the bushing.

This is a very serious problem even when there are two circuits per phase. The excess heat will build up quickly and cause rapid deterioration of the groundwall insulation, as well as melting of the copper itself, depending on the extent of the damage. In a water-cooled winding with more than one circuit per phase, boiling will probably occur, blocking off the flow of cooling water, leading to a runaway condition of very high temperatures. Time to failure for this problem is not long; action should be taken immediately.

If the lead actually breaks completely, arcing will result. The presence of the ionized gases that are a byproduct of the arcing will most likely cause other faults. This is particularly true in winding designs that employ uninsulated series, phase, or other connections. The probability is high that the consequence will be an immediate failure requiring a shutdown for repair.

This is a very difficult problem to deal with once it has begun. The best solution is prevention. Great care and evaluation are essential when the connections are made to assure that surfaces are properly prepared and bolts tightened. Expert examination of critical areas should be made during each maintenance outage, making every effort to detect it at an early stage and to repair it. As can be surmised from the preceding description, this is a potentially serious problem. Diagnostic Box 4.4 describes the problem.

■ DIAGNOSTIC BOX 4.4 ■

PROBLEM: DAMAGED OR BROKEN CONNECTORS TO CONNECTION RING

SENSOR INDICATION:

- **Outlet coolant temperatures** for conductor-cooled bars will be high for any parallel circuits that are not damaged. They will likely be lower for the circuit containing the damaged connector. The bars in the phases that are not affected by the damaged connector may be at or near normal temperatures. The algorithms discussed in Section 4.2.9 may detect this as a deviation.
- **Slot RTDs** will follow the same pattern as the outlet coolant indicators.

- **The radio frequency monitor** may detect the arcing caused by the damaged lead. However, this may not be sufficiently early to avoid damage.
- **The generator core/condition monitor** may detect the combustion residues in the gas from the unavoidable burning that will take place in the groundwall insulation. This, too, may not be sufficiently early to avoid damage.

CORRECTIVE ACTION: This is a difficult condition to detect and still have enough time to react to minimize damage. Nevertheless, assuming that it is detected, trip the unit from the system immediately. Reduction of load may solve the problem, but that is not likely unless the fault is in its very early stages. The first course of action is preferred. Contact the manufacturer for more specific recommendations for the unit in question.

4-3.2 Problem Area #2: Individual Bare Armature Bars

This section describes problems that relate to the individual bare armature bars. These problems are internal to the armature bar and are the type for which the algorithms described in Section 4-2.9 were designed. Following are the most common problems in this area:

- Strand short circuit
- Broken strands
- Cooling circuit blockage because of foreign objects
- Cooling circuit blockage because of gas leaks into water
- Stator cooling water leakage into the generator

All of these potential problems are covered in the following sections.

Strand Short Circuit

Short circuits between strands, between cooling ducts, or between combinations thereof, within the groundwall insulation of an armature bar are potentially serious problems. Strand short circuits are caused by the breakdown of the thin insulation that isolates the individual copper strands from one another. The cause of the short circuit is probably mechanical rubbing induced by the vibrations of the armature winding. These short circuits permit circulating currents that perform no useful work, yet generate excessive heating in the coils. The heating may take place at localized points, as well as throughout the length of the bar, causing degradation and ultimate failure of groundwall insulation. Additionally, local destruction of the bar may take place at the point where the circulating currents flow between the strands.

Unfortunately, a short circuit can be quite difficult to detect on-line. When the symptoms generated by this problem are great enough to be registered by sensors, they can resemble other problems that are more common. Therefore, a diagnosis of strand short circuit may be possible only through the process of elimination of other potential problems.

■ **DIAGNOSTIC BOX 4.5** ■

PROBLEM: SHORT CIRCUIT BETWEEN STRANDS

SENSOR INDICATION:

■ **The outlet coolant temperature** for the affected bar of a conductor-cooled winding will increase in proportion to the square of the circulating currents produced. Any one of the algorithms discussed in Section 4.2.9 should work if the circulating currents are high enough. The key is that only the temperature of the affected bar will deviate. In the case of a water conductor-cooled generator, the situation is straightforward because most water-cooled units have all their bars individually monitored. However, detection is somewhat more complicated for a gas conductor-cooled unit. If only one sensor is used per parallel circuit, and if the short circuit happens to involve the bar that is monitored, then the same discussion applies as to the water-cooled. However, if the short circuit does not involve the monitored bar, the problem will likely never be detected through this means.

■ **Slot RTDs** may be successful in detecting this problem if there is an instrument in the local area of the short circuit. Otherwise, they may not be helpful in this case.

■ **The radio frequency monitor** may detect any arcing that might take place, but the nature of the problem is such that the presence of arcing is not assured. Thus, RFM indication should be used only as supporting evidence and not as a necessary condition for the presence of the problem. In any case, it does not indicate where the problem is.

■ **The generator core/condition monitor** may detect any pyrolysis residue particulates in the cooling gas. As with the RFM, it should not be considered necessary evidence for the existence of the problem.

CORRECTIVE ACTION: If only one of the above indications is received, then the validity of the sensor should be verified. The damage can be minimized by reducing load on the machine until the temperature returns to within normal range, but a preferable course of action would be to bring the unit off line and repair the problem at once. Prompt action could minimize the extent of the damage and the extensiveness of the repair. The manufacturer should be contacted for specific recommendations pertaining to the unit in question.

Broken Strands

The problem of broken strands has many points of similarity to that of a broken connector to a connection ring, although it does not deteriorate as rapidly. If a strand or group of strands break, the current in that strand or strands will be reduced or stopped, depending on the amount of electrical contact remaining between the mating faces of the break. The current, therefore, will be redistributed among the remaining strands in the bar. Because the phase current is constant, depending on the generator load, the current density in the remaining active strands of the affected bar rises, giving off excess heat.

In addition to the overall heating of the affected bar or circuit, local heating because of arcing at the strand fractures can be intense. In a hydrogen-cooled generator, this heating can lead to hydrogen embrittlement of the affected copper, which cannot be observed without special mechanical tests. Unfortunately, the damaged strands are often repaired without taking this embrittlement into account, leaving unreliable brittle strands in place.

In gas conductor-cooled generators, which have the group transpositions connected only at the ends of each circuit, the break will cause higher current densities in the entire circuit. If more than one parallel circuits are involved, the increased resistance of the damaged bar may cause load current to be shunted to the healthy circuit, also causing its temperature to increase. Therefore, if there is an outlet gas sensor for at least one bar in that circuit, it will pick up the higher temperature, regardless of which bar is being monitored. In a water-cooled machine, however, because there are typically no group transpositions, the outlet water sensors have to be placed on every coil in order to detect the temperature rise. Fortunately, this is done by many manufacturers.

Additionally, the remaining unbroken strands may be subjected to additional stresses from the vibrations that caused the break in the first strand. This can cause the problem to progress rather quickly until a significant number of strands bread, resulting in a thermal runaway condition and, ultimately, failure of the unit.

■ DIAGNOSTIC BOX 4.6 ■

PROBLEM: BROKEN STRANDS

SENSOR INDICATION:

■ **Outlet coolant temperature measurements** may detect this condition if the algorithms are implemented correctly and if the number of broken strands is large enough to cause a significantly higher current density. The affected bar should deviate from all the others, as well as from its normal reading. By the time this method can detect a problem, it will probably have progressed so that a significant number of the strands are affected.

■ **Slot RTDs** will work if there is one in the slot of interest. Otherwise, they will not detect this condition in a water-cooled machine. In a gas conductor-cooled unit with group transposition, the entire circuit will have excess heat, so a slot RTD may be able to pick it up. Nevertheless, these instruments are less sensitive than the outlet coolant temperature sensors because the groundwall insulation acts as a thermal barrier. Additionally, the other coil in the slot may help average out the reading of the RTD.

- **The radio frequency monitor** will pick up the single spike sparking caused by the initial breakage and separation of the strand from its connection or the continuous arcing resulting from proximity between the broken ends. This is a good early indication, because even a single break will cause the initial sparking, whereas the temperature measurement requires additional strands to break before the temperature deviation registers. One way to determine the validity of the RFM reading is to vary the load. Because this is a current-induced RF emission, lowering the load should reduce the RF activity. If the RF level does not decrease, then it is probably detecting something else.
- **The generator core/condition monitor** will also detect pyrolysis residue particulates in the gas resulting from badly overheated insulation. This, however, may not happen until later in the degradation process, when enough strands have broken to cause severe overheating.
- **The hydrogen leak detector** is useful if the hollow strands in a water-cooled machine break, causing the hydrogen to leak into the water and partially block water flow. This break can be detected either through a direct measurement of the hydrogen pressure in the water tank (see Chapter 10), a measurement of the hydrogen gas vented to atmosphere, or by monitoring the frequency of operation of a relief valve in the water system that will open when the hydrogen pressure exceeds a certain value.

CORRECTIVE ACTION: Either remove the unit from the system, or, at the very least, reduce load so that the RF signal is eliminated from alarm level and/or the temperatures are reduced to a normal level. A prompt unit shutdown is more desirable because of the risk of a winding failure. Consult the manufacturer for specific recommendations for the unit in question.

Cooling Circuit Blockage Because of Foreign Substance

The problem of cooling circuit blockage addresses the integrity of the coolant passage for one particular bar or bars in the hydraulic circuit of an inlet or outlet water hose. The cause can be corrosion from within the bar itself or from some foreign substance in the water. It can even result from a blockage in the water circuit left during a maintenance outage. Regardless of the cause, this blockage can lead to excess heating, which can seriously damage the bar if the blockage is severe enough. Small blockages will probably not greatly affect the performance of the bar. Nevertheless, depending on the cause, the blockage may be a progressive problem that requires attention. This is more likely to happen in a water conductor-cooled machine as opposed to a gas conductor-cooled unit because foreign substances are less common with hydrogen cooling. Blockage has happened in gas conductor-cooled generators when the entire vent tube stack is left blocked with tape after a maintenance outage.

The blockage can be of two different types:

- Partial blockage of the bar
- Complete blockage of the bar

Although this separation may seem trivial and unnecessary, the indications given are considerably different and need to be discussed separately. This is done in Diagnostic Boxes 4.7 and 4.8.

DIAGNOSTIC BOX 4.7

PROBLEM: **PARTIAL BLOCKAGE OF WATER COOLANT CHANNEL**

SENSOR INDICATION:

- **Coolant outlet temperature measurements** will be higher than normal for the blocked bar(s) because the reduced water flowing will pick up heat with less water. A well-implemented algorithm should pick up the deviation.
- **Slot RTDs** may be able to detect the problem if it is in the bars in question, but they will be much less sensitive to the problem.
- **The generator core/condition monitor** will pick up residues if the situation has progressed to the point where insulation is pyrolyzing. In the presence of the temperature indications described above, the operator should realize that the situation has reached a dangerous point. Nevertheless, corrective action should be taken before this point is reached. Pyrolysis will not occur until the insulation is well above the boiling point of the "cooling" water.
- **The radio frequency monitor** should register normal readings. This is not an electrical fault, at least not initially. Thus, arcing would normally not be expected. Arcing could take place, however, if the high temperatures cause insulation damage or strand melting. This is more likely to happen at the later stages.
- **The hydrogen leak detector** into the water system will not detect anything for water-cooled machines if the blockage is due to a foreign substance.

CORRECTIVE ACTION: If only one of the indications is received, then the validity of the sensor should be verified. If valid, remove the unit from the system. Reduction in load will help, but may not be sufficient to avoid a complete failure. Contact the manufacturer for specific recommendations for the unit in question.

DIAGNOSTIC BOX 4.8

PROBLEM: **COMPLETE BLOCKAGE OF WATER COOLANT CHANNEL**

SENSOR INDICATION:

- **Coolant outlet temperature measurements** may *not* be higher than normal. In fact, they may read either the average outlet temperature or the ambient hydrogen gas temperature, both of which will be lower than the true water temperature in the blocked channel. A complete blockage will not allow any flow, thus thermally isolating the sensor that is typically located at the end of the effluent hose, a point quite remote from the source of the heat, which is in the starved bar itself.

If the blockage is near the inlet end of the hydraulic circuit, the following events may occur. The water in the manifold will fill the hose as the heat is transferred from the water trapped inside the coil, which may boil. As it boils, the steam exits through the outlet, driving the temperature read by the outlet water thermocouple to a high level. This will continue until the space is replenished by cooler water from the manifold, which may result in a normal temperature reading for a period of time before it also boils. This cycle may repeat continuously. Therefore, the most significant indication of this problem may be a cycling of the coolant outlet temperature. The same process may take place in a gas conductor-cooled bar, but the swings will probably not be as pronounced, and thus may be altogether unnoticed.

If the blockage is near the outlet end of the hydraulic circuit, the outlet temperature sensor will simply read the average water temperature, which is totally unrelated to what is happening within the starved armature bar. Thus, blockages near the outlet end of the bar may not be detected through this technique.

- **Slot RTDs** may be able to detect the problem if an RTD has been installed in the slot of the bar affected, but it will be much less sensitive to the problem than the coolant outlet temperature.
- **The generator core/condition monitor** will pick up pyrolizates if the situation has progressed to the point at which insulation is being pyrolyzed. In addition to the temperature indications described above, the presence of pyrolizates should alert the operator that the situation has approached a serious point. Nevertheless, corrective action should be taken before this point is reached; otherwise, it may be too late to prevent damage.
- **The radio frequency monitor** should register normal readings. This is not an electrical fault, at least not initially. Thus, arcing would not normally be expected. Arcing could take place, however, if the high temperatures cause insulation damage or strand melting. This may happen at the later stages and may indicate that the situation is reaching a serious stage.

CORRECTIVE ACTION: If only one of the above indications is received, then the validity of the sensor should be verified. If the indication is valid, remove the unit from the system. Reduction in load can help, but may not be sufficient to avoid a complete failure. Consult the manufacturer for specific recommendations pertinent to the unit in question.

Cooling Path Blockage Because of Gas Leak into Water

In water-cooled generators, the stator cooling water pressure is typically set a few lbf/in^2 below the hydrogen gas pressure. Thus, in the case of small breaks in the hollow strands, the hydrogen will tend to leak into the water, rather than vice versa. In such cases, the leakage of hydrogen gas into the water system may cause a partial blockage of the water flow in that hydraulic circuit. This may cause a phenomenon very similar to the one discussed in the previous section. Also, if the break in a hollow strand is severe enough, water will get into the groundwall insulation and depending on its chemistry, may cause sufficient deterioration to result in an electrical failure. Polyester binding resins are especially susceptible to damage from warm or hot water.

■ DIAGNOSTIC BOX 4.9 ■

PROBLEM: COOLANT PATH BLOCKAGE BECAUSE OF GAS LEAKAGE INTO THE WATER

SENSOR INDICATION:

■ **The hydrogen leak detector** will detect the hydrogen leakage into the water system, either through a direct measurement of the hydrogen pressure in the water tank (see Chapter 10), through a measurement of the hydrogen gas vented to atmosphere, or by monitoring the operation of a relief valve in the water system, which will open when the hydrogen pressure exceeds a certain value. This sensor is the key indication, and it separates this problem from blockages because of foreign objects discussed in the previous section.

■ **Abnormally high hydrogen consumption** may be a clue to this problem if the hydrogen leak is significant.

■ **Outlet coolant temperature measurements** may detect this condition if the algorithms are implemented correctly and if the blockage is large enough to cause a significantly higher temperature. The affected bar should deviate from all the others, as well as from its normal reading.

■ **Slot RTDs** will also work if there is one in the slot of interest. Otherwise, they will not detect the condition. Furthermore, these instruments are less sensitive than the outlet coolant temperature sensor because the groundwall insulation acts as a thermal barrier. Additionally, the other coil in the slot may help average out the reading of the RTD.

■ **The radio frequency monitor** should not detect any signals because this is not an electrical problem in its initial stages. If the situation is allowed to progress, then deteriorating insulation because of overheating may cause it to become an electrical problem, possibly causing RF signals at a later stage.

■ **The generator core/condition monitor** will also detect pyrolysis residue particulates in the gas resulting from badly overheating insulation. This, however, may not happen until later in the degradation process.

CORRECTIVE ACTION: If only one of the above indications is received, then the validity of the sensor should be verified. If the indication is valid, either remove the unit from the system, or, at the very least, reduce load to the no-water-flow capacity of the unit and remove water from the cooling circuit using a vacuum. A prompt unit shutdown is more desirable because of the risk of a winding failure. Consult the manufacturer for specific recommendations for the unit in question.

Stator Cooling Water Leakage into Generator

If a break in the stator cooling water path inside the generator is large enough, a significant amount of water may leak into the generator, regardless of the pressure differential between the water and the gas. The breaks can take place in different locations; these are discussed below.

One problem to be considered is a partial or complete break of a hose connection as it connects either to the armature bars or to the manifold. The hose may pull out of its fitting, or the fitting may fatigue. These hoses experience vibration throughout their useful lives because at one end, they are attached to the winding, which has a controlled level of vibration, whereas at the other end, they are held to the manifolds, which are typically quieter. On a heavy-current transient, the winding can see a significant motion briefly. This puts a dynamic loading on the hoses and the connections on both ends.

In addition, there may be a built-in strain if there is a mismatch between the configuration of the hose and the positions of the parts it connects. These strains, along with thermal motions, can place nonvibratory strains on the hoses and their connections. Thus, the hoses and the connecting hardware at both ends are subjected to a combination of alternating stresses and a unidirectional bias stress during their entire operating lives.

Another potential source of a leak is where the hollow strands are brazed into the strand headers at the ends of each bar. Fatigue breaks of hollow strands can also result in large enough leaks to cause water to leak into the generator. Any such water leaks would reduce or eliminate the water supply to the connected bars. They could also flood the generator with water, which may be sprayed over the end winding. If the machine's frame liquid detectors are not functioning properly, this leakage may go undetected for a period of time and lead to an electrical breakdown of the hydrogen bushings and to overheating of the affected armature bars. Water on the insulation may also cause unwanted electrical activity on the surface of the end winding.

■ DIAGNOSTIC BOX 4.10 ■

PROBLEM: STATOR COOLING WATER LEAKAGE INTO GENERATOR

SENSOR INDICATION:

■ **Liquid leak detectors** are likely to register an indication. This should be the main means of detection. These detectors are covered in Chapter 8.

■ **The hydrogen leak detector** may detect the leakage because some hydrogen is bound to leak into the water as well. See the description in the previous section.

■ **Abnormally high consumption** may be an indication of this problem if there is significant hydrogen leakage into the water.

■ **Outlet coolant temperature measurements** may detect this condition if the algorithms are implemented correctly and if the leak is large enough to cause a significant reduction in the flow of water. The affected bar should deviate from all the others, as well as from its normal reading. However, this indication may take relatively long to develop to the point that it noticeable.

■ **Slot RTDs** will also work if there is one in the slot of interest. Otherwise, they will not detect the condition. Furthermore, these instruments are less sensitive than the outlet coolant temperature sensors because the groundwall insulation acts as a thermal barrier. Additionally, the other coil in the slot may help average out the reading of the RTD.

■ **The radio frequency monitor** should not detect any signals because this is not an electrical problem in its initial stages. If the situation is allowed to progress, then deteriorating insulation because of overheating may cause it to become an electrical problem, thus possibly causing RF signals at a later stage.

■ **The generator core/condition monitor** will also detect pyrolysis residue particulates in the gas resulting from overheating insulation. This, however, may not happen until later in the degradation process and may be a sign of the severity of the problem.

CORRECTIVE ACTION: If only one of the above indications is received, then the validity of the sensor should be verified. If the indication is valid, remove the unit from the system. Consult the manufacturer for specific recommendations for the unit in question.

4-3.3 Problem Area #3: Armature Bar Ground Insulation

This area represents problems with the armature bar's ground insulation capability. These problems are not generally detectable through on-line monitoring, but are still significant enough to be included in the discussion of generator diagnostics. While the industry trend may be to avoid opening the generator periodically to inspect it, expert visual inspection is still the best means of early detection of insulation problems. The following problems will be discussed here:

- Internal damage and delamination of insulation
- Electrical failure of insulation
- Insulation Tracking
- Flashover
- Hose Tracking

For an excellent discussion on insulation of large rotating machines, see Culbert, [1989].

Internal Damage and Delamination of Insulation

The topic of concern here is a reduction in the integrity and insulating capability of the armature groundwall insulation, which would lead to its inability to hold voltage. This can be due to thermal/chemical, electrical, and/or mechanical degradation of the insulation material. These are stated in the order of increasing likelihood of occurrence, although in actuality, a combination of these factors may be worse than each acting alone. Insulation degradation can also be due to abrasion by foreign parts left erroneously on the winding, particularly in the end region.

Insulation (either groundwall or internal) of components directly cooled by water will normally not see significant thermal/chemical degradation over the machine's lifetime because the thermal capability of modern insulation systems is far beyond the operating temperature of the water. One potential exception to this is insulation that has not been properly cured in the manufacturing process.

In gas-cooled windings (either conventionally cooled or conductor-cooled), thermal/chemical degradation is not expected to happen in thermoset-resin insulation systems. It can, however, sometimes be a significant problem in older machines with asphalt-based insulation systems. Even with these machines, however, mechanically induced weakening or failure (this will be covered in the next section) is more likely than chemical degradation. Moisture within the insulation may degrade the groundwall insulation of water-cooled bars. The severity of this degradation depends on the chemistry of the insulation and, particularly, of the binding resins that give it its integrity, strength, and toughness. Insulation systems based on polyester resins have been particularly susceptible to warm or hot water.

Another cause of internal damage and delamination of the groundwall insulation is the mechanical degradation of the insulation. Cycling and normal load variations cause thermal expansion and contraction of the material. This is generally not a problem for modern thermoset insulation systems, but it has been a significant problem for relatively long generators with asphalt-based insulation in the armature bars. The first sign of this problem is normally a delamination just outside the ends of the core.

Insulation degradation can be detected only through expert inspection and selected electrical tests when the machine is shut down and disconnected from the system, including its transformers. A high-potential (hi-pot) test can also be made at a carefully selected voltage to determine the suitability for service of a questionable part. Nevertheless, the user should be aware that because it is a destructive type of test, high-potential testing should be used only after weighing all the risks involved, and perhaps having spare bars on hand. It is well to seek the advice of the generator's manufacturer as to how to make these tests and evaluations.

Because there is no way to dependably detect this problem from on-line sensors, there will not be a diagnostic box for this problem.

Electrical Failure of the Groundwall Insulation

Electrical failure of the groundwall insulation of the armature winding is the consequence of many of the problems discussed in the preceding section, if they are not corrected in time. The whole point of pursuing the detection of many of these problems is to be able to act before this electrical failure actually takes place, and thereby prevent it. Therefore, although it is currently difficult to detect and anticipate the deterioration of the groundwall directly, the incipient mechanism that leads to this failure mode may be detected, as are other abnormal conditions, using the techniques described in the previous sections. Nevertheless, electrical failure of the groundwall insulation merits a discussion by itself.

One possible way to detect this problem is with a partial discharge monitor. As described in Chapter 3, the partial discharge monitor is a passive device that is in operation at all times while the generator is in service. It detects signs of insulation degradation by analyzing small electrical discharges that occur within or adjacent to the groundwall when insulation is in the process of degradation.

Because the partial discharge phenomenon within the insulation will also emit RF signals, such activity may be detected by an RF monitor. The most highly stressed insulation in the generator is typically in the 5%–10% of coils nearest to the generator bushings, where the armature voltage is at its highest. Because the armature winding is characterized by high inductance, these high frequency emissions are heavily attenuated and may be undetectable by the time they reach the generator neutral. Because the RF monitor picks up its signal from the neutral, partial discharge activity is not always detectable through RF monitoring.

Both partial discharge and RF indications may also be caused by vibration sparking because of vibration of the bars in the slots.

Because generators generally are grounded through a high resistance or reactance, one point of failure to ground through the insulation will permit only small currents that may not do much damage. A second failure to ground, depending on its location and on the impedance of the fault path, will most likely result in heavy fault current that may cause significant damage to the involved armature bar, as well as to the core at the point of failure (if it is in the slot). Generator differential relaying is designed to protect against this type of fault, but it is effective only when more than one phase is involved in the insulation failure. This type of relay is very sensitive and designed to act almost instantaneously, recognizing the rapid escalation of damage during a heavy-current fault. See Chapter 12 for a more detailed discussion of differential relaying.

Nonetheless, the differential relay may not complete its function soon enough to avoid significant damage; that is, it is likely that at least some repair to the core at the point of failure would be required to restore the generator to health for continued reliable service. Differential relaying will 1) trip the generator from the system and 2) trip the excitation. The unit trip removes the armature current immediately. But the excitation system trip is followed by a relatively slow decay of the generator field current. The armature voltage induced by this decaying field current continues to drive the fault current through the points of failure. This may delay extinction of the fault current by as much as a second or more. While this may seem like a short interval, appreciable damage can occur during this time because of the very high level of current at the point of the fault.

■ DIAGNOSTIC BOX 4.11 ■

PROBLEM: ELECTRICAL FAILURE OF GROUNDWALL INSULATION

PERIODIC INSPECTION AND TEST PROGRAM:

Probably the most important and effective method of detection and prevention of failure of the armature groundwall insulation is through expert periodic inspection during planned maintenance outages. Subtle physical signs of degradation are sought by careful inspection of the entire winding in detail, throughout the end region and the slot section. In the slot section, inspection devices are used to examine the insulation through the cooling ducts of the stator core. In addition to mechanical, thermal, and chemical distress of the insulation, evidence of electrical tracking is sought on the surface of the insulation, especially in the end region.

Electrical tests that should be performed during periodic planned maintenance outages include 1) electrical resistance, 2) dissipation factor, 3) dissipation factor "tip-up," and perhaps 4) suitability-for-service high-potential test (which may reduce fault current and consequent damage, but is risky).

None of these test results are definitive in detecting or anticipating failure, but experts can use the information as an aid in assessing the health of the insulation. Especially significant are changes in the readings from one maintenance period to the next. Such things as operating history, including frequency of stops and starts, are also considered. Additionally, information on the performance of other machines with similar construction and operating history may be used in the assessment.

SENSOR INDICATION:

■ **The partial discharge monitor** discussed in Chapter 3 should be able to detect the partial discharges that take place within the groundwall insulation. Trends in this measurement are important in assessing the health of the insulation.

■ **The radio frequency monitor** may also detect the sparking, if any. If RF signals are detected, changing generator load is the first step in determining the source of the RF. If reducing load reduces the signal, the source is not in the insulation. But if the RF signal persists after the load is reduced to zero and the generator is disconnected from the system, then reduce voltage. If the RF signal is reduced by reducing armature voltage, the source is probably arcing external to or internal to the insulation. Normally, discharges within the insulation (partial discharges) will be so attenuated by the high inductance in the stator windings as to be difficult to detect through an RF monitor located at the neutral. This is why the more sensitive partial discharge detector is used. Visual inspection will be important in pursuing the findings of the electrical tests.

CORRECTIVE ACTION: Ideally, a deteriorating condition can be detected early enough to allow planning for correction of the problem, if more than superficial efforts are required. This planning would consider assembling required materials, including replacement bars. Making local patches to the insulation is to be avoided except in emergency situations, and then it should be recognized as being only temporary.

It would be well to seek the advice of the generator manufacturer in pursuing findings of groundwall deterioration and in bringing the generator back to an acceptable condition for continued reliable operation.

If the first evidence of trouble is when the insulation fails in service or during a maintenance hi-pot test, the machine should not be returned to service until the failed component is replaced or repaired. The failed component—a bar or connection ring, or lead, or bushing—should be removed and replaced. If a bottom bar has failed, a significant number of top bars will have to be removed to gain access to the failed bar. There is great risk of damaging the healthy bars in the process, and they may also have to be replaced in order to return the generator to as-new, or at least serviceable, condition.

Insulation Tracking

High electrical stresses can develop in parts of the armature winding, particularly when the voltage grading system has degraded or is not within the necessary range. The consequence of this on the end region of the armature winding is considered here. Higher than desired voltage stresses between adjacent bars of different phases can lead to local corona activity. In the presence of contamination, this can lead to carbonization of the material in the region, which can cause electric currents and overheating.

■ DIAGNOSTIC BOX 4.12 ■

PROBLEM: INSULATION TRACKING

PERIODIC INSPECTION AND TEST PROGRAM:

Probably the most important and effective method of detection and prevention of tracking is through expert periodic inspection during planned maintenance outages.

SENSOR INDICATION:

- **The partial discharge monitor** discussed in Chapter 3 should be able to detect the corona discharges by analyzing the partial discharges that take place within the groundwall insulation. Trends in this measurement are important in assessing the health of the insulation.
- **The radio frequency monitor** may detect the sparking.

CORRECTIVE ACTION: It is recommended that the advice of the manufacturer be immediately sought whenever tracking in the insulation is suspected, whether from on-line monitoring or from off-line inspections. Off-line inspections should be made periodically to increase the probability of finding this condition before it endangers the operation of the unit.

Flashover

In many gas conductor-cooled designs, the spacings at the ends of the bars are left bare in order to receive or discharge the cooling gas. The creepage distances introduced there can result in bypassing the insulating properties of the armature insulation system of gas conductor-cooled armature bars. Tracking and electrically conducting contamination in the creepage path can reduce the capability of the geometry to withstand and contain the armature voltage. Also, ionization of the hydrogen in the vicinity can lead to breakdown and a failure from the armature conductors to ground or to the conductors of the adjacent armature bars.

This has usually been the consequence of certain deteriorating conditions in this region. Therefore, monitoring and prevention should be directed at the cause. If, for example, there is a progression of a fracture of a bar-to-bar connection, perhaps of one of the groups of strands in the transposition system, monitoring for such things as RF emissions (RF monitor) or pyrolysis particles (gccm), as covered elsewhere in this book, may prevent a serious failure of this type.

When such a failure occurs, deterioration will proceed too quickly for operator response to prevent extensive damage (especially since he will probably have no additional evidence that something is wrong). If the failure includes currents to ground, the ground detection relay will trip the machine and remove excitation, but not quickly. If more than one phase becomes involved, differential relaying will limit the duration of the damaging current.

In the past few years, *clip caps* constructed from insulating material have been designed that prevent this problem by providing a measure of insulation to the ends while allowing the gas to freely enter and exit the stator bar.

Hose Tracking

The armature groundwall insulation system of a water conductor-cooled armature winding can also be bypassed by a cooling hose that has lost its ability to isolate the armature conductors electrically from ground. The hose can "track" on its inside surface and thus become a semiconductor that fails at armature voltage. Water chemistry can be the cause. Also, a stagnant or slow-flowing water column in the hose while the armature winding is at voltage can lead to increased temperature and tracking. This type of phenomenon will usually develop slowly over weeks, months, or even years. As currents build up in the tracking, it may be possible to detect the situation by a number of monitors.

A partial discharge detector may be able to detect these currents, although there may not yet be sufficient experience to substantiate them with this newly available device. An RF monitor may be able to detect the currents, but, again, there may not be enough experience with this technique either.

Measuring insulation resistance during an outage when the water has been completely removed will indicate a reduced value if this tracking has progressed far enough. This should be pursued during each periodic maintenance outage.

4-3.4 Problem Area #4: Armature Voltage Suppression System

The armature voltage gradient suppression system is designed to prevent destructive corona discharges and sparking in the generator by draining off electrical charges on the bar surface and thereby providing gradual changes in the surface voltage. It consists of a semiconductor or possibly high-capacitance elements. Two different designs of the grading system are used on each armature bar: one in the slot section and the other in the end region.

The electrical characteristics of the slot-section grading system are selected for the special needs of the slot section where the surface of the groundwall insulation is in contact with the grounded surfaces of its slot, at random points along core length; elsewhere, it is close to that grounded surface. The job of the voltage grading system in the slot is to keep the bar insulation at grounded potential over its entire surface, so there is no place at which there would be a voltage difference between the slot surfaces and the bar insulation.

The voltage grading system for the end region is of much higher impedance than that for the slot section and is designed for the situation where the bars emerge from the grounded core. Without grading, they would experience a very sharp change in voltage along the surface of the insulated bars. Without voltage grading, this would be an area of very intense electrical surface activity.

If the impedance of the grading system is too high, it will not do its job adequately, and the electrical activity at the surface of the insulation will have its destructive effect in the slot or in the ends, whichever system deviates from the specified characteristics.

If the impedance of the voltage grading system is too low, the electric current in it will experience extra losses, resulting in higher than desired operating temperature. Overheating the voltage gradient suppression system will cause damage to the bonding of the semiconducting particles. Excessively high operating temperatures of the grading system can cause the binding resins to lose their mechanical integrity; thus, the electrical impedance will change until the grading system will not do the job for which it was intended, resulting in damage to the groundwall in the vicinity of the trouble.

Mechanical abrasion, because of the relative motion of components in contact with each other, can remove portions of the suppression system. The reduction in effectiveness of this system because of abrasion and overheating will lead to increased voltage gradients and therefore corona discharges and sparking at the surfaces of the insulated bars. This will reduce the life of the groundwall insulation. Diagnostic Box 4.13 describes the detection of this problem.

■ DIAGNOSTIC BOX 4.13 ■

PROBLEM: OVERHEATING AND ABRASION OF VOLTAGE GRADIENT SUPPRESSION SYSTEM

PERIODIC INSPECTION AND TEST PROGRAM:

Probably the most important and effective method of detection and prevention of voltage grading system failure is through expert periodic inspection during planned maintenance outages. Subtle physical signs of degradation are sought by careful inspection of the winding in detail, with particular attention to the slot section and the end region near the ends of the core. In the slot section, inspection devices are used to examine the insulation through the cooling ducts of the stator core, looking for evidence of vibration sparking, the most serious consequence of in-slot electrical surface activity.

SENSOR INDICATION:

■ **The partial discharge monitor** discussed in Section 3.5 will generally detect the surface corona discharges. Comparison with previous readings should determine if there is a trend toward higher readings and therefore increased degradation.

■ **The radio frequency monitor** will probably detect sparking in the slot section. The sparking in this case will be a voltage-driven phenomenon. Thus, the RF signal level will not change with changes in load. However, changes in voltage will cause changes in the signal. Comparison with previous readings should determine if there is a trend toward higher readings and therefore increased degradation.

CORRECTIVE ACTION: Confirmation of the problem has to be done by visual inspection. The damaged system should be repaired at the first possible moment. It is generally not necessary, however, to force an outage to carry out this repair. Consult the manufacturer for specific recommendations pertinent to the unit in question.

4-3.5 Problem Area #5: Armature Winding Slot Section Support System

Alternating forces constantly act on the armature bars throughout their lengths while the generator is in service. Heavy-current transients, however, cause forces many times as high as these normal forces, but only for brief periods of time. Conductor-cooled windings, with higher current densities, experience much higher force densities than the conventionally-cooled windings. Therefore, while all windings require a suitable support system in the slots and in the end regions, the conductor-cooled windings require much more effective restraints.

The failure of the armature winding support system becomes the mechanism for other modes of failure, which have greater consequence for the generator than the damage done in the process to the support system itself. In general, the looser the support system, the more serious the influence that the vibrations, the transient forces, and the thermal motions will have on the integrity of the winding. Such looseness may be the direct cause of problems such as strand breakage and insulation erosion. Thus, whereas the electrical failure of the groundwall insulation may be the ultimate manifestation of many problems, the degradation of the winding support system, both in the slot section and in the end-turn region, may be the root cause of many of these failures. Experience has shown that a conductor-cooled armature winding, with its high electromagnetic forces, cannot survive when its support system permits vibratory motion relative to the components adjacent to it. In the slots, no bar motion can be permitted. In the end region, the bars must not move relative to their blocking and supports. It stands to reason that if looseness can be detected either on-line or during periodic maintenance outages, then a host of other seemingly unrelated problems can be averted.

The two problems pursued at this point are

- Bar vibration
- Vibration sparking

Bar Vibration in Slot Section

Bar vibration within the slot section is typically associated with the heightwise and sidewise clearances. If the related packing for these two areas is not tightly set, or if it degrades, the resulting clearances will allow for vibratory motion. But, for conductor-cooled windings, avoiding clearances is not enough. Sufficient sidewise pressure is needed in these windings to provide additional frictional restraint against the heightwise vibratory forces on the bars. Once bar vibrations are allowed to begin, they will progressively worsen, although often rather slowly.

These bar vibrations are destructive forces on the bar and its groundwall insulation. Conventionally cooled bars are less affected by these forces because of the lower level of the related forces. Such is not the case for conductor-cooled bars, however, on which the vibratory forces may be three to six times higher because of their larger current density. If the vibrations are left uncorrected, the damage will manifest itself in such problems as cracked strands, blocked cooling channels, and groundwall insulation erosion.

There is no way to detect slot-section bar vibration on-line, other than placing vibration monitors on the bars themselves. However, this is not normally done because of difficulty in installation within the slot section and because any such detector would of necessity be local, which means that it would detect vibration only in its immediate vicinity.

Vibration Sparking

Vibration sparking is one of the manifestations of bar vibration in the slot section. If there is a looseness in the slot, or insufficient restraint, so that the bars can vibrate, they certainly will vibrate under the large alternating forces imposed in the slot section. The bar vibrations in the slot section may cause the voltage grading system to wear and be gradually removed. When this happens, the vibrations may also cause a voltage to appear on the surface of the insulation because of capacitive coupling with the copper of the bar. Each time the bar makes or breaks contact with the grounded slot surface, a spark will occur as the surface of the insulation is forced back to the electrical potential of ground. These sparks will occur at each point of intermittent contact 120 times a second. Thus, even though only a very small amount of erosion or material removal would occur on each spark, severe damage can and would take place over a period of time at this rate of reoccurrence.

Vibration sparking can lead to the gradual machining away of the strand insulation and, changes in the electrical and mechanical characteristics of the groundwall insulation. It can also change the electrical characteristics of the voltage suppression system. All of these occurrences can lead to electrical failure of the bar and result in an electrical fault within the generator. When vibration sparking occurs, it is frequently found in many places throughout the armature winding, probably because the same condition of the restraints has developed in many slots. It would be expected, therefore, that a complete armature rewind would be required if vibration sparking has had a chance to inflict serious damage on many of the armature bars. This problem has been more widespread in hydraulic turbine-driven generators, but the results of vibration sparking have been found with turbine generators as well.

■ DIAGNOSTIC BOX 4.14 ■

PROBLEM: VIBRATION SPARKING

PERIODIC INSPECTION AND TEST PROGRAM:

Probably the most important and effective method of detection and prevention of armature groundwall insulation failure by vibration sparking is through expert periodic inspection during planned maintenance outages. Subtle physical signs of degradation are sought by careful inspection of the winding in detail in the slot section. Devices are used to examine the insulation through the cooling ducts of the stator core, looking for residue and other evidence of vibration sparking.

SENSOR INDICATION:

- **The radio frequency monitor** will pick up RF signals generated by vibration sparking.
- **The partial discharge monitor** will detect any partial discharges emitted.
- **A frame vibration monitor,** mounted on the stator frame, will detect vibrations if the problem is sufficiently widespread. This will be more effective for a four-pole generator that does not have spring mounting of the core.

This vibration is load sensitive. Reduce the load and observe whether the frame vibration is reduced.

For normal healthy generator operation, twice-frequency frame vibrations will be sensitive to generated voltage and will change only a little with changes in load. Vibrations may actually increase as the load is reduced to zero.

CORRECTIVE ACTION: If the problem is suspected, the machine should be thoroughly inspected and tested during the very earliest scheduled maintenance outage. Also the problem can probably be detected by the dissipation factor, although a high dissipation factor will not necessarily identify vibration sparking as the problem. Inspection by an individual trained in detecting this problem should be undertaken as the primary path to identifying and then correcting this problem. The things to look for are wear deposits and erosion as well as degradation of the insulation. If continued operation is required without correction of the problem, the loads should be reduced to slow down the progress of the deterioration. The problem should be repaired at the earliest opportunity. Contact the manufacturer for specific recommendations pertinent to the unit in question.

4-3.6 Problem Area #6: Armature End-Winding Support Region

Like slot-section vibrations, end-winding vibrations are associated with looseness of restraints. The magnitudes of these vibrations progress relatively slowly. They can be affected by the vibrations of the bars within the slot section as well. Nevertheless, end-turn vibrations are the root cause of various other problems within the stator winding system. Therefore, end-winding vibration should be corrected at the earliest opportunity after detection.

Relative motion between an armature bar and the blocking adjacent to it will cause wear of both the block and the insulation, causing additional looseness, which will be followed by greater relative vibratory motion and more wear. This is an accumulating phenomenon. Abrasion has been known to lead to insulation failure of conductor-cooled windings where the winding forces are much greater than for conventionally cooled bars.

Whereas the force densities in the end region are about one-third of those in the slots, the geometry and voltages in the end regions impose special requirements. Bracing against these forces is a demanding design challenge. As a result, the end winding deserves special attention in regard to the effects of vibration and consequent mechanical degradation of the armature winding and its support system.

When the bar end-region vibration gets severe enough and affects enough of the bars in the armature winding, it may cause sufficient vibration of the stator frame to be picked up by a frame vibration sensor (or even heard by the human ear in the power station near the frame). This is more likely in a four-pole generator where the core is not generally spring-mounted. If the bars in all three phases are vibrating, a six-times frequency is felt by the frame; that is, in addition to the expected 120-Hz vibration of a 60-Hz generator, a 360-Hz vibration of the stator frame has been detected. The cause of this six-times frequency is that the bars in the three phases are "bouncing" out of phase with one another. Cracked strands are typically a result of this vibratory motion in the end regions of the armature bars.

Monitoring the vibrations in the end-turn region is somewhat easier, and monitors exist that can do this. The end-turn vibration monitoring system, for instance, is covered in detail in Chapter 3. Impact hammer tests may be necessary prior to installation of this monitor to determine the resonant frequencies.

■ DIAGNOSTIC BOX 4.15 ■

PROBLEM: END-TURN VIBRATIONS

PERIODIC INSPECTION AND TEST PROGRAM:

Probably the most important and effective method of detection and prevention of armature end-winding vibration sparking is expert periodic inspection during planned maintenance outages. Subtle physical signs of motion and degradation are sought by careful inspection of the end winding in detail. It is difficult to be specific because the different designs of the various manufacturers require attention to different details and different evidence of motion.

SENSOR INDICATION:

■ **The end-turn vibration monitor** should detect the problem directly, as well as provide an indication of the severity of the problem. The severity is directly proportional to the magnitude of the vibrations.

CORRECTIVE ACTION: Inspect the end turns during maintenance outage, and repair any looseness, greasing, or damage at the earliest opportunity. If continued operation is required without correction, then reduction in load may be necessary in order to decrease the vibration magnitude. This will slow down the damage being cause by the vibratory forces. Consult the manufacturer for specific recommendations pertaining to the unit in question.

4-3.7 Problem Area #7: Armature Leads and Bushings

The following abnormal conditions in armature leads and bushings can arise that merit discussion here:

- Fracture of flexible lead connectors
- Reduction of coolant flow to bushings

Other failure modes that have been known to occur are loss of bushing sealant and subsequent thermal failure, and arcing from the bushing cap to the stator frame. However, as these are quite rare, they will not be covered in this section. Loss of bushing sealant is alluded to in Diagnostic Box 4.17, but it is not covered as a separate problem. The reader is referred to the manufacturer for further information.

Fracture of Flexible Lead Connectors to Bushing

There needs to be a point in the circuitry of the armature winding at which mechanical isolation is provided between the winding and the bushings. The winding is susceptible to constant vibration throughout its life and to thermal expansions and contractions. The bushings are solidly mounted to the frame. They may not, however, be absolutely stationary if there are vibration and thermal motions in the frame, but these motions will be unrelated to those of the armature winding. It is common to provide a flexible connection between the connection rings and the lower leads that go to the bushings, where there is a concentration of the effects of relative motions.

As with any conductor, the flexible lead connectors are susceptible to breakage because of the heat and vibrations to which they are subjected. Their flexibility reduces, but does not eliminate, this problem. Local overheating can occur at the point of fracture, and if severe enough, copper metal can be vaporized, thus further reducing the current-carrying capability of the connector. If left unmitigated, the situation can result in a complete breakage of the connector with its consequence of catastrophic electrical failure to ground.

DIAGNOSTIC BOX 4.16

PROBLEM: FRACTURE OF FLEXIBLE LEAD CONNECTOR

SENSOR INDICATIONS:

- **The radio frequency monitor** would detect RF signals because there would probably be some arcing. The arcing would result from the breaking and separation of the parts of the leads. The arcing may be continuous in nature, or it may be intermittent. In any case, because this is a current-driven type of arcing the magnitude of the RF signal will vary proportionally to changes in the generator load current.

CORRECTIVE ACTION: Fractured leads can be confirmed only through a visual inspection while the unit is off-line. This is a serious problem. When the problem is suspected, the unit should be removed from service immediately. A reduction in load will mitigate the situation somewhat by slowing the rate of fracture, but it may not be sufficient to eliminate the problem until the next planned outage. Consult the manufacturer for specific recommendations pertinent to the unit in question.

Bushing Coolant Flow Reduction

There are three methods for cooling the bushings that transmit armature current through the hydrogen containment to the world outside.

For machines with relatively low current per phase, conventional cooling is used, whereby the heat generated within the bushing is "washed off" the outer surface of the bushing by the hydrogen gas environment around it on the inside of the generator and by the air surrounding the bushing outside the generator. This is commonly done by way of natural convection. Some generators with conductor-cooled armature windings have conventionally-cooled bushings.

For generators with an intermediate level of armature phase current, gas conductor-cooled bushings are used. Each bushing is forced-cooled over its entire length by the passage of hydrogen within its cylindrical copper conductor. This method is used for the bushings of generators with conductor-cooled armature windings that are either hydrogen-cooled or water-cooled.

For generators with high armature phase current, liquid-cooled bushings are used for the most effective means of heat removal. Here, water is passed within the cylindrical conductor of the bushing to cool its entire length. The water is supplied by the same external water treatment system that is used for the armature winding itself.

Loss or reduction of coolant flow to the bushings can result in their overheating and subsequent failure. The reduction in coolant flow can be due to an overall coolant deficiency within the entire generator, in which case other parts of the unit will exhibit severe symptoms. Other causes can be leakage or blockage of coolant, which can be determined only through off-line inspection. In such cases, it is not so important to know what the cause of the problem is, but rather, that the unit has to be removed from service and internally inspected. Nevertheless, Diagnostic Box 4.17, describing this situation, is included in this section.

■ **DIAGNOSTIC BOX 4.17** ■

PROBLEM: LOSS OF COOLANT FLOW TO THE BUSHINGS

PERIODIC INSPECTION AND TEST PROGRAM:

An external visual inspection of the area underneath the lead box, where deposits of leaking sealant will accumulate, can provide indication of this problem.

SENSOR INDICATION:

- **Bushing temperature sensors** for either the water or the gas passed through the inside of the bushings can indicate the presence of this problem. The indication would be of higher temperatures than normal for the current load levels.
- **General temperature sensors** for the stator winding may also indicate abnormalities if the problem is a general deficiency in the flow of water or gas.
- **Liquid leak detectors** underneath the lead box can serve to identify a water leak in the bushings.

CORRECTIVE ACTION: The unit should be removed from service and the problem corrected. Alternatively, a reduction in load may be sufficient to maintain the temperatures within a tolerable range. However, with no direct measuring device, there is no way of knowing whether or not that is the case. Consult the manufacturer for specific recommendations pertinent to the unit in question. If the problem needs to be corrected by replacing a bushing, a spare bushing should be on hand.

SUMMARY

This chapter has described the stator windings and the various means of cooling them. It also included a description of the various sensors that can be used to detect problems within them. Lastly, some of the problems that can affect the operation of this important component have been described in terms of the sensor readings that can identify the problem, as well as what course of action to follow when the condition is identified.

REFERENCES

[ANSI, 1990] "American National Standard for Rotating Electrical Machinery—Synchronous Machines," American National Standard Institute Standard C50.10–1990, New York, 1990.

[Culbert, 1989] CULBERT, I. M., H. DHIRANI, AND G. C. STONE, "Insulation Condition of Large Rotating Machines," Chapter 16, *Power Plant Electrical Reference Series*, Electric Power Research Institute, 1989.

CHAPTER 5

The Generator Rotor

The rotor is the second major component of the generator. It is the rotational element that serves as the interface between the electrical output of the generator and its mechanical input derived from the prime mover.

More specifically, the purpose of the rotor is to provide the rotating magnetic flux field (the *excitation*) required to generate the needed voltages in the stator winding. For very small generators, such as pilot exciters, a rotating permanent magnet is usually sufficient to create the magnetic field. However, this is not possible for larger generators because of the large magnetic fluxes required from the rotor. To achieve this with permanent magnets would require unacceptably large magnets. The only way to achieve such high fluxes, therefore, is to pass a (relatively) large current through a conductor wound repeatedly around a magnetic core. By rotating this conductor, we can create a rotating magnetic field. This rotating winding is called the *field winding*, or the *rotor field*.

Moreover, being able to control the magnetic field in the generator is important in the successful operation of the electric power system of which the generator is a part. Perhaps this can be emphasized by stating that there are two main means of controlling the operation of a power system through the generator: 1) by adjusting the magnetic flux generated in the rotor field of each of the generators to control the terminal voltage and reactive power, and 2) by controlling the valve position of each of the turbines to control the real power.

In the first case, such control cannot be carried out with a permanent magnet because the flux generated by the magnet is constant and cannot be varied. This control function thus requires varying the current in the excitation (field) winding as a means of controlling the flux.

5.1 DESCRIPTION OF THE GENERATOR ROTOR

Compared to the armature winding discussed in the preceding chapter, the field winding carries direct current at relatively low driving voltages. Therefore, the electrical duty on the field winding is not as severe as that of the armature winding. However, because of the space constraints and the mechanical duty on all of the rotor components, there are some significant challenges in accommodating the electrical demands of the field winding design.

Very severe mechanical demands are made on the field winding design and on the rotor forging and the other mechanical components that support the winding. Thermal expansion and centrifugal forces are the two principal concerns. The centrifugal forces on the surface of the rotating element can exceed 6000 g's. The burden of the rotational duty falls upon the *rotor forging* and the *retaining rings,* as well as on some other less important mechanical components. Therefore, the description of the rotor will be divided into three sections.

- Rotor forging
- Field winding
- Retaining rings

5-1.1 The Rotor Forging and Mechanical Components

The function of the rotor forging is to supply drive power to the generator. This is the rotational element required to produce electricity. The power comes from the prime mover, namely, the steam or gas turbine. As such, the forging must be able to withstand the normal speed and a momentary overspeed, as well as a normal, constant load torque applied by the stator magnetic field and any transient torques induced by electrical disturbances in the power system. Lastly, the rotor forging carries the field winding and provides a magnetic path for the useful flux in the machine.

The rotor is made up of one or more forgings, prepared from steel of carefully selected metallurgy and forged to precise specifications for high quality and performance. For a small generator, one forging may be all that is needed. Separate shrunk-on couplings are frequently needed, and these are also made of high-performance materials. Some rotors have been assembled from a few major forgings and held together by carefully designed means.

The term "forging" will apply here to the assembly of forged components that have a large diameter rotor body and a relatively slender spindle at each end. Its key properties are

- Ultimate strength
- Yield strength
- Ductibility and toughness
- Ability to withstand high-cycle fatigue
- Ability to withstand low-cycle fatigue
- Magnetic properties

The rotor has radial slots (called the *rotor* or *field slots*) milled out parallel to its axis to accept the field conductors. The section of the forging in which the slots are located is called the *rotor body*. The *pole faces* in many rotors are smooth surfaces in the body through which the magnetic flux flows across the air gap. Axial slots may be machined in the pole face to carry amortisseur conductors.

Cross slots at frequent intervals are cut into the pole faces of a two-pole rotor in the transverse (or circumferential) direction to equalize the stiffness in all directions and thus reduce double-frequency vibrations. This can be clearly seen in Figure 5.1, which shows a machined generator forging.

Some manufacturers have used axial slots in the poles to equalize the bending stiffness in the two perpendicular axes of their two-pole rotors. These slots have been filled with magnetic material to carry the flux. Unfortunately, cracking of the forging material in the vicinity of the ends of the wedges in these slots has occurred in some generators, with dire consequences.

Because mechanical power is transferred to the generator from the prime mover through the rotor, the rotor forging must also accommodate the prime mover coupling on one end and the rotating exciter coupling on the other (if one is used). The end of the rotor shaft that is connected to the prime mover is referred to as the *turbine end*, or the *drive end*. The end of the rotor shaft that is coupled to the rotating exciter, or the point at which the collector assembly for connection to the

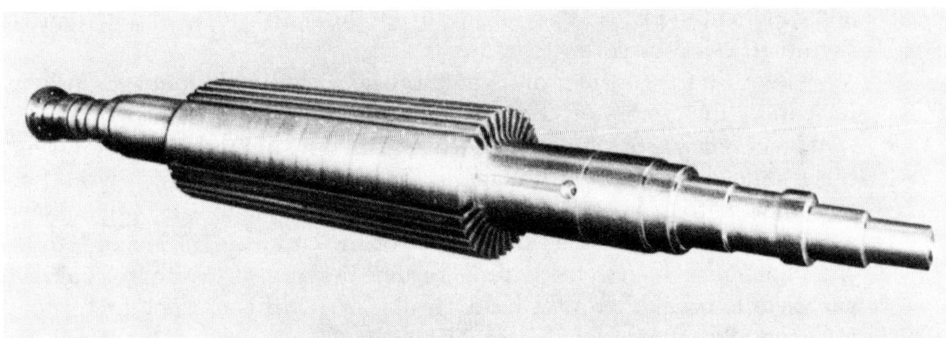

Figure 5.1 Machined generator forging. (Used by permission from Westinghouse Power Generation, Orlando, Florida.)

exciter is located, is called the *exciter end,* the *collector end,* the *connection end,* or even the *opposite the drive end.* The rotor is supported at each end by a journal or sleeve-type bearing. These bearings are, in turn, supported by *end shields* or bearing brackets, or by separate pedestals external to the generator.

Whereas the rotor body is of full diameter, the bearing journals and the couplings are part of smaller diameter integral *shaft extensions* on each end of the forging. This is done to reduce bearing sizes and, therefore, bearing losses. However, the consequence of the reduced diameter in the shafts is that the rotor is limited in withstanding transient torque loads. These shaft extensions also provide the surfaces used for the hydrogen gland seals.

One significant problem in generator rotors is that electrical currents flowing through the close tolerance gaps in the bearing journals, the thrust bearing, and the hydrogen seals would typically result in arcing if the current were not prevented. This arcing can cause these precision surfaces to pit severely, and ultimately to fail. The voltage level required to cause the pitting of these surfaces will vary with the journal-bearing oil gap, the degree of surface irregularity in the journal, and the degree of oil contamination. Nevertheless, typically 12 volts is all that may be required to break down the oil film and initiate the destructive arcing.

There are two common approaches to the problem of relieving the voltage buildup in the rotor shaft. A third alternative has appeared in the marketplace recently and will also be discussed. The three methods are

- Insulation of the bearing or pedestal. Some manufacturers insulate the exciter end only; others insulate both ends.
- Passive shaft grounding through a brush or copper braid.
- Active shaft grounding system.

Insulation is used in the bearings or pedestals for the generator and exciter to electrically isolate their bearings from the frame's ground potential. This is generally sufficient to protect the bearings from the shaft end-to-end voltages caused by asymmetries in the core.

Because the turbine rotors are not significantly magnetized and are not near the core of the generator, they do not build up an end-to-end voltage to a significant level because of magnetic effects. Steam flow in the turbine, however, can produce static electricity and voltage in the turbine rotors, which would be uniform over the length of the turbine rotors. This static voltage is usually taken care of by a grounding brush at the "front end" of the turbine. The maximum end-to-end shaft voltage as seen in the various turbine bearings is less than 1 volt, which is insufficient to break down the bearing oil film. For this reason, the turbine bearings are generally not insulated.

The situation is different within the generator. Inevitable magnetic dissymmetries in the stator core will generally cause a small alternating voltage

(60-Hz and higher harmonics) to be produced from one end of the generator rotor to the other. By having the turbine-end bearing grounded and the bearing at the other end of the generator rotor insulated, destructive current because of this end-to-end voltage is prevented. The shaft of the exciter may be isolated from that of the generator by an insulated coupling. Alternatively, the bearings of the exciter would be insulated. The shaft grounding brushes keep these voltages from the turbine shaft.

The problem with passive grounding of the shaft is that the current densities experienced are much too low to induce self-cleaning of the brush-shaft interface. The result is that a semiconductive film frequently forms in this interface, which inhibits the current flow. This film creates a buildup of voltage and periodic discharge through the interface in the form of arcing, or worse, through the oil films in the bearing or in the gas seals. The problem can be relieved considerably through the use of copper braids instead of carbon brushes, but the basic problem may still remain, albeit to a lesser degree.

The third device, called the *active rotor shaft grounding system* (ARSG), overcomes the charge transfer voltage drop at the brushes or braids by actively counteracting the voltage buildup in the rotor. It does this by impressing an oppositely charged voltage on the shaft that is equal in magnitude to the shaft potential in order to achieve zero voltage on the latter. It works in the following fashion.

The system requires three brushes in order to carry out its function. One of the brushes measures the shaft potential continuously. This brush is connected to a high-impedance measuring circuit, so even if the semiconducting film forms, its drop will be insignificant compared to this high-input impedance. The parameters measured by this first brush provide an error signal for the control of the second brush, which impresses the oppositely charged potential onto the shaft in order to reduce the built-up voltage to zero or thereabouts. If a semiconducting film develops in the second brush, the control system will automatically compensate for its voltage drop. The third brush is used purely for monitoring purposes to indicate when the system is not performing adequately. It will alarm when it senses that the brushes are in need of service. Additionally, if the system detects a fault within itself, it automatically fails "safe" to a passive grounding system. It will also do that when the bearing insulation fails and the impression of external voltages causes a high current to flow in the bearing.

5-1.2 The Field Winding

The field winding is made up of a group of concentric coils centered on each pole. The coil sides of each coil are located in the rotor slots in corresponding positions on opposite sides of a pole, with the conductors of a single coil in each slot. Each coil is made up of a selected number of series-connected turns of solid (not stranded) copper. In some generators, the turns may each be made up of more than one full-width conductor. The conductors are bent or brazed in the end region to form the coils. While copper is used universally in modern generators, a few older rotors contain field windings of an aluminum alloy such as condal.

Coils are also connected in the end region, in series, usually all in one electric circuit. The coils for each successive pole are connected with opposite polarity to produce alternate north and south poles of flux entering and leaving the air gap.

Another mechanical component associated with the rotor are the *rotor wedges.* The function of a rotor wedge is to carry the centrifugal load of the coil conductor (bar) in the slot section. The wedges are fitted into the *wedge slots* and locked against axial movement. This means that they have to be designed to withstand the centrifugal force mentioned above, which may be as high as 6000 g's, or even higher. They may also contain radial vent holes for gas conductor-cooled field windings. Figure 5.2 is a cross-sectional view of the field slot with the bar and the wedges shown mounted inside the slot.

Because of the different coefficients of expansion of the copper and the rotor forging, a generator startup will place a high stress on the wedges as the copper expands more rapidly than the slot teeth. This subsides as the forging warms up somewhat later. Unit cycling (numerous starts and shutdowns of the generator) can ultimately cause weakening of the wedges as they are repeatedly placed under stress.

In a conductor-cooled field winding, there are cooling passages in the conductors, which may be axial, radial, or arranged in a direction angled from the radial. High-velocity gas passing through these passages provides intensified cooling in the slot and end regions. This is shown in Figure 5.3.

The field winding induces a magnetomotive force (mmf) that will produce a magnetic flux that will flow from the pole face across the air gap and into the stator magnetic core to generate the electromagnetic force in the stator wind ings. The mmf is expressed in *ampere-turns;* therefore, the area available within each rotor slot for copper (the coil) is subdivided into series-connected *turns.* The number of turns per slot is a function of the rated-load mmf and the desired field voltage.

Most manufacturers insulate the individual turns in a coil from each other by separating strips. Another, less common method is to wrap every other turn with insulation tape. Ground insulation is typically provided by armor insulation on the sides, and perhaps the bottom, and by creepage block insulation on the top and perhaps the bottom of each rotor slot. This can be seen in Figure 5.2. Because the voltage in the field winding is typically low (150–750 volts), there is no need for heavy groundwall insulation, even though the rotor body is at ground potential. The individual turns of each coil are connected in series in the end region.

The end region is located at both ends of the rotor body, and the coil sections located there are held in place by the retaining rings. Heavy rugged insulation is placed on the outer surface of the field winding end turns to isolate them from the inner surface of the retaining rings that are shrunk onto the rotor body. This insu-

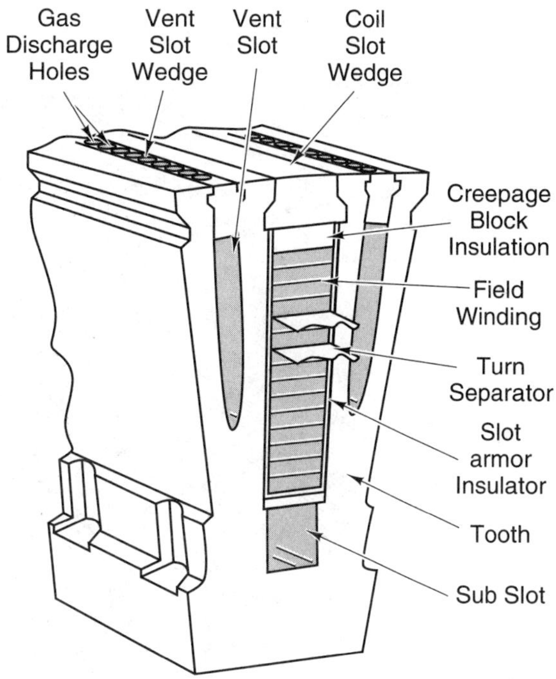

Figure 5.2 Cross-sectional view of field slot. (Courtesy of General Electric Co., Schenectady, New York.)

lation is also known as the retaining ring liner in some places. The retaining rings themselves are discussed in Section 5.1.3.

In addition to the normal excitation voltage levels, the field winding must also be able to handle the ceiling excitation current for brief periods. Application of ceiling excitation to the field winding for a certain duration of time is thus not only acceptable, but also necessary. By boosting the excitation during certain types of system disturbances, voltage levels in the power system are supported, and system stability is enhanced. Figure 5.4 depicts a table showing the generator field short-time thermal capability as specified in ANSI American National Standard C50.13-1989, "Requirements for Cylindrical Rotor Synchronous Generators," American National Standards Institute, New York, 1989. It relates the per-unit field current application to the amount of time such an application is allowed to persist. The values shown are based on a constant increment of heat added per incident and can be approximated by the equation

$$(V_f{}^2 - 1) * t = 33 \tag{5.1}$$

where V_f is in per-unit of rated field voltage and t is in seconds. Figure 5.5 shows a continuous plot of the values contained in the table.

Figure 5.3 Turn connections in the field winding end region. (Used by permission from Westinghouse Power Generation, Orlando, Florida.)

Per Unit of Rated Field Voltage	Permissible Time in Seconds
1.12	120
1.25	60
1.46	30
2.08	10

Figure 5.4 ANSI C50.13–1989 table of allowable transient overvoltage of field winding.

The field winding should also be designed to withstand normal voltages as well as transient overvoltages, such as those present during *out-of-step operation.* Out-of-step operation was covered as a system problem in Chapter 2 and will be discussed again in more detail in Chapter 12. Lastly, the field winding should also withstand the 6000 g's or more of centrifugal force discussed above.

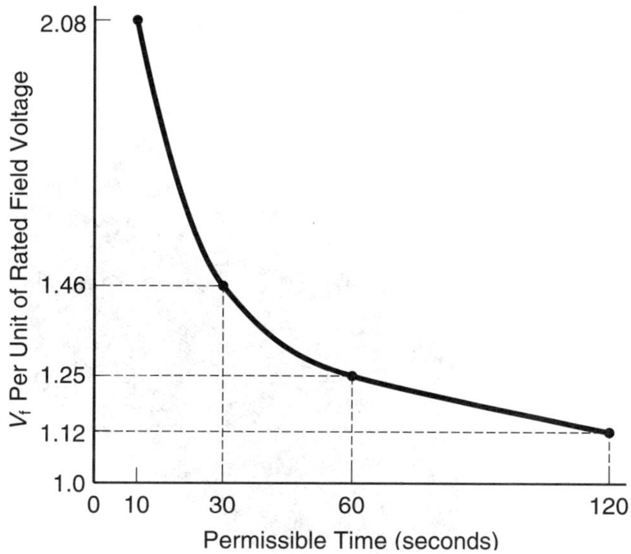

Figure 5.5 Curve of ANSI values of allowable overvoltage of field winding.

Cooling of field coils is also an important consideration. The coils can be cooled either indirectly or directly. Indirect cooling means solid copper turns, whose heat is removed from the outside ground insulation. Cooling takes place at the teeth and at subslots. This *conventional* method of cooling is generally used for medium and small generators. Conductor cooling of the field winding is used in medium and large generators, both air- and hydrogen-cooled. Here, the conductors are cooled internally by the flow of the cooling gas.

In an *axially cooled design,* the gas passes through axial passages in the conductors, being fed from both ends and exhausted to the air gap at the axial center of the rotor. Figure 5.6 shows an example of conductor-cooled field turns in a rotor slot with axial cooling passages.

In another design, radial passages in the stack of conductors are fed from subslots machined along the length of the rotor at the bottom of each coil slot. This can also be seen in Figure 5.2.

In a *zoned* or *gap-pickup* design, cooling gas is picked up from the air gap and used to cool a relatively short length of the coil, and then discharged back to the air gap.

Some European manufacturers make water-cooled rotors for the larger ratings. These require hollow conductors, a water connection to the rotor, and the ability to reliably accommodate the high water pressures caused by the centrifugal

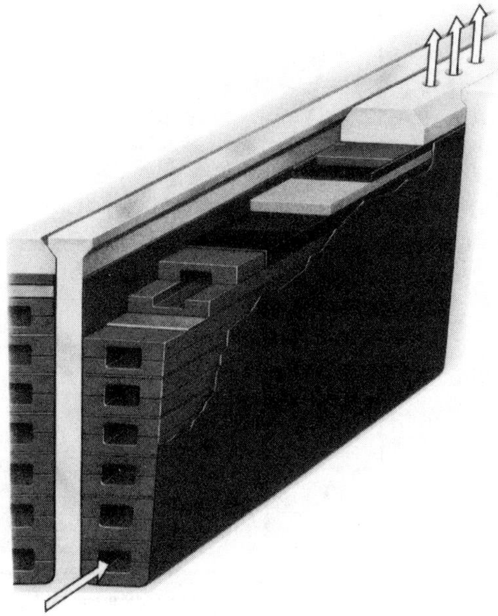

Figure 5.6 Axial cooling of field turns. (Used by permission from Westinghouse Power Generation, Orlando, Florida.)

forces involved. Foreign-made rotors with field windings of this design, however, are rare in the United States and will not be discussed further.

5-1.3 The Retaining Rings

Keeping the coils in the slots of the rotor field is more complicated than keeping them in the slots of the generator stator. Not only must the forces due to current and flux interaction be restrained, but the effects of the centrifugal forces must also be addressed. The best way to support the rotor field end turns at synchronous speed is with a pair of high-strength retaining rings.

The retaining rings are probably the most highly stressed part of the rotor. They are mounted to support the rotor end winding, as the turns exit the slots in the rotor forging and cross over to complete the coil connections. Their function is to carry the field winding centrifugal load at the end regions. Therefore, they must also be able to withstand the 6000 g's or more of force. But the field winding end region does not enjoy the natural support provided by the wedges and teeth. Therefore, the retaining rings are required to carry all of the end-turn forces. Moreover, they must also withstand a high axial thermal thrust load because of the thermal expansion of the field winding ends. The retaining rings are usually shrink-fit

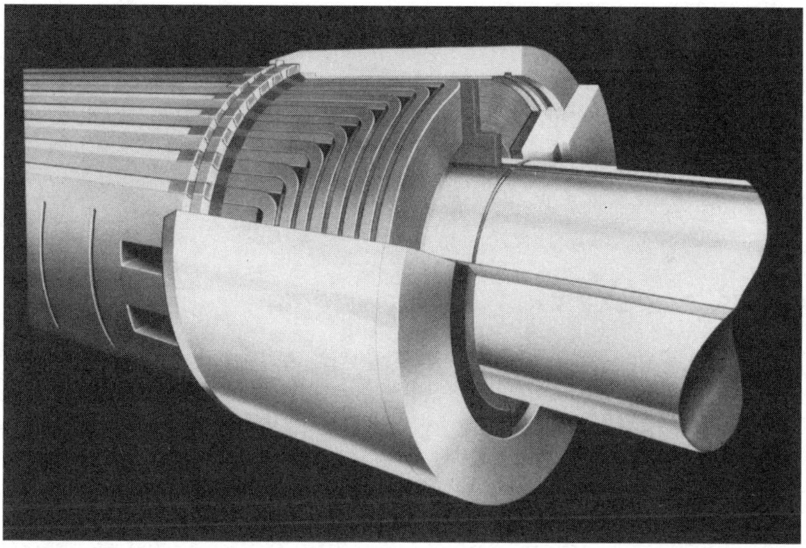

Figure 5.7 Cutaway photograph of retaining rings on the rotor body. (Used by permission from Westinghouse Power Generation, Orlando, Florida.)

and locked against the thrust loads of the field winding. Figure 5.7 depicts a cutaway view of a retaining ring on a rotor body.

Alloy chemistry of these retaining rings varied widely during the early years of synchronous generator development. Some of the different kinds of rings extant are listed below. They are usually identified by their noniron constituents and alloy chemistry.

- Magnetic
- 8Mn–8Ni–4Cr
- 9Mn–6Ni–4Cr
- 10Mn–6Ni–4Cr
- 13Mn–5Ni–3Cr
- 18Mn–5Cr

An 18Mn–5Cr retaining ring is composed of austenitic steel with 18% manganese and 5% chromium.

Following World War II, nickel had become difficult to obtain; therefore, by the late 1950s, the industry had generally standardized on the 18Mn–5Cr nonmagnetic material for retaining rings and other strength parts such as rotor-mounted air gap ventilation baffles. The retaining rings normally contact the rotor field forging at either the inboard end or the outboard end of the retaining ring (through a centering ring), but not both. This is to minimize the possibility of wearing or fretting at the contact areas because of bending or sagging of the rotor field that could occur if both ends of the retaining ring were in contact with the rotor forging. Because the centrifugal loading on the retaining ring during normal operation at synchronous speed tends to expand the retaining ring diameter, the ring is designed with an interference fit with the rotor at rest. This prevents the retaining ring from lifting completely off the rotor during operation. Nevertheless, most newer retaining rings have either a bayonet fit or a snap ring slot key to minimize axial movement of the retaining ring.

5.2 PARAMETERS, SENSORS, AND MONITORING

The rotating nature of the rotor makes it a difficult component to monitor. Therefore, there are relatively few sensors available to determine its state. This is especially true for field windings supplied by *brushless exciters*. Because of their design, the electrical quantities in the generator rotor cannot be directly measured. Also called *rotating rectifier* exciters, their access to the generator field winding is directly through the rotating shaft. Brushless exciters are covered in Chapter 7. Fortunately, the number of different modes of failure that can exist in a rotor are somewhat fewer than in the stator.

5-2.1 Field Voltage and Current

Knowing the field voltage and current values at all times is quite important. Various problems can be detected by comparing the sensed values with the expected values. For a static or rotating exciter with slip-ring access to the field, this is fairly easy to do. A voltmeter and an ammeter at the proper location will suffice.

That is not so, however, for generators equipped with brushless exciters, for which the voltage and current values have to be indirectly calculated from the (rotating) exciter field current, using a saturation curve for the exciter. Figure 5.8 shows a typical exciter saturation curve. It contains the no-load saturation curve as well as the constant-resistance saturation curve, which describes the operation of the exciter load. Like the generator saturation curve described in Chapter 2, the exciter saturation curve relates exciter field current to exciter output direct voltage, which is also the generator field voltage. In order to calculate the generator field current, however, its field resistance must be already known. For example, at an exciter field current of I_{fe} amperes, the exciter output voltage, which is the same as the generator field voltage, E_{fg}, is E_{fd} volts. Assuming a generator field resistance of R_b ohms at t_b° C, the generator field current can be calculated as follows:

$$I_{fg} = \frac{R_f}{E_{fd}} \text{ Amperes} \tag{5.2}$$

5-2.2 Generator Terminal Electrical Characteristics

Most of generator terminal electrical characteristics were described in Chapters 2 and 4. They include the generator MW, MVA, power factor (PF), armature phase

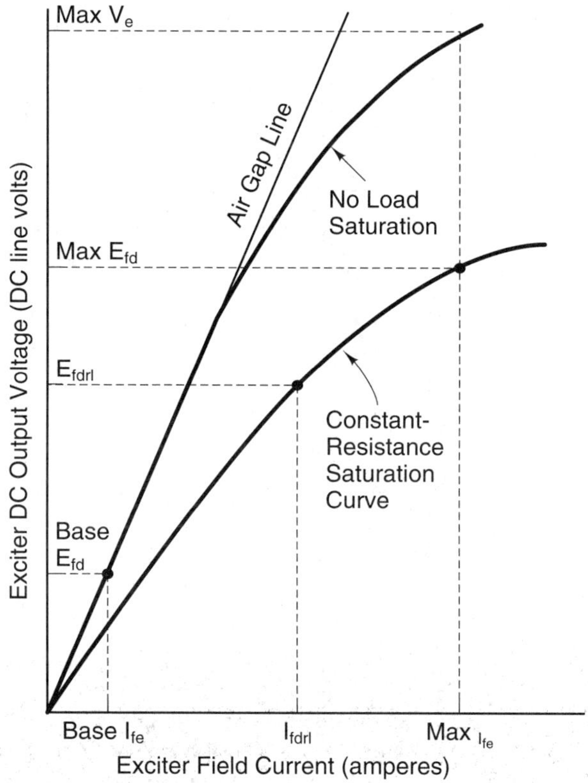

Figure 5.8 Typical exciter saturation curve. (Used by permission from Westinghouse Power Generation, Orlando, Florida.)

currents I_a, I_b, and I_c, and the generator terminal voltage E_t. E_t is the line-to-line output voltage of the generator at its terminals, typically expressed in kilovolts.

The above electrical terminal quantities can be used to determine the field current for the generator operating conditions. One method commonly used to calculate the field current value required to supply a certain load at a certain PF and terminal voltage is through the load-excitation curves described in Chapter 2.

If a specific curve is not explicitly drawn for a particular combination of V_t, MVA, and PF, then the *Potier method* can be used. The Potier method is the most simple known technique to analytically compute the field current required for specific generator terminal conditions. It can be quite useful to detect shorted turns and/or overheating of the field winding.

Figure 5.9 illustrates a generator saturation curve, previously described in Section 2.3.2. (For the convenience of the reader, Figure 5.9 is a reprint of Figure 2.18). As can be seen in Figure 5.9, one additional feature of the saturation curves

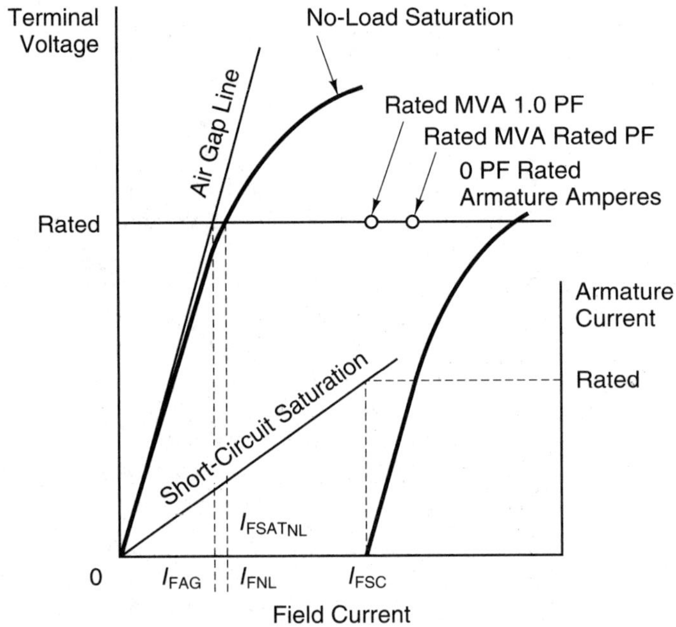

I_{FAG} = Generator field current required to produce rated voltage on the air gap line.
I_{FNL} = Generator field current required to produce rated voltage at no load.
I_{FSC} = Generator field current required to produce rated armature current with generator terminals short-circuited.

Figure 5.9 Generator load saturation curves.

is the demarcation at rated voltage of the unity PF, rated power operating point, and the rated PF lagging, rated power operating point.

The Potier method is based on a unique property of the saturation curves. A triangle can be constructed as shown in Figure 5.10 that can be slid along the saturation curves such that the right corner will always contact the rated current curve, whereas the top corner will always contact the no-load curve. If this triangle is not shown on the saturation curves, it can always be constructed by measuring the distance *O–t* at rated voltage from the rated current saturation curve, and then drawing a line parallel to the air gap line from *O′* until it contacts the no-load curve.

For the remainder of this section, it will be assumed that the stator winding AC resistance is negligible, so it will not appear in any of the calculations. The *Potier reactance drop*, X_p, can be defined as follows:

$$E_p - E_t = IX_p \tag{5.3}$$

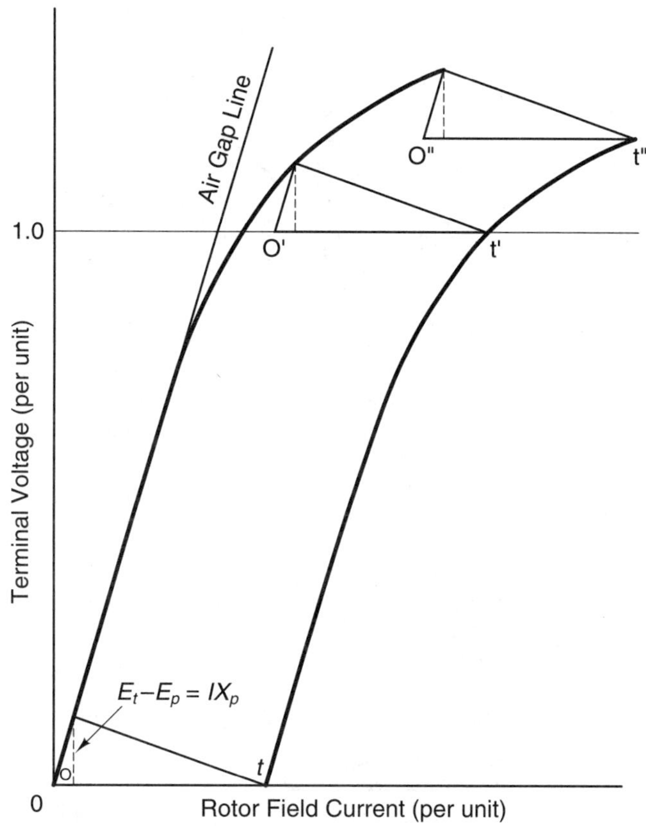

Figure 5.10 Potier triangles.

At zero PF lagging and rated stator current, the X_p drop can be represented by the dotted line in Figure 5.10.

The Potier method employs a graphical technique to determine the saturation at the voltage behind Potier reactance in order to determine the rotor field current needed to operate at a specific stator terminal voltage, PF, and stator current. There are different Potier methods for different synchronous machine designs. The one that will be demonstrated below is the round rotor Potier voltage method, which gives the most accurate results for most two-pole and four-pole synchronous generators [Westinghouse, 1964].

The *saturation correction factor, s,* is shown in Figure 5.11. To construct it, a line is drawn from point *a* at the PF angle. The Potier reactance drop is scaled off from *a* to *c*. Note that this line will be shorter than the dotted line in the Potier triangle if the stator current is less than rated. The distance *O–c* then represents the voltage behind Potier reactance for the operating condition specified. Using a compass, E_p

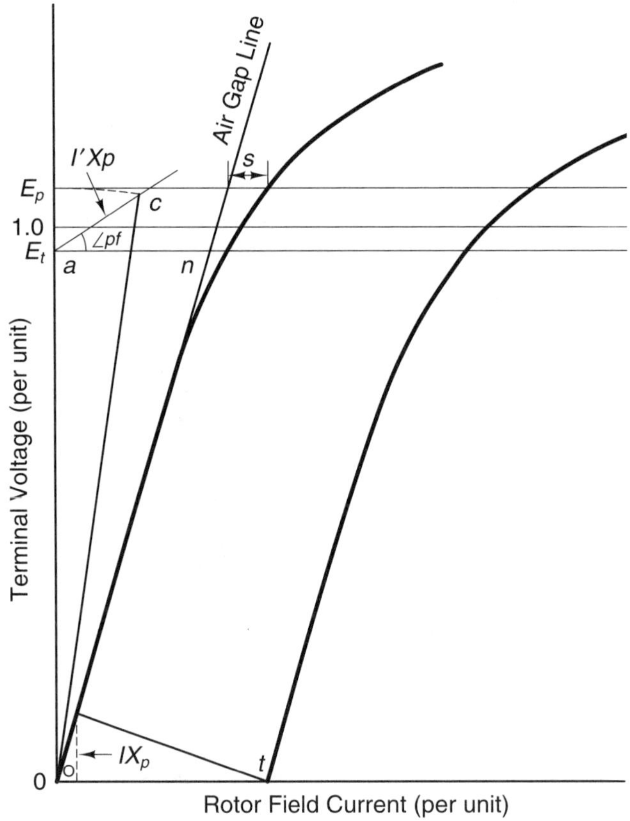

Figure 5.11 The saturation correction factor.

is scribed back to the ordinate. The distance from the air-gap line to the no-load saturation curve at the voltage, E_p, represents the saturation correction factor, s, for the operating condition specified.

The final step is shown in Figure 5.12. The rotor field current required to excite the no-load voltage at the air-gap line, a–n, is scaled off on the ordinate. As was done in the previous case, a line is drawn from point j at the PF angle. The rotor field current required to support the stator current is scaled off from j to d'. Note that this line will be shorter than O–t if the stator current is less than rated. A line is then drawn from O to t' and extended an additional amount to m to accommodate s, the saturation correction factor. O–m is now scribed down to the abscissa. O–m' is the rotor field current required to operate at the specified stator terminal voltage, PF, and stator current.

To improve the accuracy of the graphical technique, a computer program can be written that employs the Potier method. A simple FORTRAN program that uses

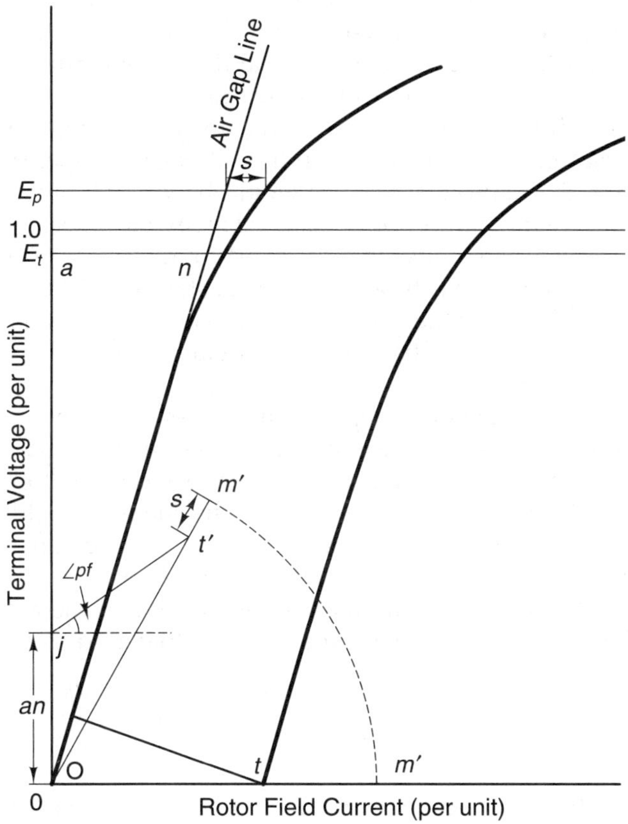

Figure 5.12 Determination of rotor field current.

an exponential model for saturation and includes all the coding steps is available in the literature [Nilsson, 1993]. This technique is being used by at least one electric utility to monitor synchronous generators for the onset of rotor field shorted turns by comparing the calculated rotor field current requirements with the measured rotor field current. Section 5.3.1 describes the problems that can arise if a rotor field has a significant number of shorted turns.

5-2.3 Average Field Winding Temperature

The average field winding temperature is not an instrument but rather a value that can be calculated from the field current and voltage. The resistivity of the field winding, as that of any electrical conductor, is related to its temperature. The average temperature of the field winding can be easily determined on-line for a machine equipped with slip rings if a baseline of field temperature, t_b, and resistance,

R_b, have been previously measured accurately, either in the factory or in the field, when the rotor had been stationary long enough for the temperature to be at a steady value. Temperature of the field during operation is deduced from a measured change in resistance from these baseline values.

One drawback to this technique, however, is that short-circuited turns will lower the resistance of field winding at every temperature. If turns are short-circuited, the reduced resistance would be wrongly interpreted as a change (reduction) in temperature. One of several methods of detecting short-circuited turns needs to be used to detect that the short circuits exist, a condition that may disappear when the rotor is brought down to standstill. Thus, short-circuited turns in the rotor will render the temperature measurement meaningless until a correction is made—either a new baseline measurement (which may not be possible) or repair of the rotor.

As will be seen in Chapter 7, brushless exciter-equipped generators do not provide direct access to the winding. Therefore, the actual resistances of the winding cannot be calculated from field voltage and current measurements. In fact, to calculate the field winding current in such generators, field current is calculated by first assuming a particular temperature and, therefore, resistance.

IEEE Standard 115–1983 [IEEE, 1983] *IEEE Guide: Test Procedures for Synchronous Machines* presents the method of calculating the average temperature of a winding from its resistance. Equation 36 of that guide follows:

$$t_t = t_b + \frac{(t_b + k) * (R_t - R_b)}{R_b} \tag{5.4}$$

where

t_t = total temperature of winding when R_t was measured, degrees Celsius

R_t = resistance measured during test, ohms

R_b = reference value of resistance previously measured at known temperature t_b, ohms

t_b = temperature of winding when reference value of resistance R_b was measured, degrees Celsius

k = 234.5 for pure copper
225 for aluminum based on a volume conductivity of 62% of pure copper. (A few old machines may have aluminum alloy field windings.)

One weakness of this method is that hotspot temperatures cannot be detected. Additionally, as mentioned above, this calculation is limited to directly accessed field, such as those accessible through a slip ring.

5-2.4 Negative-Sequence Relay

A *reverse-phase* or *phase-balance current* relay functions when the polyphase currents are of reverse-phase sequence, or when the polyphase currents are unbalanced or contain negative-sequence components above a given amount. The nature of generator phase current unbalance and the resulting negative-sequence currents have already been discussed in Chapter 4, their effect on the rotor will be covered here.

Steady negative-sequence currents are reverse-phase currents that flow in the armature winding as a result of current unbalance in the loads connected to the generator phases. Such unbalances can also exist as a result of the opening of one phase of a set of single-pole circuit breakers. They can also result from unbalanced faults (e.g., single-phase-to-ground or phase-to-phase) that are allowed to persist, possibly because of the failure of a relaying scheme or a stuck circuit breaker pole in the power system. The reverse-phase sequence currents flowing in the armature winding will produce twice-rated frequency (120-Hz) currents in the rotor. These currents tend to flow on the surface of the forging and in the slot wedges and retaining rings. Continuous flow of these currents will cause significant heating in these areas and, in severe cases, even lead to melting of these surfaces if the negative-sequence currents are allowed to persist for a long time.

One can determine the negative-sequence component of three-phase currents by using the Law of Cosines and the equations for symmetrical components. Refer to Figure 5.13. Three unequal phase currents may be represented by the triangle of this figure. Let x equal the X phase current, y the Y phase current and z the Z phase current. Then the angles of the triangle may be determined by the Law of Cosines as follows:

$$z^2 = x^2 + y^2 - 2 * x * y * (\cos Z) \tag{5.5}$$

Likewise,

$$x^2 = y^2 + z^2 - 2 * y * z * (\cos X) \tag{5.6}$$

Having determined the angles between the three currents, the negative-sequence component is then determined by the following equation:

$$I_2 = \frac{(I_X + a^2 * I_Y + a * I_Z)}{3} \tag{5.7}$$

in which a is the operator that rotates a phasor counterclockwise 120° and a^2 rotates a phase 240° counterclockwise, and I_x, I_y and I_z are vector quantities.

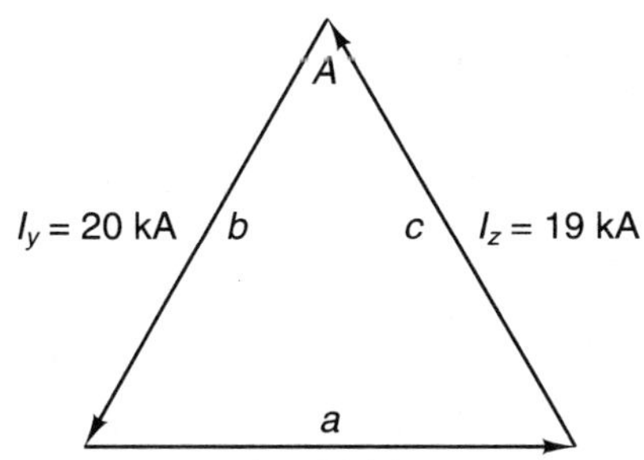

Figure 5.13 Determination of negative phase sequence component of unbalanced currents.

An example would help to explain the concept.

The operator of an 800-MVA, 21-kV, 60-Hz, 36000-r/min, 0.90 PF generator observes the following phase currents of the generator:

Phase X current (x) = 21.0 kA
Phase Y current (y) = 20.0 kA
Phase Z current (z) = 19.0 kA

Is there cause for concern about the continuous negative-sequence capability of the generator being exceeded?

The three current values represent the sides of a triangle as shown in Figure 5.13. The magnitudes of the currents are known, but the angular relationships are not. The angles can be determined by the law of cosines as follows from Equation 5.4 above.

$$z^2 = x^2 + y^2 - 2 * x * y * (\cos Z)$$

Rearranging Equation 5.4 can be expressed as follows:

$$\text{Cos } Z = \frac{(x^2 + y^2 - z^2)}{2 * x * y} \tag{5.8}$$

which results in the following:

$$\text{Cos } Z = (21^2 + 20^2 - 19^2)/(2 * 21 * 20)$$
$$\text{Cos } Z = 480/840 = 0.5714$$
$$Z = \text{Cos}^{-1}(0.5714)$$
$$\mathbf{Z = 55.15°}$$

Similarly, the same can be done for angle Y:

$$\text{Cos } Y = \frac{(x^2 + z^2 - y^2)}{2 * x * z} \tag{5.9}$$

$$\text{Cos } Y = (21^2 + 19^2 - 20^2)/(2 * 21 * 19)$$
$$\text{Cos } Y = 402/798 = 0.5038$$
$$Y = \text{Cos}^{-1}(0.5038)$$
$$\mathbf{Y = 59.75°}$$

resulting in

$$X = 180° - 55.15° - 59.75°$$
$$\mathbf{X = 65.1°}$$

Refer to Figure 5.14. The three phase currents are

$I_X = 21$ kA @ 0.0°
$I_Y = 20$ kA @ −124.85°
$I_Z = 19$ kA @ 120.25°

Using Eq. 5.7

$$I_2 = \frac{(Ix + a^2 * I_y + a * I_z)}{3}$$

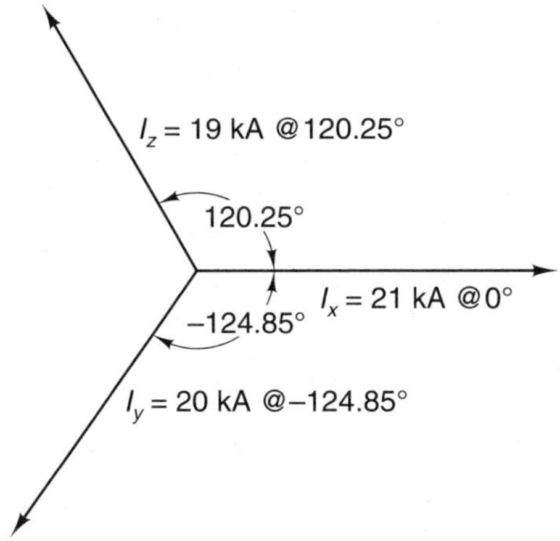

Figure 5.14 Unbalanced phasors defined.

$$a^2 I_Y = 20 \text{ kA @ } 115.15°$$
$$a I_Z = 19 \text{ kA @ } 240.25°$$
$$I_2 = \frac{(21 - 8.5 + j18.10 - 9.43 - j16.5)}{3}$$
$$I_2 = \frac{(3.07 + j1.6)}{3}$$
$$I_2 = (3.46 \text{ @ } 27.53°)/3 = 1.154 \text{ kA}$$
$$\%I_2 = 100 * (1.154/21.994) = \mathbf{5.25\%}$$

According to ANSI C50.13–1989, a generator of this rating has a continuous negative-sequence current capability of 8%; therefore, there is no hazard to the generator rotor.

The negative-sequence current monitor is not a sensor, per se, but rather, a monitor that calculates this value from the measurements of the armature phase currents I_a, I_b, and I_c.

Transient negative-sequence currents are also of concern. The parameter $I_2^2 * t = K$ is the key parameter for determination of damage to a rotor for negative-sequence currents of short duration. The term I_2 is the negative-sequence current in terms of per unit of base machine armature current; t is the time, in seconds, of duration of the negative-sequence current episode. As mentioned in Chapter 4, ANSI C50.13–1989 specifies the allowable value for K for different types and ratings of turbine generators. K ranges from 5 to 30. The negative-sequence relay will monitor the $I_2^2 * t$ parameter and be set to either alarm or trip the machine upon reaching the K limit allowed. For monitoring and diagnostics, this sensor is self-contained. It is basically a protective relay and does not provide warning that the situation is present, only that a limit has been reached. As such, its utility as a monitor is rather limited.

5-2.5 Overexcitation Protection Relay

The overexcitation protective device in the excitation control equipment is an inverse time element, the setting of which can be coordinated with the short-time thermal capability of the generator field winding. It is generally set to alarm or to trip the unit when the ceiling excitation thermal limits have been reached or exceeded. It can be applied to generator field voltage either directly when access to the field is available through slip rings or indirectly through the exciter field current for brushless exciters. This is discussed further in Chapter 7.

5-2.6 Field Ground Detection System

This monitor will be discussed in detail in Chapter 12.

5-2.7 Field Interturn Short-circuit Field Detector

The reader is referred to Chapter 3 for detailed information and discussion regarding this type of monitor.

5.3 FAULT MONITORING AND DIAGNOSIS

This section will cover some of the problems that can arise in the rotor as a result of operating the turbine generator. The section will be divided into three parts:

- The field winding
- The forging
- The retaining rings

As with all diagnostic situations, the validity of sensors should not be unquestionably assumed. Uncalibrated or malfunctioning sensors can cause a misdiagnosis and its attendant costs. The reader is advised to ensure to the greatest degree possible that all readings are accurate before any diagnosis is made.

5-3.1 Problem Area #1: The Field Winding

Problems in the rotor field winding are typically traced to 1) electrical problems in the field circuit, 2) heat removal problems, 3) effects of the very high centrifugal force, or 4) thermal expansion and contractions, primarily of the field winding conductors. These problems are covered in the sections below.

Short-circuited turns

If the insulation fails between the turns in a rotor slot, a short circuit will cause a portion of the coil in the slot to conduct no current. This will have the following consequences:

- The short-circuited turn causes the number of turns in the field winding to be effectively reduced. Because the generator requires a certain amount of mmf (measured in ampere-turns) to produce the magnetic flux needed to generate rated terminal voltage, the field current then has to be increased in order to maintain the desired ampere-turns. The *voltage regulator* will increase the field current automatically in response to a measurement of the generator terminal voltage. If, however, the limit of the normal excitation boundary is reached, then a reduction in maximum machine output can result.

- The random nature of short-circuited field turns will lead to an uneven pattern of short circuits. This will, in turn, lead to uneven heating and uneven conductor expansion, and, therefore, thermally sensitive rotor vibrations. Because the heating of the rotor is (partially) dependent on the field current, the rotor vibrations will be a function of field current.
- In a four-pole rotor, the uneven pattern of short-circuited turns will result in an unbalanced radial magnetic pull between the rotor and stator, resulting in rotating transverse vibrations at a frequency of once per revolution. These vibrations would be felt by both the rotor and the stator.
- The shorting of turns can sometimes lead to grounds in the field winding.

A generator can tolerate a certain number of short-circuited turns in the field winding without significant problems. This is especially true for two-pole rotors, and less so for four-pole rotors. However, erosion in lagging capability can result if the generator has a significant number of shorted turns in its rotor field. Figure 15.15

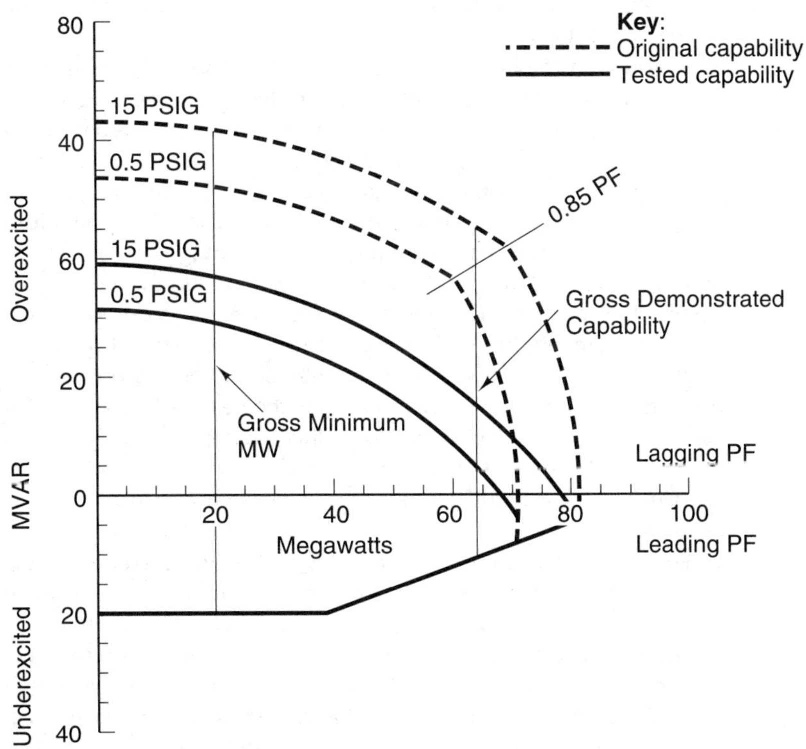

Figure 5.15 Loss of capability because of rotor field shorted turns.

depicts a capability curve for a machine having a rotor with a significant number of shorted turns in the rotor field winding. Diagnostic Box 5.1 will detail the detection means and the corrective course of action.

■ DIAGNOSTIC BOX 5.1 ■

PROBLEM: SHORT-CIRCUITED TURNS IN THE FIELD WINDING

SENSOR INDICATION:

■ **Lateral rotor vibrations** will probably be the first means whereby the presence of short-circuited turns in the field winding is made apparent. This is because the uneven heating of the rotor will result in uneven thermal expansion of the conductors and, therefore, in rotor vibrations detected at the bearings.

■ **The short-circuited field turn detector** will likely provide definite indication of the problem. Described fully in Chapter 3, this detector uses a search coil placed in the air gap to detect the flux waveform produced by the mmf of each rotor slot. The output of this sensor requires expert human interpretation of the waveforms, and it is typically not continuously monitored. This monitor can be retrofitted on existing generators, or it can be provided on new machines by some manufacturers.

■ **Correlation of output conditions** (i.e., MW load, PF, and terminal voltage) with the generator field current measured during operation can be useful in determining the amount of field current that would be required for a specific combination of load, PF, and terminal voltage. (See Section 5.2.2 for the details.) In reality, the field supplies the mmf to drive the needed magnetic flux through the generator's magnetic circuit. The mmf is measured in ampere-turns. Because the number of turns is constant (determined during design), the current is the only variable involved. However, if the number of turns is reduced through short circuits, the regulator has to increase the current in order to achieve the same number of ampere-turns. Thus, for a given set of terminal conditions, the field current measured will be greater than would otherwise be required. This is a good indication of short-circuited turns in the field, but it may be a difficult procedure to execute. The best way to do this is to repeat exactly a set of load conditions for which careful readings were made when the field winding was in a "healthy" condition, and compare the values of field current required to supply that load.

■ **Rotor field resistance calculations** using the measured values of field current (I_f) and voltage (V_f) where $R_f = V_f/I_f$ can be made for field windings with slip rings or other access to the terminals of the field winding. Short-circuited turns will result in *decreased* field winding resistance, which can be used to directly determine the presence of short-circuited turns in the field winding. Using this method requires the assumption of the rotor temperature so that a comparison with a previous measurement can be made. This method may miss some of the short-circuited turns because they may disappear at standstill.

CORRECTIVE ACTION: If only one of the above indications is received, then the validity of the sensor should be verified. Short-circuited turns in the field winding are not likely to cause a major failure unless there are a significant number of them (which causes the temperature limit to be exceeded or vibrations to be beyond safe limits). Because they could lead to more serious problems such as field grounds or bearing vibrations, however, the short-circuited turns should be corrected as soon as convenient. The manufacturer should be consulted for specific recommendations if there is reason to suspect short-circuited turns in the field wiring.

Broken Field Winding Turn

Broken field winding turns are found almost exclusively on older rotors that have retaining rings mounted at their outboard ends, on the shaft. Relative movement between the body and the retaining ring imposes high-cycle strains in the winding conductors, especially those at the tops of the coils. This can lead to fatigue failures, resulting in fracture of the affected turns. It has been found that this inevitably leads to short-circuited turns as the field current finds its way through the damaged winding. Therefore, the effects of a broken field turn are very similar to those of short-circuited field winding turns, and as such, are really a variation of short-circuited field winding turns, although the repairs are likely to be quite different.

■ **DIAGNOSTIC BOX 5.2** ■

PROBLEM: BROKEN TURNS IN THE FIELD WINDING

PERIODIC INSPECTION AND TEST PROGRAM:
Probably the most important means of detecting the onset of broken field turns is by expert examination of the field winding under the retaining rings during periodic maintenance shutdowns. If perhaps the top turn in a coil has already broken or is in the advanced stages of fracture, repair can be planned and carried out before the failure has progressed further.

SENSOR INDICATION:
The sensors used to detect short-circuited field turns apply also to broken turns. (See Section 5.3.1.)

CORRECTIVE ACTION: Repair of a field winding with broken turns is probably best carried out by the manufacturer. It is common to introduce, but not connect into the winding, a selected number of "turns" of special material that has a very high endurance to high-cycle strains. These are not connected into the electrical circuit because this material has high electrical resistivity. These mechanical turns act as a protective bridge over the remaining active turns of the coil. Consult the manufacturer for specific recommendations.

Insulation Degradation and Grounded Field Winding

Degradation of the ground insulation in the rotor field is typically due to the thermal expansion and contractions of the insulation when the unit is cycled through shutdowns and startups. Ground insulation degradation is not normally caused by electrical stresses, as these are relatively low because of the low field driving voltages. The discussion of field ground degradation is similar to that of the stator groundwall insulation found in Section 4.3.3.

The result of rotor field insulation degradation is a field ground. However, field grounds can also result from the failure of any part of the circuit, comprising also the exciter, rectifiers, collector (rotating and stationary parts), and leads connecting these components.

Because the circuit containing the field winding and the exciter is essentially ungrounded, it can operate unperturbed in most situations, with a ground at a single location. Nevertheless, one utility has reported a failure of the rotor as a result of only one ground.

A second ground in the field winding or anywhere else in its electrical circuit, however, can result in high currents through both ground points and probably cause serious damage at the location of the grounds. If one or both of these are in the rotor, the forging may sustain serious damage, which may make it necessary to replace the rotor.

The basic technique for protecting against this kind of failure, therefore, is to look for the first ground before damage is done and repair the unit. This topic is covered in more detail in Chapter 12.

■ DIAGNOSTIC BOX 5.3 ■

PROBLEM: GROUNDED FIELD WINDING

SENSOR INDICATION:

■ **The field ground detector** is especially designed to pick up the first ground fault and alarm the operator of its presence. In most cases, it will detect grounds in the rotor field winding as well as in the excitation supply circuit. See Chapter 12 for a detailed description.

CORRECTIVE ACTION: Unlike short-circuited turns, a second field ground can generally cause severe damage to a rotor and should be considered a very serious problem as indicated above. Upon detection of the first ground, the unit should be taken off-line immediately and the ground repaired as a precaution against serious damage. It is recommended that the unit manufacturer be consulted for recommendations.

There have been cases in which grounds in cables to regulating and other electrical equipment have registered as rotor field grounds. The operator may want to attempt to locate any grounds in the excitation circuit. If the resistance of the ground fault is sufficiently low, a single fault can be located by electrical means when the rotor is stationary. If a direct voltage is impressed across the field winding from an ungrounded source and the voltage is measured from each terminal of the winding to the forging, the location of the ground as a percentage of the distance through the winding from either terminal can be calculated by proportions. Then if a direct voltage is connected from one end of the forging to the other with no current in the field winding, and if the voltage from each end of the rotor to the field winding is measured, the axial location of the fault can also be calculated by proportions.

Blocked Ventilation Path

The self-pumping of the rotor can pull any debris through it that is being carried by the cooling gas of the generator. Thus, foreign material can become lodged in the ventilation paths of the rotor and impair the effectiveness of the cooling. Alternatively, the vent holes in the wedges or in the turn insulating strips can be misaligned, thus blocking the flow of coolant through the coils. The result can be overheated individual conductors or an entire coil-side.

■ DIAGNOSTIC BOX 5.4 ■

PROBLEM: BLOCKED VENTILATION PATH

SENSOR INDICATION:

■ **Field temperatures** can be calculated from field voltage and current readings (i.e., E_f and I_{fg}) using the formula defined in Section 5.2.3. Because added heating will make the resistance increase, a higher-than-normal field resistance can indicate blocked ventilation paths. In coming to this conclusion, it needs to be recognized that any short-circuited field turns would have had the effect of reducing that resistance.

■ **Rotor lateral vibrations** as detected at the bearings can also indicate a blocked ventilation path problem. If this is the problem, these vibrations should be sensitive to load changes. However, because rotor vibrations can indicate a myriad of other problems, this indication should be used only as confirmation, rather than as primary evidence of blocked ventilation path.

■ **Terminal conditions** (i.e., MVA, PF, and E_t) should indicate a near-normal level of field current. This should be used to separate a belief in the presence of blocked ventilation paths from that of short-circuited turns in the field winding.

CORRECTIVE ACTION: If only one of the above indications is received, then the validity of the sensor(s) should be verified. If the problem is allowed to persist, resulting in severe overheating of the field, or part of it, extensive damage can be done to the winding and forging. Such overheating can, in extreme situations, lead to melting of the unventilated copper sections and general failure of the rotor winding. The machine should be removed from service and repaired at the earliest possible time. The manufacturer of the equipment should be consulted for specific recommendations.

Broken Field Winding Lead

A broken field winding lead occurs when one of the leads from the slip rings or the brushless exciter breaks. This can occur without warning, and the resulting arcing within the lead conductor will proceed without detection (except perhaps by an RF monitor) until the insulation is overheated and fails to the forging. Then it will be detected as a ground fault by the field ground detector. Unfortunately, this is likely to proceed very quickly, and the lead will be destroyed before corrective action can be carried out by an operator.

■ DIAGNOSTIC BOX 5.5 ■

PROBLEM: BROKEN FIELD WINDING LEAD

SENSOR INDICATION:

■ **The field ground detector** can be used to detect this problem. If the lead breaks, it will likely cause a ground.

■ **The radio frequency monitor** can be useful in detecting the problem if arcing is present.

CORRECTIVE ACTION: This is a serious problem that can lead to extensive damage. However, the time to action may be so short that once it is detected, it may already be too late. Nevertheless, immediate tripping of the machine and removal of field excitation is imperative to prevent extending the arcing (material removal) and heating damage of the forging. The preferred corrective action to take, then, is to shut the unit down and inspect for damage. (Note that at this time in the incident, it is probably not known why there is a ground fault in the field winding.) The location of the resulting fault to the forging can be verified by electrical means (see Section 5.3.1). One problem is that the location of the failure may be quite difficult to reach and verify visually. Consult the manufacturer for specific recommendations.

Prolonged Overexcitation of the Field

Prolonged application of ceiling conditions in the field winding can cause significant damage to the conductors. The results of such a malfunction are high field current levels and, consequently, high field winding temperatures. The armature winding could also possibly be overloaded because of the higher voltages induced. The problems are typically with the control of the excitation system, but the overheating and any consequent damage are suffered by the rotor field winding and the armature winding. While the causes will be discussed in Chapter 7, the means of detection and the corrective action to take will be discussed here.

As mentioned in Section 5.1.2, the capability for accommodating a certain amount of transient overexcitation is specified by ANSI C50.13–1989. Furthermore, because power system transients are of very short duration, continuous application of the ceiling conditions is generally not necessary. But the heat buildup resulting from prolonged ceiling conditions in both the field and armature windings can be quite damaging and should be avoided.

It should be noted that overexcitation of the rotor field winding usually cannot occur under normal conditions if the automatic voltage regulator (AVR) is in service and is equipped with an overexcitation limiting device (called *Maximum Excitation Limiter*, or MXL by some manufacturers). This device will work through the regulator to automatically limit the level of excitation current in accordance with ANSI C50.13–1989 so as to not allow the thermal capability of the field to be exceeded. However, if the regulator is not equipped with a limiter or if the limiter fails to operate properly, a separate overexcitation protection relay can serve as a backup protection to either alarm or trip the regulator or the unit if the thermal capability curve limits of the rotor field are exceeded. This device is external to the regulator and has a characteristic similar to the thermal capability curve for the rotor, but it is typically set to allow the limiter to operate first. An overexcitation event could also take place when the excitation control system is in manual operation if an electronic component of the system were to fail, resulting in an overvoltage condition.

■ DIAGNOSTIC BOX 5.6 ■

PROBLEM: PROLONGED CEILING EXCITATION

SENSOR INDICATION:

■ **Field current and voltage** can be used to directly detect this condition, although not its cause. Nevertheless, its cause may not be important here—only what to do about it when it happens. Calculating the field current and voltage, however, can be difficult in brushless exciter-equipped generators.

■ **The overexcitation protection** relay, if one is used, will alarm when the limits are surpassed. Nevertheless, when this happens, it may already be too late, so the machine and voltage regulator may be set to be tripped when this takes place. Generally, there is not much time to react to this problem.

CORRECTIVE ACTION: This is probably a problem with the voltage regulator controls. It is important to eliminate the source of the overexcitation quickly, but it is not always possible to respond quickly enough because of the rapidity with which the field and armature winding temperatures will be increasing. Perhaps the fastest way to respond to this problem is to immediately trip the voltage regulator as a way of possibly removing the cause of the problem. Manually adjusting the excitation level may alleviate the problem temporarily. If not, and the problem cannot be otherwise identified, then the unit should be tripped to avoid immediate damage to the field and armature windings. If the problem is not readily resolved, the manufacturers of the voltage regulator and the generator should be consulted regarding their specific recommended course of action for their equipment. In any case, a study of the consequences of the event should be pursued to assure that the reliability of the generator has not been impaired.

Copper Abrasions (Copper Dusting)

Some large four-pole rotors have subdivided conductors that result in copper-to-copper contacts between conductors. At very low speeds, such as when the machine is on turning gear, this contact can lead to rubbing and result in copper erosion particles (copper dust). If this is allowed to go on, the deposits of copper particles may be large enough to cause a short circuit of the ground insulation at the top of the slot. It can also result in a short circuit of the turn insulation in the rotor bars.

This problem cannot generally be detected with on-line instrumentation. Expert visual inspection of the rotor after the unit is opened is typically the preferred means of detecting this problem. For this reason, there will not be a diagnostic box covering this problem.

Not all generators are sensitive to the copper dusting problem, as it generally applies only to some four-pole generators. A generator's propensity to copper dusting is a function of its design. Those that are subject to this problem, however, should be inspected regularly and have their slots cleaned of copper dust during outages. Furthermore, time on turning gear should be minimized for these units.

Out-of-Step Operation

Out-of-step operation of a turbine generator can induce high voltage levels in the field winding. These abnormal voltages are concentrated in the upper turns in each coil. Also, with exciters with rectified output, high voltages between the winding and the forging can result. Such voltages are not expected to exceed the capability of healthy insulation. However, insulation that has been weakened by a number of factors during service may be inadequate for this duty and may fail, resulting in short-circuited turns or failed ground insulation, or both. High torque pulsations are also imposed on the generator stator and rotor and are felt in the entire rotor shaft system.

When pole slipping is occurring, the switchboard meters of armature current and voltage will be swinging wildly. Because fast action is required, an out-of-step relaying scheme is typically available for protection against this condition. This is described in Chapter 12. No diagnostic box will be included for this condition in this chapter.

5-3.2 Problem Area #2: The Rotor Forging

The rotor forging is the second major area of problems in the rotor. The following sections include discussions of some of these problems.

Flaws in the Forging

Over the years, the methods of steel making, testing, and appraising have greatly improved. As a result, there are many older rotors in service whose forging quality would not be accepted if they were made today. Two aspects of the quality of older forgings will be considered: One is their flaws, and the other is their material mechanical properties.

Older rotors may have flaws (*porosity* and *inclusions*) that can be detected by modern sonic testing. These flaws may be tightly clustered or of a sufficient size to be of concern for the reliability of the generator. Present methods of analysis are commonly used on these old rotors to determine their suitability for service, including a prediction of the effects of future start-stop cycling. Many old rotors have been replaced or the generators retired because of the high risk of failure if continued in service. Of particular concern are rotors in generators that are now used as peaking units with frequent starts and stops. This cycling can cause flaws to grow and combine, putting the rotor at further risk. Also, older rotors may have a high *fracture appearance transition temperature* (FATT); that is, the material may be brittle at low temperatures, and not become ductile until a temperature (the FATT) that is well above room temperature is reached. The rotor may well come up to speed before this change from brittle to ductile has been achieved, with an added risk of failure. However, many of these old rotors can be successfully used if certain precautions are taken. Prewarming is used at a reduced low-stress speed so that the rotor is not brought up to full stresses until its temperature is above the FATT.

Older rotors commonly had bore holes. This provided the opportunity to remove flaws that, in the steel making process, tend to concentrate near the centerline of the forging. Bore-sonic testing has been developed into an advanced state of the art, whereby carefully designed signals are sent from the bore into the forging and received in a transmitter-receiver unit placed in the bore of the rotor forging. By recording the received signals and analyzing them by computer, the entire volume of the rotor forging can be evaluated as to its present and anticipated future condition. The work is done as the rotor is turned slowly in a lathe, with a long arm slowly fed in along the length of the shaft, from the end opposite the drive. The leads of the field winding that are in the bore hole are removed before this test and replaced afterward.

It is difficult to detect this problem through on-line monitoring. Therefore, a diagnostic box will not be included.

Shaft Torques

Transient disturbances in the power system can cause severe torques to be imposed on the generator rotor and consequently on the whole shaft system of the turbine-generator set. If the disturbances are severe enough to cause damage to the rotor shaft system, it will occur in the smaller diameter sections of the shafts of the generator, exciter, and prime mover, in the regions of the bearing journals, hydrogen and steam seals, and slip rings, and not in the generator or turbine rotor body masses.

There are various circumstances that cause these transient torques, and they are not all equally damaging. The susceptibility of a shaft to these torsional loads depends heavily on the shaft system's natural torsional frequencies and how close they are to the frequencies of the components of the electrical torque felt by the generator rotor. The latter depends on the type of disturbance that is felt. The unidirectional torques are of particular concern because as they are suddenly imposed on the shaft system, they stimulate every natural frequency and contribute disproportionately to the total shaft torque stresses.

A rotor shaft has two parameters of significance: the *high-cycle fatigue stress limit*, and the *yield stress* of the shaft material. These will be different for each of the shafts in the turbine-generator set because of the different materials that are used.

The high-cycle fatigue limit is the peak equivalent tensile stress at which no damage will occur on the shaft, no matter how many torsional cycles are applied to the shaft.

The yield stress is the highest equivalent tensile stress to which the shaft material can be exposed without sustaining a permanent deformation. This deformation may cause not only a permanent twist in the shaft, but also bending which, if sufficiently severe, could make the affected rotor inoperable because of vibration. The design margins that are built into the rotor shaft system are such that, to the authors' knowledge, there have been no purely electrical events that have caused such a single peak torque that resulted in a shaft being rendered unusable. But, in theory, the most extreme out-of-phase switching against a very stiff system may exceed the yield strength of some of the shaft system components.

At stress levels above the high-cycle fatigue limit of a shaft and below its yield strength, or with more severity, its ultimate strength (which will be higher than the yield strength), damage will accumulate if the rotor system is exposed to torque peaks within that region. Each incident of such torsional torque application will reduce the remaining fatigue life of the shaft until it fails or otherwise becomes unusable, probably because of vibration. Loss-of-life for a single episode could range between being negligible to as high as a few percent of total life. Chapter 3 contains a more detailed discussion of the torsional vibration monitor.

The categories of incidents that can contribute to accumulated shaft loss-of-life are the following, listed roughly in decreasing order of severity:

- Subsynchronous resonance[1]
- Out-of-phase synchronization
- Three-phase sudden short circuit, subsequent clearing and reclosing onto a bolted fault, resulting in 60-Hz and unidirectional torques
- Single-phase sudden short circuit and subsequent clearing and reclosing onto a bolted fault, resulting in 60-Hz and 120-Hz as well as unidirectional torques. This will be the equivalent of a phase-to-phase fault at the terminals, except that the generator step-up (GSU) impedance would diminish the severity of the fault current and thus of the resulting electrical torque. If the shaft natural torsional frequencies happen to be at or near to 120 Hz, then this type of fault could in fact be more severe than the three-phase fault.
- Transmission line opening and reclosing
- Out-of-step operation

A three-phase fault will cause high electrically induced torques on the generator rotor shaft. These torques have two decaying components; one is at 60 Hz, the other is unidirectional. The effect of the 60-Hz torque is magnified if any torsional resonances are close to that frequency. The step-function nature of the unidirectional torque causes a response from every torsional resonance of the shaft system. The result is that the unidirectional electrical torque can be responsible for approximately 30% of the total torque on the shaft. The closer the fault is to the generator, the more severe are the torques that result. But reclosing on the fault can cause additional torque stresses that build on those from the first fault.

Single-phase-to-ground faults in the transmission system are more common than three-phase faults. The electrical torque for this case, which is equivalent to a phase-to-phase fault at the generator terminals except through an impedance, also includes the 60-Hz and unidirectional components, but in addition, because it is an unbalanced fault, it has a 120-Hz component. This can cause a severe problem for rotor shafts that have a natural frequency at or near 60 or 120 Hz. Rotor shaft designers should ensure that the shaft system's natural frequencies are sufficiently distant from both of these frequencies in order to reduce the ensuing shaft torques and stresses.

[1]While destructive subsynchronous resonance has been experienced, techniques have been developed to avoid it, or at least to reduce it to a nonhazardous level. But if these techniques fail or are not in service, this is a major hazard for turbine generators in power systems that have long AC transmission lines with series capacitive compensation. See Chapters 2 and 12 for more detailed discussions of this topic.

Reclosing onto bolted faults can have an effect on the severity of the transient shaft torques. The generating unit design is based on withstanding a single specified fault on the generator terminals. Although transmission system faults are not as severe as generator terminal faults because of the impedance of the GSU, the torque imposed on some shaft elements as a result of the reclosing action may amplify an already existing torque oscillation and exceed the design criterion for the unit. In general, the timing of the reclosing action is based on the time required for the arc path to deionize. It is consistent with the circuit breaker and relay operating times, with no consideration for the torques occurring on the generating unit. For this reason some utilities do not employ high speed reclosing on transmission lines that have a terminal at a generating station.

Out-of-step operation can also cause high oscillating torques with each passing of any of the poles of rotor flux past a pole of stator-produced flux, the latter being synchronous with the currents of the power system. Essentially this is the application of full-load torque of alternating plus and minus polarity at the frequency of pole slipping. Once again, discussion of out-of-step conditions is delayed until Chapter 12.

When a generator is connected to a power system and the voltage of the generator is not in phase with the voltage of the power system, there is potential for damage to the generator. The peak torque caused by out-of-phase synchronization depends on the degree of mismatch of magnitudes and phase angles of the voltages on the two sides of the circuit breaker before closing, as well as on the reactance (or electrical stiffness) of the system into which the synchronizing is being done.

Assume that the generator frequency is slightly above that of the power system just prior to synchronization. The voltage that will appear across the open contacts of the circuit breaker at any instant will be the phasor difference between the generator and power system. The effective impedance in the generator/system loop at the instant of closing of the circuit breaker will be the sum of the subtransient reactance of the generator and the power system impedance, which will be basically reactance. The current that will flow in the loop at the instant of closing will be the instantaneous voltage difference divided by the loop impedance. This current will lag the *difference voltage* by 90° because of the predominantly reactive nature of the impedance. The electrical torque will be proportional to the product of the generator voltage and the in-phase component of the current. This torque will have a maximum value when the current lags the generator voltage by 45°, that is, when the generator and system voltages are at 90°.

The maximum *current* in the loop will occur when the voltages are 180° out of phase. The current may exceed that which would result from a three-phase short circuit on the generator terminals if the system impedance is less than the subtransient reactance of the generator. The effect on the generator shaft is related to the torque, but the effect on the winding is also related to the current magnitude. Thus, the worst-case voltage angle difference for a combined effect on the generator shaft and windings is somewhere between 90° and 180°, typically approximately 120°. Discussion of the reasons for this, however, are beyond the scope of this book.

Generators are not typically designed to withstand the shaft torques generated by such worst-case incidents. To do so would increase the bearing diameters significantly with a large increase in bearing losses and a detriment to the heat rate of the turbine-generator set. Manufacturers provide an upper limit of voltage angle and magnitude mismatch that is recommended across the circuit breaker during synchronization.

Other situations in the turbine-generator set can cause shaft stresses, as well as the torques that have been discussed. Severe misalignment, or torque jolts from component failure, impact the same areas of the shaft system as the torques from electrical causes. If, for example, misalignment and torque were to occur together, they might drive the shaft stresses beyond the endurance limit or yield strength and cause damage in cases where either one by itself would not.

A study was performed in the early 1980s to determine the actual effect of power system transients on turbine generator shafts [Bowler, 1994]. The study placed a number of torsional monitors in different generating units to record any disturbances over a period of time that might result in loss of shaft life. The results of the study showed that, with one exception, no loss-of-life was incurred in any of the units monitored. The exception was a unit electrically located near a DC converter station. This unit did indicate a small, yet measurable, loss-of-life.

■ DIAGNOSTIC BOX 5.7 ■

PROBLEM: SHAFT TORQUES

SENSOR INDICATION:

■ **A torsional vibration monitor** is the only sure way to detect this problem and calculate the resulting loss of life over a machine's entire history. This instrument covers torsional episodes from all the different causes and calculates the loss-of-life for a particular rotor whose design parameters have been input. Chapter 3 contains a detailed description of a torsional monitor. If the details of the episode are known, then a computer simulation can be made with an appropriate model to calculate the loss-of-life. However, that is beyond the scope of this book.

■ **The negative-sequence relay** may be able to detect torsional vibrations indirectly as a result of the unbalanced faults (single-phase-to-ground) that give rise to the transient torques. But that is a rather ineffective means because it does not address the loss-of-life question and is limited only to episodes that induce negative sequence currents in the stator winding.

■ **The bearing lateral vibration monitor** may pick up the problem when a shaft has been bent or cracked as a result of its exposure to torques (and other causes, such as misalignment).

CORRECTIVE ACTION: If the transient torque took place because of severe out-of-phase synchronization, then the rotor should be inspected at the next possible opportunity and loss-of-life calculated. If there is reason to believe that the rotor shaft is reaching the end of its useful life, then serious consideration should be given to replacing it. If it is suspected that the shaft has been bent in any of the rotors of the turbine-generator set (from any cause), this can be measured by uncoupling the rotors and measuring the lateral and axial runout. The manufacturer should be consulted for specific recommendations.

Negative-Sequence Rotor Surface Heating

The means of identifying the presence of the destructive negative-sequence currents have already been discussed. The potential damage to the generator rotor will now be described.

A large conductor-cooled generator is specified by ANSI C50.13–1989 to have a steady capability of 8% negative-sequence current. An 8% negative-sequence armature current can cause rotor surface currents of several thousand amperes per pole pitch flowing in the rotor. Likewise, a momentary single-phase fault in the system can cause hundreds of thousands of amperes to flow in the rotor per pole. These may seem to be unexpectedly large currents.

Potentially severe local heating can take place as a result of such currents. There are several areas where the current concentrates and where the electrical contacts are not necessarily excellent: 1) over the entire length of the rotor body, at the corners of the ends of each coil slot wedge, where axial current enters or leaves the wedge coming from or going into the teeth; 2) at the ends of the wedges if they are touching; 3) at the ends of the rotor body where the current turns in the peripheral direction and crosses all of the teeth and wedges; and 4) between the rotor body and the retaining ring. In a two-pole generator, potentially severe heating will also be likely in the pole surface at the ends of the cross slots. Additionally, the wedges themselves can overheat.

■ DIAGNOSTIC BOX 5.8 ■

PROBLEM: NEGATIVE SEQUENCE CURRENTS IN ROTOR

SENSOR INDICATION:

■ **The negative-sequence current relay** will protect the rotor from excessive steady negative-sequence currents (I_2), as well as transient currents that exceed the ANSI-specified I_2^2t limits, discussed earlier in this chapter.

■ **The phase currents** will be significantly different, but the amount of unbalance can be deceiving. This can be calculated as shown above, but it is best to implement this algorithm in the plant computer or other monitoring device to continually calculate these values.

■ **The bulk armature coolant outlet temperatures** may not indicate anything out of the ordinary, because the other two phases will be carrying less current and running cooler. Thus, this sensor should probably indicate normal temperatures.

■ **The coil outlet coolant temperatures**, whether gas or water conductor-cooled, will indicate a current unbalance problem, in which case all the readings in the same phase will be close to each other, but the average values for the different phases will deviate from each other significantly. The algorithms discussed in Section 4.2.9 may indicate a problem, because one of the phases will be running hotter for the load level, whereas another one will be running cooler than normal. Additionally, if the deviation from the average is used, that may also exceed the normal values. The high-low temperature differential limit may also be exceeded. The ability to detect this may depend on the machine design; therefore, not all situations may exhibit this indication.

■ **The slot RTDs** may show a similar pattern to that described above for the coolant outlet temperatures. This may also depend on the design of the machine.

CORRECTIVE ACTION: An excessive steady current unbalance must be corrected, or damage may be incurred on the rotor. If the steady I_2 has exceeded its limit or if a transient negative-sequence current has exceeded the limiting value for I_2^2t, then the generator should be removed from service at the earliest possible opportunity and given expert inspection for possible damage. Consult the manufacturer for specific guidelines and recommendations.

Shaft Voltage Buildup

Voltage in the rotor shaft can be damaging to the bearings. It can break down the oil film electrically with shaft currents between the journal and bearing. Shaft currents across hydrogen seals can also be produced because of this buildup. Whereas the bearings are insulated against damage from this voltage, dirt or previously existing damage can compromise the integrity of the insulation and allow current to flow through it. Shaft grounding brushes or braids are typically installed to eliminate the buildup of static voltage, but can either make poor contact or remain unconnected, thus allowing the buildup to occur. Therefore, it is quite important to monitor this voltage.

■ DIAGNOSTIC BOX 5.9 ■

PROBLEM: SHAFT VOLTAGE BUILDUP

SENSOR INDICATION:

■ **The active rotor shaft grounding system,** a voltage suppression device described in Section 5.1.1, is the most direct means of detecting this problem. It directly measures the voltage in the shaft forging, and it is analog in nature. It can also remedy the problem automatically by impressing an equal and opposite voltage on the shaft.

■ **The radio frequency monitor** detects shaft voltage buildup indirectly by detecting arcing in the shaft grounding brushes. Arcing implies poor contact between the shaft and the grounding brushes and, therefore, inadequate grounding action. The RF level should be insensitive to load changes, and the source can be easily confirmed by visual inspection and corrected while the machine is on-line. Nevertheless, the brushes will not always be arcing when there is a voltage buildup. In fact, if there is no contact with the brushes, there will be no arcing.

CORRECTIVE ACTION: This problem generally does not require a unit shutdown. Corrective action includes a visual inspection of the grounding brushes and their replacement if they are inadequate. Additionally, a check of the integrity of the bearing insulation can be made on-line. At the first planned outage after a voltage buildup episode, the journals and the bearings as well as the hydrogen seals should be visually inspected for pitting damage. Consult the manufacturer for recommendations.

Shaft Unbalance

The rotor shaft must be closely balanced in order to minimize lateral vibrations. This is typically done at the factory and again at the initial startup, as well as when modifications are made to the rotor, such as the reassembly of the retaining rings. Nevertheless, the migration of rotor parts because of high centrifugal forces and thermal expansion effects can cause an unbalance in the shaft, which will result in increased vibrations. Also, load sensitive unbalance may be present because of the causes discussed earlier in this chapter ("Short-circuited Turns" and "Blocked Ventilation Path"). Bearing vibrations can also be caused by gland seal oil that is out of the range of recommended operating temperatures, particularly temperatures that are too low.

When attempting to diagnose these types of problems, it is more important to monitor the trend in bearing vibrations than it is their absolute value. This is discussed in Diagnostic Box 5.10.

■ **DIAGNOSTIC BOX 5.10** ■

PROBLEM: SHAFT UNBALANCE

SENSOR INDICATION:

■ **The bearing vibration monitor** should be able to detect this condition if severe enough. This is the most direct means of detection. The problem is that the vibration sensors can detect many other problems as well, and there is no easy way to distinguish among the many problems that cause vibrations. The best indicators are bearing vibrations that have been steadily increasing over time, rather than a relatively sudden increase, when all other indications for field and rotor shaft problems are normal. If bearing vibrations increase dramatically after a shaft balance operation during an outage, then the balance procedure should be the prime suspect.

CORRECTIVE ACTION: In addition to the consequences for the rotor itself, the potential for damage from excessive rotor vibration comes from high vibration levels at the bearings, which would be felt by the structure of the generator and power station. Nonetheless, as long as the vibration levels are not beyond the manufacturer's limits and there are no other problems, the unit can be operated without concern. The unbalance is, in and of itself, generally not a problem for the rotor body or the shafts, as long as the vibration does not exceed the manufacturer's recommendations. However, if the levels of the rotor vibrations exceed the manufacturer's recommendation for safe operation, the turbine generator should be shut down at the earliest opportunity, and the manufacturer should be consulted for further recommendations.

Improper Cooling of Rotor During Shutdown

The generator rotor gets considerably warm during normal operation. After it is taken off-line, most manufacturers will recommend that the rotor be placed on turning gear. This will prevent bowing of the rotor that would take place if it were allowed to remain still while warm. Failure to do this may result in high levels of bearing vibrations upon return to operation, as the bowed rotor comes up to speed.

5-3.3 Problem Area #3: The Rotor Field Retaining Rings

As indicated in Section 5.1.3, retaining rings are mounted on both ends of the rotor field body in cylindrical rotor generators. They are the most highly stressed part of the rotor system. The following sections include a discussion of the stress corrosion cracking phenomenon and inspection techniques.

Stress Corrosion Cracking

There are two conditions required for stress corrosion cracking in those retaining rings whose material is susceptible to such stress cracking:

- High hoop stresses on the retaining ring
- Presence of moisture

Most large generators operate with high hoop stresses, so the only condition that can be controlled is the moisture content of the internal environment of the generator. Unfortunately, moisture can come from a number of different sources, including

- Generator hydrogen cooler leaks
- Generator water-cooled stator winding leaks
- Water entrained in the seal oil
- Water vapor in makeup hydrogen gas
- Water vapor in purging gases
- Atmospheric humidity (during outages)

Although there has not been an epidemic of retaining ring failures because of stress corrosion cracking, there have been a few failures over the years. When a retaining ring fails, for whatever reason, it has severe consequences for the rest of the generator. There is normally stator core iron and stator winding damage when this occurs, not to mention damage to the rest of the rotor field. Accordingly, most generator manufacturers are encouraging generator owners to replace their 18Mn–5Cr retaining rings with 18Mn–18Cr retaining rings. This new material does not appear to have the susceptibility to stress corrosion cracking that the 18Mn–5Cr material has. On the other hand, there is no need to replace the retaining rings if they do not show any indications or cracks denoting stress corrosion and there is no problem with moisture accumulation.

There are no on-line monitors for measuring stress corrosion on retaining rings. This is why no diagnostic box is included with this problem description. However, there are methods for minimizing moisture in the machine, and most manufacturers have systems that can help achieve this result. Typical systems include a hydrogen gas moisture separator or dehumidifier (gas dryer). To measure the effectiveness of these systems, dew point monitors can be applied at strategic locations in the gas path. Dew point monitors were covered in Chapter 3, and gas dryers are covered in Chapter 8.

The other major source of moisture exposure occurs during turbine generator maintenance outages. Both the rotor field and the stator should be maintained in enclosed areas with sufficient heating to prevent moisture condensation. Corrosion pitting and stress corrosion cracking occur in spots where retaining rings have been wet. Typical locations are shown in Figure 5.16.

Any inspection, even when the retaining rings are to be removed, should be made first with the retaining rings on the rotor. Some of the indications will be missed if the inspection is made with the retaining ring removed from the rotor because the indications will tend to close up when the ring is not under stress. It should be remembered that the retaining ring is under great stress at rest because of the interference fit with the rotor forging.

Surface inspection can be performed using high-frequency eddy current testing techniques and dye penetrant techniques. A borescope can be used for visual inspection of some otherwise inaccessible retaining ring locations and can be helpful in finding corrosion. To locate indications below the surface, ultra-

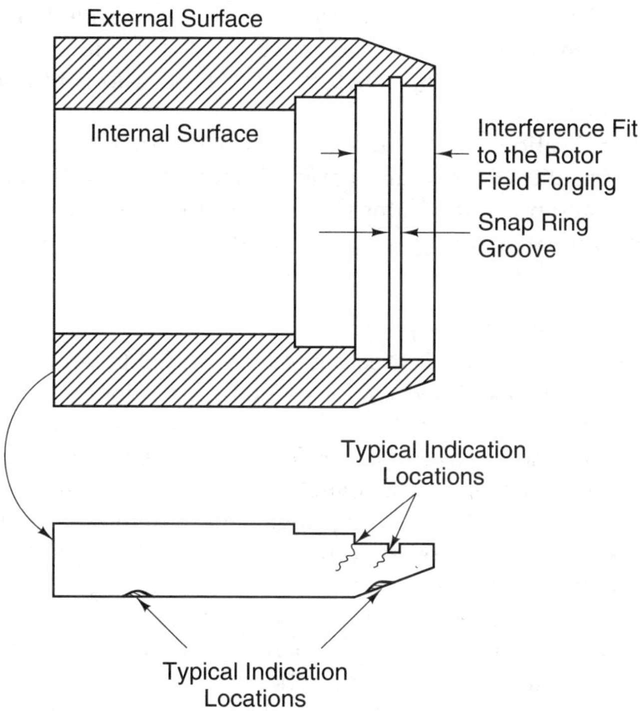

Figure 5.16 Generator rotor field retaining ring stress corrosion cracking characterization.

sonic testing (UT) techniques are required. Personnel with the proper qualifications should perform the UT examination. A couplant is needed to guarantee good contact between the UT equipment and the retaining ring. The couplant should be nonhygroscopic (not water based and not water absorbing) and noncorrosive.

A classification system has been developed to quantify retaining ring UT inspectability. Cold-worked retaining rings are usually made of low structural noise materials known as Class A Materials. The materials can be UT examined at frequencies between 2.5 and 5.0 MHz. Retaining rings with higher structural noise are denoted as Class B retaining rings, and they must be tested at lower UT frequencies. Warm-worked retaining rings fall into Class C because of high structural noise. These cannot be reliably inspected through the UT method. Whereas these designations are specific to one company, they clearly point out the fact that not all retaining rings can be practically analyzed through UT.

Directing ultrasound circumferentially around the retaining ring can detect radial-axial indications. Directing the ultrasound toward the nose of the retaining ring can locate radial-circumferential indications. Because of the shape of the rotor field forging teeth and the interference fit of the retaining rings, ultrasonic information can be difficult to interpret. It is recommended that areas be investigated using advanced techniques, such as the *mode converted longitudinal wave* technique and the *time-of-flight diffraction* technique. Discussion of these, however, is beyond the scope of this book.

If indications are located, an informed decision needs to be made regarding the repairability of the retaining ring. Minor surface indications can be removed by grinding and polishing. Consideration should be given to replacing 18Mn–5Cr retaining rings with 18Mn–18Cr retaining rings if several large indications are located and verified.

Retaining Ring Liner Slippage

Quite often when inspecting older rotor fields, vent holes in the retaining ring are partially blocked because of retaining ring liner slippage. This problem normally creates heat-affected zones in the rotor field and results in high vibration. Either the vent holes have to be drilled out, or the retaining rings have to be pulled and the liner replaced. Fortunately, only retaining rings made of magnetic steel have vent holes in the retaining rings. Few generators are of this type. If the holes are to be drilled out, however, the repair would have to be done *very* carefully to avoid damaging the top turns of the field winding. It is recommended that the manufacturer be consulted prior to carrying out this repair.

There have been cases in which insulation has migrated out between the rotor body and the retaining rings where the latter are mounted at its outboard end (only on old rotors). Also, there have been cases where the slot armor has migrated up along the sides of the rotor slots (only on old rotors), but it cannot get past the wedge. This can lead to a ground failure of the field winding.

SUMMARY

This chapter briefly described the generator rotor and the field winding. It discussed the various sensors and monitors that are used to identify problems with it and listed some of the problems that are common to this component of the generator. There is considerable overlap between this chapter and Chapters 2, 7, and 12 because control of the field winding is significant in the operation of the power system.

REFERENCES

[ANSI] American National Standard C50.18-1989, "Requirements for Cylindrical Rotor Synchronous Generators", American National Standards Institute, New York, 1989.

[Bowler, 1994] Bowler, C., personal communication, May 1994.

[EPRI, 1993] "Generator Retaining Ring Moisture Protection Guide," EPRI TR–102949, Project 2719 Final Report, EPRI Nondestructive Evaluation Center, North Carolina, September 1993, p. A–4.

[IEEE, 1983] IEEE Standard 115-1983, "Test Procedures for Synchronous Machines," Institute of Electrical and Electronics Engineers, New York, 1983.

[Nilsson, 1993] NILSSON, N.E., and J. MERCURIO, "Evaluating the Service Degradation of Large Hydrogen Cooled Generator Rotor Fields," *IEEE Transactions on Power Apparatus and Systems,* Vol. PAS–102, No. 9, September 1983, pp. 2950–2958.

[Westinghouse, 1964] *Electrical Transmissions and Distribution Reference Book,* Central Station Engineers of the Westinghouse Electric Corporation, East Pittsburgh, Pennsylvania, 4th ed., 1964, pp. 149–152.

CHAPTER 6

The Generator Core and Frame

The generator core is the third major component of the generator. It is the main magnetic path for the useful flux in the stator. The frame is the fourth and last of the major components. It provides the support for the many elements in the stator, as well as serving as the enclosure for the core, the stator winding, and the rotor.

6.1 DESCRIPTION OF THE GENERATOR CORE AND FRAME

This section will provide a descriptive overview of these components.

6-1.1 The Generator Core

The stator core of a large turbine generator may contain more than 200,000 thin punched laminations of electrical grade silicon steel (called *punchings*), which are approximately 0.014 in. thick. Each punching is insulated with a thin coating of varnish, enamel, chemical treatment, or other form of insulation to block the flow of destructive currents between punchings. Figure 6.1 shows a photograph of a single punching.

With the frame positioned vertically, these punchings are stacked onto stacking bars located at the back of the core, side-by-side, to form a complete 360° ring around the stator bore where the rotor is inserted. This is referred to as *segmenta-*

Figure 6.1 Sketch of single punching.

tion of the punchings; between 12 and 24 of these are required to form a complete ring around the stator bore. Each of these rings forms one layer of the core in the axial direction. Successive core layers are staggered relative to the locations of the gaps between punchings to ensure an integral mass of steel. Sufficient layers are stacked onto stacking bars supported by the frame and tightly clamped together axially to form the full-length solid mechanical body known as the core. If the core is to have radial gas cooling, then the punchings are stacked in packages separated at intervals by spacers that are inserted to provide radial cooling ducts.

After the core is completely stacked, final clamping pressure is applied, and the core is permanently clamped with end flanges and clamping fingers. Some manufacturers use additional insulated axial through-bolts to apply the clamping pressure. Core lengths can vary from less than 100 in. up to 330 in. for medium and large turbine generators. Figure 6.2 shows the core of a generator before the stator bars are wound within it.

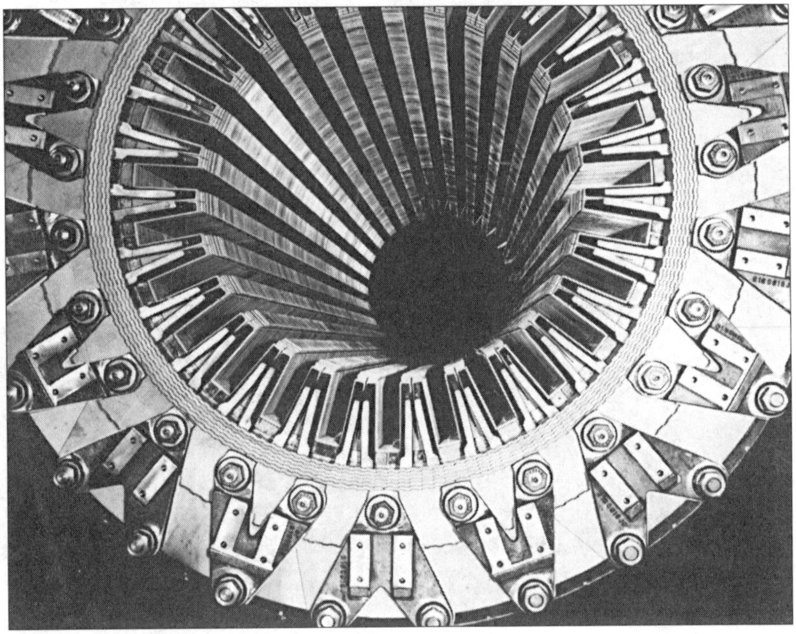

Figure 6.2 Stator core before placement of stator bars. (Used by permission from Westinghouse Power Generation, Orlando, Florida.)

The core in a two-pole generator experiences a rotating inward magnetic pull at the two diametrically opposite locations of maximum air-gap flux density. These forces are called *ovalizing forces* because they cause the core to deflect a few thousandths of an inch from a circle to an oval. A mechanical engineer would call this a "four-node pattern of deflection" because, at any instant, there are four locations around the periphery of the stator core where the deflection goes through zero (has a node). The result is a vibration at a frequency twice that of the power system (120 Hz), which must be isolated from the frame and foundation so as not to transmit it to the earth and building. For this reason, the stacking bars in two-pole generators are commonly isolated from the frame by spring mountings.

The mode of deflection of the stator core of a four-pole machine comes from radial inward pull at four locations, 90° apart, giving a deflection pattern with eight nodes. The core is much stiffer to eight-node deflection, and the level of the 120-Hz core vibration is about one-sixth of that of a two-pole generator. It is not necessary to isolate this smaller magnitude of core vibration from the frame and foundation; therefore, the stacking bars are welded directly to the structure of the frame.

Most generator designs provide for gas cooling of the core (using hydrogen or air). Space blocks between the punchings in selected pairs of segmented layers are used to form radial vents (called *ducts*) for the gas to flow radially from the back of the core towards the bore, or vice versa. These radial ducts subdivide the core into packages of laminations. Each package is cooled on both sides by the radial flow of cooling gas in the ducts.

Other designs include water-cooled cores in which the coolant flows from end to end through axial passages within the core. This type of cooling is used on a very small portion of the turbine generators in the United States.

The top ends of the punchings, as can be seen in Figure 6.1, contain teeth, which, when stacked next to other punchings, will combine to form the slots in which the armature bars are placed. Each punching in the example of this figure has two teeth, which define one slot section in the center and two half-slots on each side. On each side of the punching, the teeth will combine with those of the neighboring punchings in the same layer to form two more slots. Because the segmentation is staggered, every slot along its length will be contained within some punchings that are integral, as well as others that are made up of two half-slots of neighboring punchings.

The most susceptible areas in the core for overheating are its ends. The end laminations must carry the stray magnetic flux coming from the end region in addition to the useful flux that comes from the air gap. The end flux impinges on the broad face of the punchings and produces eddy currents therein and consequent losses. The teeth are commonly slit to reduce these currents and the consequent heating. The yoke of the end laminations and the clamping flanges and press pieces may be shielded to reduce the losses that would result from end fluxes, especially in generators with conductor-cooled windings. In these machines, the end-flux densities are substantially higher than those in conventionally cooled generators.

The flux field in this location is quite complex in nature and is affected by the magnitudes and angular positions of the currents in the armature and field windings. At constant armature current and voltage, as the field current is reduced and the power factor (PF) changes from lagging to unity to leading (also called going from an *overexcited* to an *underexcited* condition), more loss is developed in the core ends, and they necessarily become hotter. Therefore, it is common to reduce the maximum allowable armature current for PFs lower than .95 leading. This is reflected in the underexcited region of the generator reactive capability curve shown in Section 2.2.

The basic functions of the core, therefore, are

- To carry the generator's useful magnetic flux
- To support the armature windings in the slots

- To carry the electrical torque produced by the interaction of the armature current and the air-gap flux, this torque being equal and opposite to that imposed on the rotor shaft system

6-1.2 The Generator Frame

The stator frame is all that one sees of a generator in a power station. It tends to be long and cylindrical in shape. Deviations from the cylindrical shape are introduced to accommodate coolers. One or two terminal boxes are commonly welded onto the bottom of the frame to house and support the bushings of the armature winding, if the generator is hydrogen-cooled. Separate end shields are bolted onto the ends of the frame to contain the hydrogen (or air ventilation circuit) and, in many designs, support the rotor through bearings and house the hydrogen seals. Once attached, the end shields become a functioning part of the frame.

The stator frame is the shell structure upon which the core is mounted. It serves the following functions:

- Support for the stator core
- Containment for the cooling gas, air or hydrogen
- Housing for the coolers (gas-to-water heat exchangers)
- Baffling for machine ventilation

There are generally two types of frame designs, an *integral stator* frame and a *frame with a separate cage to support the core.* In an integral stator, the frame holding the core also acts as the casing, and is one complete structure. In the caged core, the core is mounted into a cage made up of a weldment of annular plates and axial key bars. In a two-pole generator, the cage is then spring-mounted from the outer frame. For a very large generator, the cage is removable, axially, from an outer frame, called the *casing,* so that it can be shipped separately from the frame. The casing is also used to support the coolers and is ducted and baffled to form specific paths for the flow of the cooling gas (air or hydrogen).

The frame includes feet for mounting the generator on the foundation. These feet must accommodate not only the weight of the generator, but also the upward and downward loads created by the overturning torque of high-current transients, such as nearby sudden short circuits in the power system. The frame stiffness and natural frequencies must be such that a natural resonance with the main stimulating frequency (twice the power system frequency) is avoided.

Additionally, the frame must avoid natural resonances with the mechanical frequency of rotor vibration. These mechanical frequencies are power system frequency (60 Hz) for two-pole machines and half-power system frequency (30 Hz) for four-pole generators.

Whereas the cross-sectional shape of the frame is typically round, the frames are often flat-sided in order to better accommodate the hydrogen coolers. Figure 6.3 is a photograph of a stator frame.

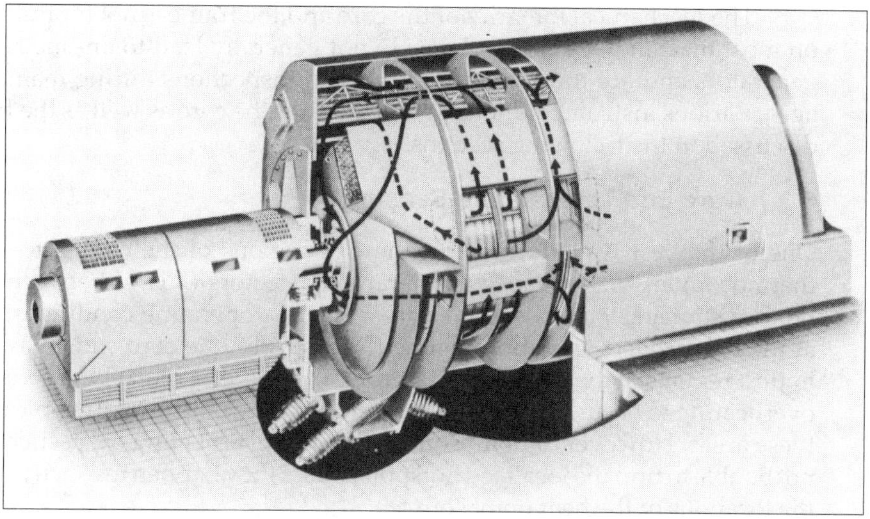

Figure 6.3 Stator frame. (Used by permission from Westinghouse Power Generation, Orlando, Florida.)

6.2 PARAMETERS, SENSORS, AND MONITORS

Some generators have a well-instrumented core, while others have no instrumentation at all. The level of instrumentation may well be determined by the experience the owner and/or the manufacturer has had with the core of machines of similar design. Historically, major core failures have been few; however, the outage time required to restack a core can be rather lengthy.

The most common way to monitor the core is indirectly through the generator core/condition monitor (GCCM) described in Chapter 3. Any pyrolysis of the insulation between the punchings or of the paint on the core surface will produce products that are released as very small particulates, called *condensation nuclei,* into the hydrogen gas. These particulates are detected by the GCCM. While the GCCM will alarm on pyrolysis products from any other source, it was originally developed for application to the core, hence its name. Samples of the particulates can be collected and analyzed, and in some cases this can help locate their source. In some machines, the various components are located with different identifiable materials (tagging compounds) to aid in this process.

Unfortunately, indirect monitoring of internal heating of the core through the GCCM may not always be possible because the inorganic insulation material used by some manufacturers has such a high heat resistance that it will pyrolysize only at very high temperatures. Such temperatures are higher than typically present in the core, except at advanced stages during electrical failures. High surface temperatures on the core of one of these generators, however, may be detected through pyrolysis of surface paints, as was indicated earlier.

The mechanical integrity of the core and the frame is not typically monitored on most machines. Such problems do not generally lead to on-line failures of the generators and are thus delegated to visual inspections during maintenance outages. Various instrumentation used to monitor the core as well as the frame will be discussed in the following sections.

6-2.1 Core-End Temperature Sensors

One of the best ways to monitor the end regions of the core is to install either thermocouples or resistance temperature detectors (RTDs) between the laminations in strategic locations in the core. For some operating conditions, particularly at the overexcited operating regime (leading PF), the core ends may become the hottest regions of the core, and therefore may be the most likely general location for overheating to occur. Because of this, temperature detectors are typically placed in these areas. However, if point sources of local overheating exist, such sensors will not be able to find the localized hotspots unless they happen to exactly coincide with the locations of the heat point source.

Electrical failures of a stator core can be expected to start from locations of weakness in the punching insulation. These points may be anywhere in the core but it is near the ends of the core where the parameters that are conducive to a failure are more severe.

6-2.2 Slot RTDs

Slot RTDs used for monitoring the temperature of the armature windings can also be used to detect core problems. Because these sensors are typically not at the ends of the windings, they do not overlap the core-end sensors. Nevertheless, most generators have them only in selected slots, and others do not have them at all. Furthermore, being a temperature monitor of appreciable length, they measure an average temperature and therefore also suffer from the inability to detect local hotspots, even when the hotspots are located in the vicinity of the RTD. Lastly, their location between the top and bottom coils in the slot further isolates them from any source of the core heating.

6-2.3 Core Vibration Monitor

Temporary vibration sensors can be placed directly on the core to monitor its vibratory activity. This is typically done only if there has been a problem and additional understanding of the vibration of a particular generator is needed. At the present state of the art, it is not possible to install permanent instrumentation of this type because of the relatively short life of the sensors.

6-2.4 Volts-per-Hertz Relay

The measure of volts per hertz is an indication of the flux conditions in the core. This is derived from the following equation:

$$E_{\phi} = 4.44 * \phi * f * N_t * 0.01 \tag{6.1}$$

where

E_ϕ = armature phase voltage (volts, rms)
Φ = flux in the machine (megalines)
f = frequency (Hz)
N_t = effective number of series turns per armature phase per circuit, linking the flux

The above equation can be simplified to the following version by using a proportionality constant to combine all of the generator constants:

$$\phi = \frac{K * E}{f} \tag{6.2}$$

where

K = proportionality constant
E = line-to-line voltage of the generator, in volts rms

Thus, by measuring this ratio of voltage divided by frequency, the flux flowing through the core can be monitored. It is common to present this quantity in percent, where 100% is the value of the ratio at rated frequency and voltage. A frequency-sensitive voltage relay is generally used to monitor this condition. A curve showing the permissible volts per hertz in the core is usually available from the generator manufacturer. Typical values, however, are shown in Figure 6.4. Most manufacturers will recommend that the generator be tripped when the suggested limits are reached.

6-2.5 Generator Core/Condition Monitor

A GCCM is available for the detection of severe overheating of any part of the stator core. This is important to the surveillance of the core because there is no other means of detecting local hotspots unless the very unlikely situation arises where there happens to be an embedded point temperature detector in the location of the overheating.

This monitor evaluates the cooling gas in the generator continuously. If small particulates appear, as are produced by severe overheating of insulating materials, it records the change. If the particulates become sufficiently numerous to be of concern, the monitor issues an alarm.

The GCCM is described in Chapter 3, so its details will not be discussed here.

Percent Overflux	Allowed Time of Duration
118	2 seconds
110	45 seconds

Figure 6.4 Typical allowable overfluxing of generator core.

6.3 MONITORING AND DIAGNOSTICS

There are basically two problem areas in this chapter: the core and the frame. The core will be treated first.

6-3.1 Problem Area #1: The Core

There are various problems that can affect the core. These are

- Overloading at leading PFs
- Blocked core cooling ducts
- Loosening of the core
- Electrical failure of core insulation

Overloading of the Core at Leading PFs

As discussed previously, overheating of the core end regions can result if the generator is operated outside its design limitations in the leading PF region or if abnormal problems develop. Design operating limitations are indicated on the generator capability curve provided by the manufacturer. Abnormal problems can occur from damage to the end laminations, resulting from mechanical or thermal damage, or a combination of both.

DIAGNOSTIC BOX 6.1

PROBLEM: **OVERLOADING OF THE CORE AT LEADING POWER FACTORS**

SENSOR INDICATION:

- **Core end temperature sensors** may detect overloading once the heating has begun. If there are a number of sensors distributed throughout the core end region, they all should indicate high temperatures.
- **Electrical meters** in the control room, such as MVA and PF, can indicate whether the generator is being operated within its reactive capability curve. Most monitoring systems and plant computers have built-in algorithms that can continuously plot the operating point of the generator directly from the sensor readings and alarm if the operation migrates to a point outside of the reactive capability for the operating hydrogen pressure (if it is a hydrogen-cooled machine). This is the earliest means of detecting this problem, even before the overheating begins to happen.

CORRECTIVE ACTION: The operating point of the unit should be immediately brought back within the capability curve through one of the following means: 1) reduce MW load, 2) increase PF, or 3) increase hydrogen pressure if it is not already at maximum.

Blocked Ventilation Ducts in the Core

Ventilation in the core can be blocked by foreign material. This will result in reduced cooling of the core laminations (punchings). This condition may be difficult to detect because the heating may be localized and can occur at any location along the core's length and around its periphery. The only situation that can be detected is blockage near one of the point temperature sensors. The applicable sensors, if they exist, are typically in the end regions of the core. If the overheating is severe enough and broadly spread, it could possibly be sensed by a nearby slot RTD.

An overheated core caused by abnormal problems could occur while operating within the design reactive capability limits of the generator. Thus, there may not be any indication of a problem from normal control room electrical instrumentation. In a core with a conventionally cooled armature winding, excessive armature current will cause increased core tooth temperatures because the teeth carry the heating load away from the armature coils. In the core of a generator with a conductor-cooled armature winding, it is not likely that an excessive generator load will result in elevated core temperatures unless they are occurring in the end regions at a time when the generator is underexcited.

■ DIAGNOSTIC BOX 6.2 ■

PROBLEM: BLOCKED VENTILATION DUCTS IN THE CORE

SENSOR INDICATION:

■ **Core end temperature sensors** may detect this problem once the heating has begun if one of them is near the point of blockage. If there are a number of these sensors distributed throughout the core end region, then not all of the sensors may indicate high temperatures.

■ **Slot RTDs** may also be used to detect this problem if the armature winding is conventionally cooled. But in a core with a conductor-cooled armature winding, slot RTDs are not expected to detect excessive core temperatures because of ventilation blocking. In any case, RTDs are not a necessary piece of evidence to support the hypothesis. But if an RTD does detect a high reading, not all of the RTDs would be expected to read high, because this is typically a random occurrence. A combination of partially blocked ventilation and excessive generator load may combine to cause the affected parts of the core to have elevated temperatures.

■ **The generator core/condition monitor** will probably detect the problem in most generators if the overheating is sufficient to cause pyrolysis of the insulation and/or paint. Thus, the generator core/condition monitor acts as an indicator of the problem as well as of its severity. If no temperature sensor is near the blocked duct, the GCCM may be the only way to detect core overheating because of blocked ventilation.

CORRECTIVE ACTION: If the GCCM alarms and the alarm is verified, the situation has probably become severe. Whether or not this has been detected by other means, the generator gas should be sampled for detection of the particulates when a GCCM alarm is verified. It would be prudent to reduce the load or bring the unit off-line for inspection of the problem. Consult with the equipment manufacturer for recommendations.

Stator Core Loosening

Loosening of the core results from reduced clamping pressure, likely caused by the mechanical relaxation of the punching insulation. The aggregate length of this insulation may be several inches in a long stator.

Core loosening tends to reduce core stiffness and, if severe enough, will increase vibrations, especially the ovalizing force vibration common to two-pole machines. The level of this vibration is less in the core of a four-pole generator, but core loosening in any machine should be taken seriously and corrected at the first opportunity.

Increased vibration of the core may be felt in increased frame vibration. There are two factors that tend to impair this means of detecting core vibration: First, in two-pole machines where normal core vibration is the highest, spring mounting tends to isolate core vibration from the frame, which is precisely why spring mounting is used; second, in a four-pole generator, the level of core vibration is much less than that in a two-pole machine. However, because the core is directly connected to the frame, any change in vibration would be much easier to detect.

■ DIAGNOSTIC BOX 6.3 ■

PROBLEM: STATOR CORE LOOSENING

PERIODIC INSPECTION:

Probably the most important action that can be taken is to give the core and its mounting an expert inspection at every planned maintenance outage. Any trend toward looseness can easily be corrected at that time.

SENSOR INDICATION:

■ **The human ear** is, in fact, a very sensitive instrument. An experienced individual may be able to distinguish the noise caused by the vibrating core by standing near the frame during operation and recognize changes that took place since he observed the core on a previous visit.

■ **A core vibrometer** may be installed temporarily to explore a problem that is observed, but in a hydrogen-cooled generator, this may be very difficult. A permanent sensor of this type is not available at this time.

■ **A frame vibrometer** can be installed temporarily to gain additional information regarding a problem that has been identified or is suspected. This, however, can be used only with difficulty to detect loosening of the core. This type of sensor has the advantage, however, of being capable of identifying the frequency structure of the frame vibration. Generally the predominant vibration of the frame is at the frequency of rotor vibration, which is at power system frequency (60 Hz) for a two-pole generator or at half that frequency for a four-pole machine. If the core becomes loose and the core vibration increases sufficiently, then a 120-Hz component of the frame frequency may also be detected.

CORRECTIVE ACTION: The consequences of this condition can vary widely, from almost none to serious, depending on the design of the core and the severity of the vibrations. Therefore, generalizations cannot be made here. If the periodic inspections mentioned above are carried out, emergency procedures will probably not be necessary. If a problem has been detected or suspected, seek the manufacturer's recommendation for further course of action.

Overfluxing of the Core

Overfluxing of the core is defined as excessive excitation for the rotational speed of the unit. This is manifested as a relationship between the generator terminal voltage and its frequency. For that reason, it is also referred to as excessive volts/hertz. Refer to section 6.2.4 for a more thorough explanation.

Overfluxing of the core can take place when the generator is tripped from full load, but the excitation system does not trip, and the excitation control is set on manual control, maintaining the field current at the full-load value. Poor connections in one phase of the terminal voltage transformers (VT) while the unit is being brought on-line and the excitation is being increased manually has also been known to cause severe overfluxing. In one documented case, one VT secondary connection became completely loose, causing the secondary to act as a voltage divider. Because the burdens of the secondary legs still connected were of nearly the same value, the voltmeter read a value equivalent to approximately half of the actual voltage on the generator terminals. In such circumstances, the tendency is to increase excitation manually to reach rated terminal voltage, resulting in overfluxing.

During severe power-system disturbances, the speed of the rotor may drop substantially, perhaps because of a severe overload on the entire system relative to the remaining connected generating capacity, whereas the voltage regulator is attempting to maintain the preset voltage level in spite of the decreasing speed. It would do this by increasing the field current. Here the armature voltage would be maintained approximately constant while the frequency is reduced, increasing the volts/hertz and, therefore, also the flux in the generator. Other occurrences, such as overvoltage during tests and/or overvoltages as a result of power system transients, can cause an overfluxing condition.

Oftentimes, the failure of the core is not due to one particular incident of overfluxing, but rather to a number of lower level overfluxings. Therefore, it is important to maintain alertness for the types of operation that can place the generator at risk of overfluxing.

Along these lines, it may well be that serious overfluxing does not cause a core failure at the time of the damaging event itself. Nevertheless, the resulting high interlamination voltages can break down some of the points of insulated contacts and thus weaken the core so that subsequent overfluxing, even at lower levels, may lead to an electrical core failure.

The effect is that during overfluxing, the voltage between laminations increases significantly, which can result in local short circuits between laminations. The consequence of these short circuits is that circulating currents may develop and then persist between the punchings, which would cause localized heating. If only two laminations make contact, experience has shown that there is typically not enough power behind the voltage between the laminations to cause melting. When several laminations are short-circuited in the same area, there is more voltage behind the circulating current (and thus more power). Consequently, more loss is generated with essentially the same heat flow path. This will lead to higher temperatures. The larger the number of faulted laminations, the more likely that a runaway condition may develop.

If the fault can be located and the area is small enough, the lamination can be cut out, and no filler may be required. If a filler is used to replace a number of removed laminations, many considerations must be taken into account, including the added resistance to heat flow and the risk of increasing the temperature of the core nearby. It is strongly recommended, however, that the manufacturer be consulted before proceeding with such a repair.

A core failure represents serious damage to the generator, affecting unit availability and often requiring major repairs. Depending on the severity, core repairs can result in extended outages involving major disassembly. Restacking or partial restacking of a generator core will require partial or full winding removal, with the potential for damaging or degrading the stator insulation system. Associated damage may be severe enough to require complete rewinding of the stator.

■ DIAGNOSTIC BOX 6.4 ■

PROBLEM: OVERFLUXING OF THE CORE

SENSOR INDICATION:

■ **A volts/hertz relay** will likely indicate overfluxing and alert to the possibility of electrical failure. Any resulting damage may be due to this one particular incident of overfluxing or to a previous history of similar events.

■ **Voltage regulator on-line indication** is a simple means of detection if this, in fact, contributed to the overfluxing. If the regulator is still on-line, but the generator is rolling down, this should be an obvious alarm. If off-line, the operator should see a noticeable MVAR input to the GSU transformer in the MVAR recorder.

CORRECTIVE ACTION: If an overfluxing condition is determined to have taken place, the unit should be brought off-line and the excitation removed quickly to prevent further expansion of the failed region. The advice of the manufacturer should be sought to determine if the event was severe enough to warrant opening the machine and giving it an expert inspection.

Electrical Failure of Core Insulation

Overfluxing (discussed in the previous section) is just one of a number of potential causes of an electrical failure of the core. There are various other causes of physical damage to the core that can also cause short circuits between punchings, leading to unwanted currents and local heating. Tools can do the damage when wedges or coils are being removed or replaced. Foreign objects that find their way into the air gap can short-circuit punchings or wear and distort the punchings so that they make contact.

Also, if individual laminations have become loose, they may rub and damage the electrical insulation of the affected punchings, leading to short circuits between laminations, which may, in turn, cause heating of the affected local areas. In severe cases, melting of the laminations may result, as described in Section 6.3.1. This local damage can be difficult to detect because, unlike the case of overfluxing, there is no cause that can be detected without opening and inspecting the machine.

Because the damage may be hidden or in locations where access is difficult, the detection means tend to be indirect, thus not very accurate. Even then, there is no easy way to detect the problem, and certainly no guarantee that it will be found. Surface damage may be detected through expert inspection with the machine off-line. There exist some tests, however, that can better detect and locate failed areas of the core with the rotor removed from the machine, specifically, the ring test and the El-Cid test.

The *ring test* (also known in some areas as the *loop test*), uses a heavy exciting current to produce rated-level flux in the core yoke. The exciting current is produced by wrapping a loop of cable around the core. An area of electrical core failure will heat up under this test as it would in operation. If the points of fault are located deep within the core, a period of time is needed for the heat to reach the surface of the stator bore where it can be detected. An infrared scanner is used to detect hotspots by examining the entire bore of the stator. Installing the supports for the heavy cables used for this test presents some risks of damage to the core; therefore, extra caution should be exercised in making this test.

The *El-Cid* test uses a light exciting current to produce a low level of flux and interlamination voltages. This will sense currents that circulate between punchings. It uses a small bridging sensor coil that is moved by hand from tooth to tooth along the length and around the circumference of the stator bore, covering its entire inner surface.

Utilities and manufacturers in general prefer the El-Cid to the ring test for a variety of reasons, including ease of installation, requirement for less expensive equipment, and safety for test personnel as well as for the generator itself. Nevertheless, the high-current ring test is more discerning of failures because it is done at full flux, and the normal level of voltages between the punchings is present. This is particularly true of failures deep within the core, which may not be detected with the El-Cid test.

6-3.2 Problem Area #2: The Generator Frame

The problems that can develop in a frame tend to be mechanical. They can be due to steady stresses in a hydrogen-cooled generator or to vibration. In most areas, a frame is not subject to electrical, magnetic, or heat-related stresses. It is, however, subject to significant mechanical stresses, which were discussed previously in this chapter.

Steady stresses can cause cracks in the frame, if there is a weak area or an area of residual stresses that add to the steady in-service stresses and possibly vibratory stresses. Abnormally high vibrations may be a symptom of a local or a (less likely) general resonance condition in the frame. But vibrations are also symptoms of other core-related problems, as seen in the previous section, but it is not expected that vibrations in the core will reach severe levels in the frame, for the reasons discussed earlier.

■ **DIAGNOSTIC BOX 6.5** ■

PROBLEM: FRAME CRACK

SENSOR INDICATION:

■ **The hydrogen pressure monitor** may indicate a periodically decreasing hydrogen pressure inside the unit, which can be indicative of a continuous leak. This may be characterized by a relatively rapid drop-off of pressure after each top-off.

■ **A hydrogen detector,** placed outside the frame, may indicate the presence of hydrogen that has escaped from a crack in the frame.

■ **A tracer-gas detector** can be used for leaks that are difficult to detect. A small quantity of a tracer gas that is easier than hydrogen to detect is introduced into the frame of the generator. Then a detector designed for the tracer gas is used outside of the generator to detect the presence of that gas and, therefore, the location of the leak.

- **A hydrogen gas consumption meter** may indicate larger-than-normal usage of hydrogen gas from the supply.
- **The frame vibration monitor** may register a high vibratory level, which will probably be at the running frequency of the rotor (60 Hz for a two-pole machine and 30 Hz for a four-pole unit).
- **The seal oil system** should be operating normally in the face of these leaks. This will eliminate the possibility of a hydrogen leak through the seals as the cause of the loss of hydrogen gas from the casing.
- **Hydrogen cold gas temperature** should be normal. This is discussed here only to eliminate the possibility of a gas leak through the hydrogen coolers because of the presence of hydrogen in the water side of the cooler. This evidence, however, is not absolutely necessary for the presence of a gas leak into the cooler water. By eliminating gas leak into the hydrogen cooler as one possible cause, the indications would point toward a stator frame crack.

CORRECTIVE ACTION: If the leak is not severe, continue monitoring the hydrogen pressure to ensure that proper cooling gas is present. Also, monitor the level of hydrogen outside the generator. This can accumulate in pockets under the machine. If the leak is significant and the amount of escaping gas could cause an explosion and/or fire, then the unit should be shut down as quickly as possible and the crack repaired. In any case, the level of hydrogen consumption should be compared with the manufacturer's recommendations and, if excessive, an early shut down should be planned so that the cause of the problem can be found and repaired. Consult with the manufacturer for recommendations on the action to be taken and the repairs to be made.

SUMMARY

This chapter has described the generator core and frame and the problems that can affect their on-line operation. The core is a significant component because it serves two important purposes: It is the mechanical support for the stator winding, and it serves as the medium for the magnetic flux circuit surrounding the stator windings. The core is stressed electrically as the result of the magnetic flux flowing within it, and it is stressed mechanically by the magnetic forces placed within the slot by the high stator currents flowing in the coils therein. Core failure, whether mechanical or electrical, can result in significant outages.

The frame serves as the support for the core and as the containment for the pressurized hydrogen gas. Because the frame does not have an electrical function, problems with the frame are less likely than those in the stator winding, the core, and the rotor, as well as generally less serious.

CHAPTER 7

The Excitation System

7.1 INTRODUCTION

As stated in Chapter 2, a steam turbine generator has a rotating magnetic field consisting of a DC winding embedded in slots in a solid forging. The voltage generated in the armature winding is proportional to the strength of the magnetic field, which is, in turn, proportional to the amount of current in the field winding. The effects of the armature current on the generator terminal voltage and field current requirements are described in Chapter 2. These effects are best shown on the generator saturation curve of Figure 2.18 and reproduced here as Figure 7.1.

The excitation system must be capable of supplying the required field current for all permissible load conditions and must also be able to be controlled either manually or automatically by a control mechanism. The control system for the field current is referred to as an *automatic voltage regulator* (AVR), but modern excitation control systems include many features over and above the function of maintaining a constant generator terminal voltage. These features will be discussed in this chapter.

Some means for removing the excitation during normal and emergency shutdown of the generator must be provided in the control equipment. In the past, when generators were smaller than the major units of today, a main field circuit breaker and field discharge resistor were used. However, the heavy direct currents required for excitation of modern large turbine generators present a challenging

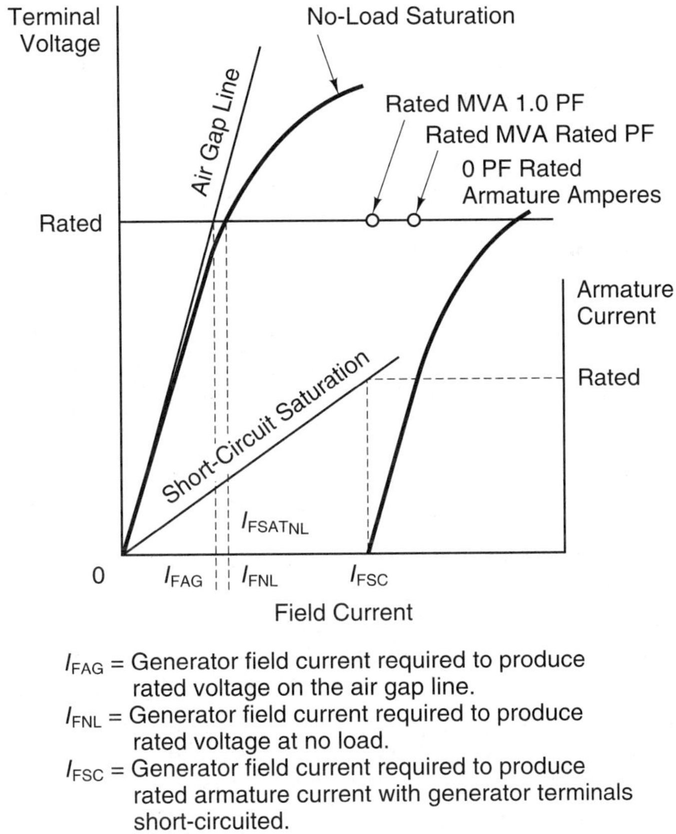

Figure 7.1 Generator saturation curve.

duty to DC interrupting technology. For this reason, and because of changes in the types of excitation systems, the large generating units in service today generally do not have main field circuit breakers. The means for excitation removal employed will depend on the type of excitation system used, as described in the following sections of this chapter.

It must be remembered that, regardless of how the excitation is removed, the magnetic flux and generated voltage will decay with some time delay. The time constant at which the flux decays depends on the generator armature current at the time of excitation removal. If the generator is at no-load, the effective time constant is the open circuit time constant of the generator, T'_{do}, or typically about 6 seconds. If the generator is feeding a fault, the time constant will be considerably less, again depending on the fault current magnitude. The fault current has a demagnetizing effect as described in Chapter 2; therefore the flux is forced to decay faster.

7.2 TYPES OF EXCITATION SYSTEMS

Excitation systems can be classified into two general categories: 1) rotating exciters and 2) static exciters. Rotating exciters, in their most basic terms, take the energy input from the rotating shaft and convert it into DC. Static exciters, on the other hand, take their energy source directly from the electrical output of the generator. These next few sections will describe the various types in more detail.

7-2.1 Rotating DC Exciters

Figure 7.2 shows a rotating DC exciter system. The exciter may be driven from the shaft of the generator, in which case it is called a direct-connected exciter, or it may be driven by a motor. The exciter must have a commutator to convert the generated alternating voltage to a direct or unidirectional voltage. The generator field must be connected to a set of collector rings to accept the output of the exciter. As generating units became larger, the size and complexity of the commutators and associated carbon brush rigging increased. In addition, the high speeds of steam turbine generators complicated the commutation problems. In the 1950s there was a trend away from the direct-connected exciter in favor of the motor-driven exciter, which could be operated at a much lower speed than that of the generator.

Figure 7.2 also depicts an exciter with three fields. The exciter main field and a rheostat in series are connected across the terminals of the exciter armature. The other two fields are called *control fields*. Most of the excitation for the exciter is supplied by the exciter itself by means of the shunt main field. The control fields are connected such that the magnemotive forces (mmfs) are in opposition; that is, one of them is connected to assist the main field, and the other one is connected to oppose or reduce the effect of the main shunt field. They are referred to as *boost* and *buck* fields, respectively.

The energy for the control fields is supplied by a *pilot exciter,* usually a small permanent magnet generator (PMG), through the AVR. A base level of excitation is set by the amount of resistance in the main field rheostat. If the resulting main generator output voltage is the same as the setting of the AVR, the currents or mmfs of the boost and buck fields cancel out, and there is no net effect from them. If, however, a change in excitation is called for, the control system can provide an increase or decrease in the amount of excitation. Assume, for example, that a boost in excitation is called for. The voltage across the boost field is increased, and that across the buck field is decreased. The effect is similar to the operation of a push-pull amplifier in an electronic circuit. This increase in excitation will cause the exciter output voltage to rise, which, in turn, causes an increase in the current in the main shunt field. The control fields then supply only the amount of excitation not supplied by the main field for any given operating condition. For the system just described, the AVR could utilize an electronic or magnetic amplifier to supply the control fields.

Some earlier systems did not use the three-field exciter described above. On older, smaller generators, the exciter had only the shunt field, and the automatic voltage control was done by varying the amount of resistance in series with the

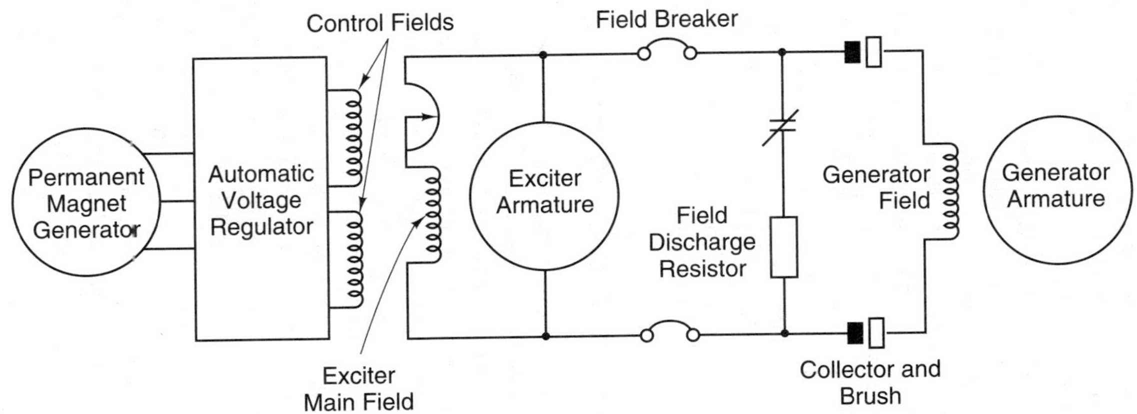

Figure 7.2 Direct current exciter system.

shunt field. Of course, some means for *field flashing* (briefly energizing the field to establish magnetic flux) was required to establish an initial buildup of voltage. In older, larger systems, the main exciter field was supplied by the output of a pilot exciter, and the alternator voltage was controlled by variation of the pilot exciter field current. As one might suspect, the older types of excitation control systems, in which the resistance in the field circuit was changed to change the alternator voltage, were very slow in responding to a call for a change in excitation. In addition to the time required to change a rheostat position, there was, and still is, the inherent time constant of the exciter. The use of magnetic or electronic amplifiers to supply the control fields of Figure 7.2 improved the speed with which the voltage could be corrected. In the magnetic amplifier and electronic systems, the output frequency of the permanent magnet pilot exciter was 420 Hz.

The system of Figure 7.2 also includes a main field circuit breaker and field discharge resistor. The excitation for the unit is removed by opening the field circuit breaker. The contacts of the field circuit breaker are adjusted so that the discharge resistor is connected across the field before the main contacts open to interrupt the field current. The resistor value is selected to hold the self-inductance voltage of the field to a value that will not be detrimental to the field circuit insulation.

7-2.2 Alternator Exciter with Stationary Rectifiers

The problems with commutation on high-speed exciters, together with the development of silicon rectifiers, prompted the development of alternator exciters. Instead of the armature of the exciter being the rotating element, the alternator exciter is a small version of the main generator; that is, the armature is the stator, and the field is the rotating element. The output of the exciter is an alternating voltage, and the current is taken from stationary terminals or bushings.

This excitation system is shown in Figure 7.3. The AC is converted to DC by a bank of stationary rectifiers. The generator terminal voltage is controlled by varying the field current of the alternator exciter. The rectified exciter output is supplied to the generator field through brushes and a collector assembly. These exciters are usually direct-connected; that is, they are connected to the main generator shaft. The excitation for the exciter field is derived from a pilot exciter. This excitation system, like the DC exciter, has the time delay of the exciter to slow down the overall response of the system during a call for a sudden change in excitation.

Because the field circuit is available, a main field circuit breaker may be employed. However, as stated earlier, most large modern units do not use a main field circuit breaker. In this case, excitation removal is accomplished by interrupting the exciter field current. The rate of generator voltage decay will be somewhat longer because of the time delay of the exciter.

A variation of this scheme uses controlled rectifiers or thyristors instead of diodes. In this system the output voltage of the alternator exciter is maintained at the exciter ceiling level, and the direct voltage applied to the main generator field is controlled by phase control of the thyristor firing circuits. When a sudden change in excitation is called for by the voltage regulator, the change in voltage at the main

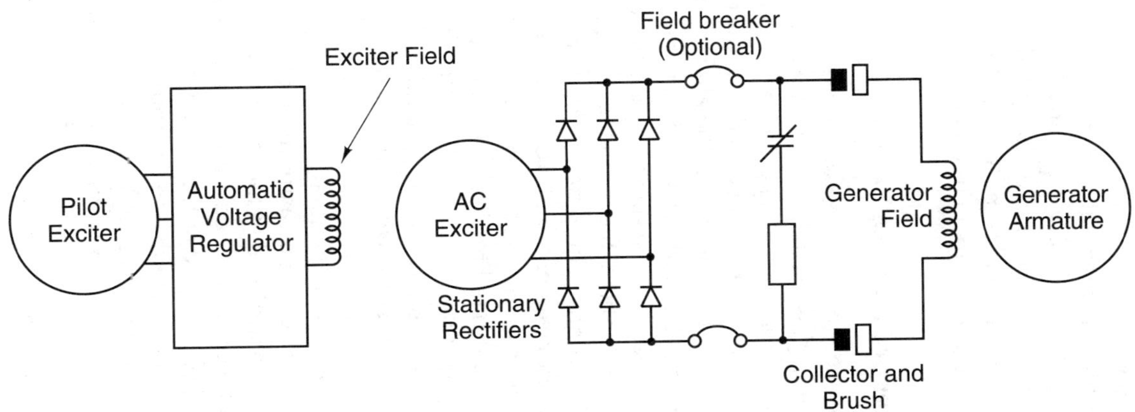

Figure 7.3 Alternator exciter with stationary rectifiers.

field terminals is almost instantaneous. The time constant of the alternator exciter is eliminated. Of course, in any system there is still the inherent time constant of the main generator. The significance of the excitation system response will be discussed later in this chapter.

In this system, excitation removal can be accomplished rather rapidly by phase control of the thyristors to apply a negative voltage to the generator field and force the field current to zero. Even though the field winding current is forced to zero, the solid rotor provides multiple paths or damper circuits for current to flow. The effect of these damper circuits is to slow down the rate of decay of magnetic flux.

7-2.3 Rotating Rectifier or Brushless Exciters

In order to eliminate the need for brushes, commutators, and slip rings completely, the rotating rectifier or *brushless* excitation system was developed. This scheme is functionally equal to the alternator exciter system described in the previous section, but instead of the bank of fixed or stationary rectifiers, the rectifiers are mounted on a wheel and rotate. In this system, however, the alternator exciter armature is the rotating element, and the field is stationary. This scheme is depicted in Figure 7.4. From this figure, it can be seen that the stationary elements of the unit are the main generator armature, the exciter field, the permanent magnetic generator (PMG) armature, and the excitation control equipment. The rotating elements are the main generator field, the alternator exciter armature, the diode wheel, and the PMG field.

The field of the exciter is supplied from a PMG or pilot exciter, the output of which is rectified by a controlled rectifier power amplifier in the excitation control equipment. The output of the (rotating) alternator exciter is rectified by the rotating diodes and supplied directly to the generator field by means of rotating connections to the main field leads, which are located in the bore of the generator shaft.

This excitation system, while eliminating the need for brushes and collector assembly, does not permit the direct measurement of the generator main field current. The field current is proportional to the exciter field current, but the relationship is not always linear because of the effects of magnetic saturation in the exciter. In addition to the slight nonlinearity of the exciter saturation curve, the units are designed to satisfactorily operate with one or more rectifiers per phase out of service. The rectifiers have fuses in series, and a failed diode will result in the fuse interrupting that leg. This may cause a further small amount of nonlinearity in the field current measurement. The rectifiers are identified by phase and location in the circuit, and the condition of the fuses can be monitored by means of a strobe light and operation indicators on the fuses.

Another slight disadvantage of this scheme is that a continuous monitoring of the field-to-ground insulation on the generator is not normally done. To determine whether a ground exists on the generator field circuit requires the use of slip rings connected to the field circuit and brushes that are momentarily connected to these rings. A detection circuit indicates the presence of a ground. Typically the

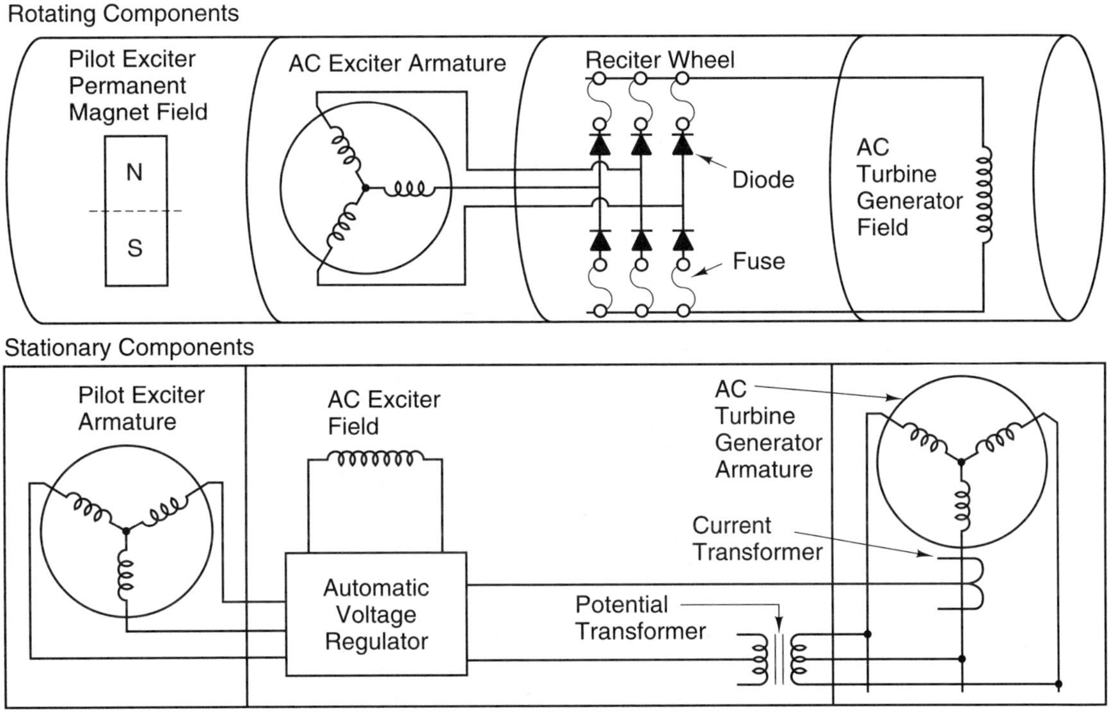

Figure 7.4 Rotating rectifier excitation system.

field-to-ground insulation is checked automatically by a sequence type of control in the excitation control equipment, or it can be initiated manually.

Because this system does not permit the use of a main DC field circuit breaker, excitation removal must be done by interrupting the exciter field current. This is accomplished by an exciter field circuit breaker. More modern control schemes include a provision for forcing the exciter field current to zero by phase control of the controlled rectifier in the power amplifier before opening the exciter field circuit breaker.

7-2.4 Static Exciters

A static excitation system is shown in Figure 7.5. In this system the excitation source is the main generator itself. The excitation power is supplied from power potential or voltage transformers (VTs) connected to the generator terminals. The AC is rectified by means of a controlled rectifier power supply. The magnitude of excitation current is controlled by phase control of the thyristors either manually or by the AVR. Some means of providing a small amount of initial excitation must be provided to "bootstrap" the system.

In the potential-only system of Figure 7.5, the excitation voltage available is proportional to the generator terminal voltage. In a large unit-connected generator,

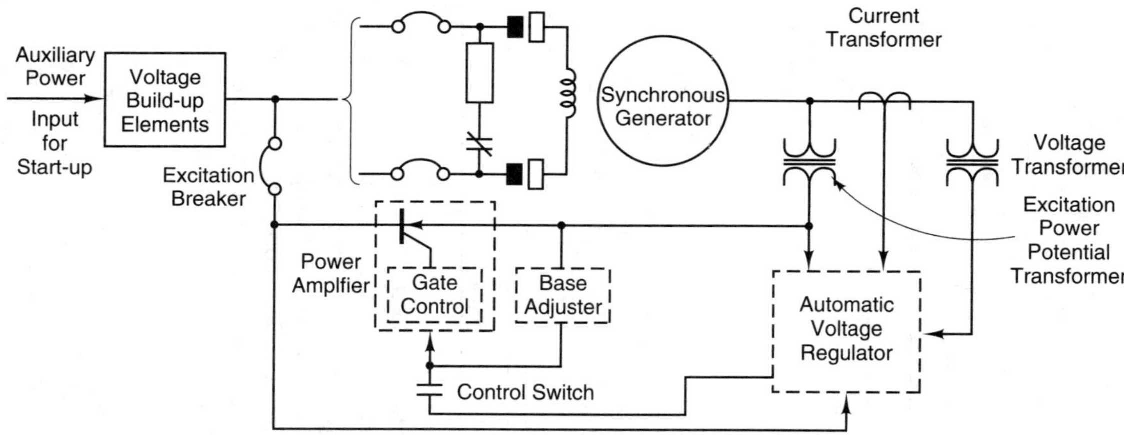

Figure 7.5 Static excitation system.

severe faults on the generator terminals are rare, but such faults may occur on the high-voltage power system. In this case, the voltage on the generator terminals will include the impedance drop of the main step-up transformer; that is, the voltage will not be zero. In specifying a static excitation system, the purchaser will usually specify the lowest voltage at which a given exciter ceiling voltage must be maintained. Typically this is about 70% of rated generator voltage.

Some static systems utilize power current transformers to provide additional excitation from the generator output current. This permits maintaining a high level of excitation even during fault conditions. These compound rectifier systems are not widely used in electric utility applications, but are used in special applications, such as naval vessel generators.

The static system requires a collector assembly and brushes. Because the field terminals are available, a DC field circuit breaker may or may not be used. The static system is inherently fast in response because switching of the thyristors is essentially instantaneous and there is no exciter time constant with which to contend. Excitation removal may be accomplished rapidly, either by phase control of the thyristor or by use of a field circuit breaker, if one is used.

A novel variation of a static excitation is shown in Figure 7.6. This may also be considered a version of an alternator-rectifier system, but with the alternator included as part of the main generator. The excitation voltage is generated by "P" bars located in the slots of the generator stator. The output of these "P" bars is supplied to the primary windings of an excitation transformer, also located within the main generator casing. The output of the excitation transformer secondary winding is converted to DC by a controlled rectifier bridge and then to the generator field winding through brushes and a collector assembly.

The system of Figure 7.6 is similar to a potential-only static system except that the excitation is not directly dependent on the generator terminal voltage. It is, however, dependent upon the magnetic flux cutting the "P" bars. The system may also use power current transformers to provide additional excitation from the generator load or fault current.

Because the excitation voltage applied to the generator field terminals is controlled by phase control of the thyristors, the response of the excitation control is almost instantaneous. The only significant time constant of the system is that of the main generator.

7.3 EXCITATION CONTROL SYSTEMS

Excitation control systems have been alluded to in the previous sections of this chapter. They will be discussed in more detail in this section. It should be pointed out that the AVRs used by manufacturers could be used on more than one type of exciter, the major difference being in the final quantity controlled by the regulator.

Figure 7.7 depicts a typical rotating rectifier excitation control system. The functions of this system will be described in some detail, but it should be noted that other excitation control systems are functionally similar.

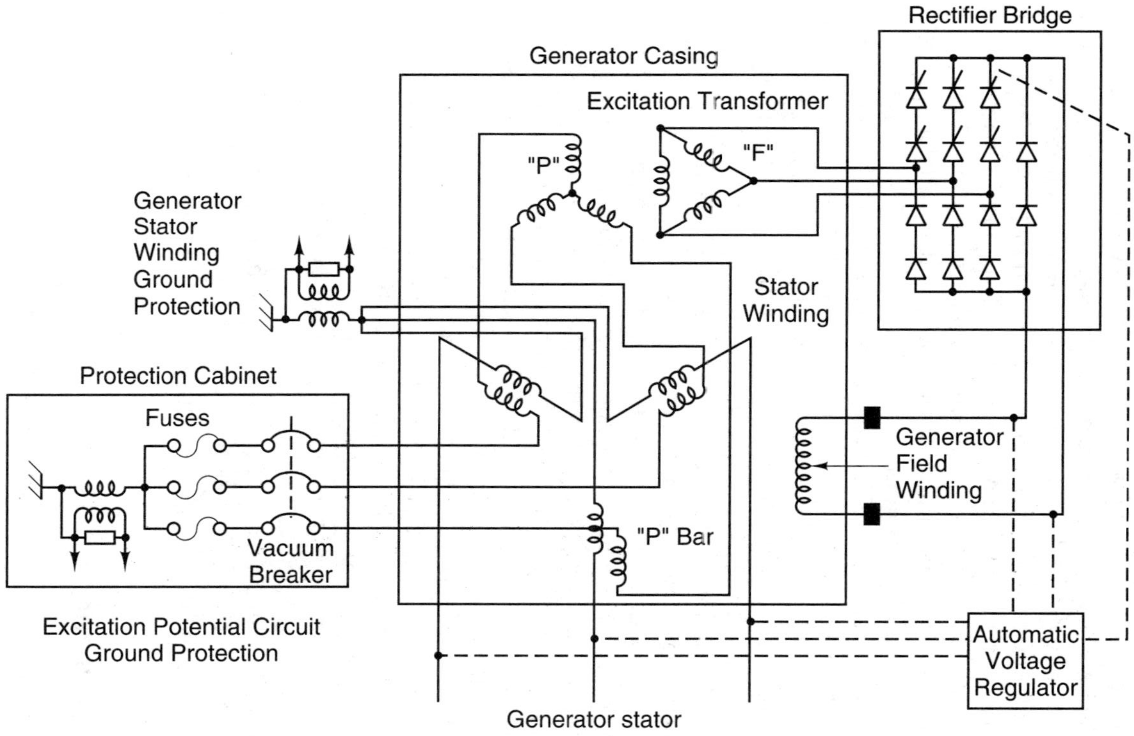

Figure 7.6 Generrex® excitation system. (Courtesy of the General Electric Company.)

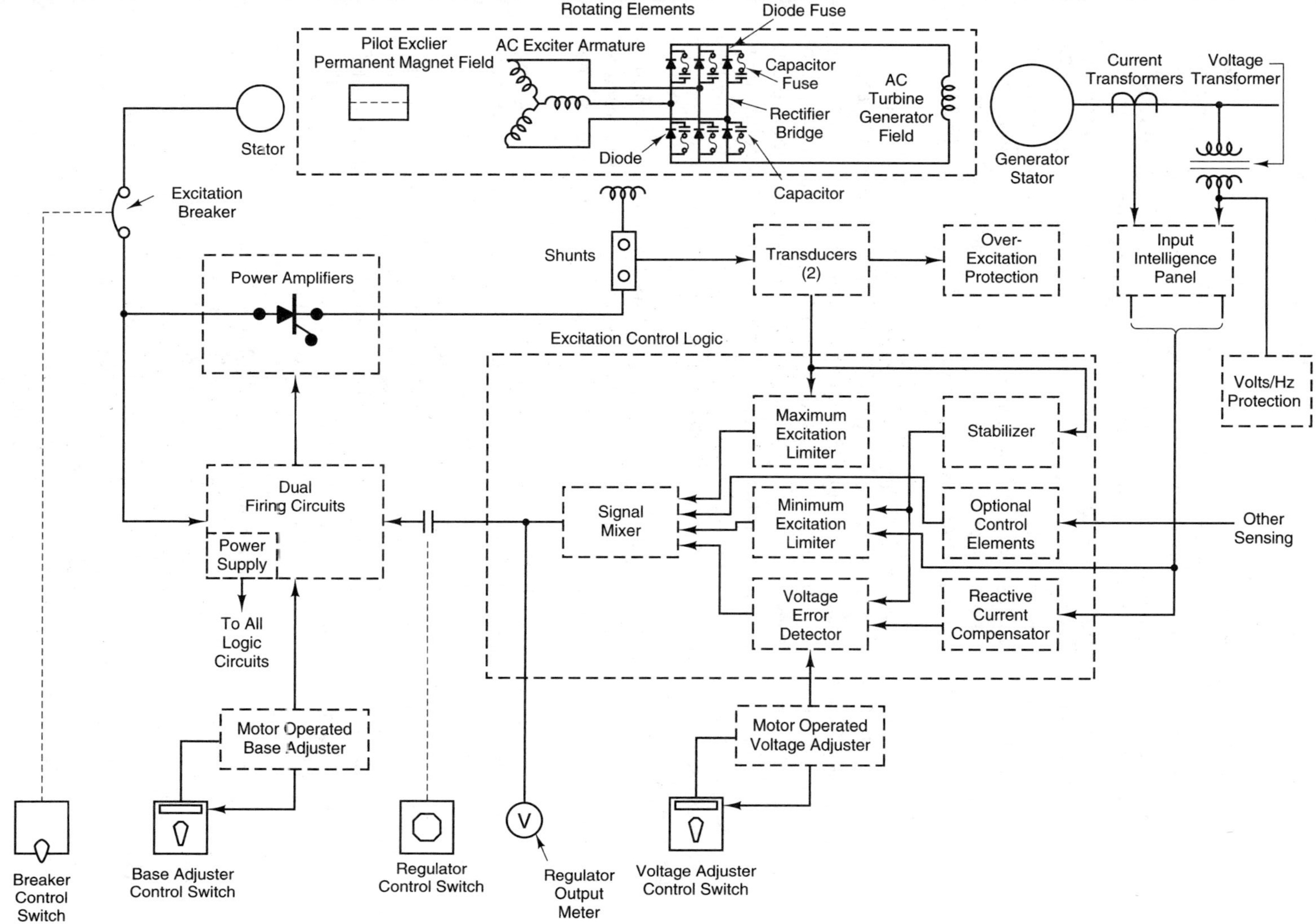

Figure 7.7 Rotating rectifier excitation control system. (Used by permission of Westinghouse Power Generation, Orlando, Florida.)

On the lower left part of the diagram is a motor-operated base adjuster. When the generator excitation is being controlled manually (i.e., the AVR is out of service), the base adjuster is used to control the generator voltage. During automatic operation of the excitation system (i.e., the AVR is in service), the base adjuster is used to set a base level of excitation about which the AVR regulates. If the regulator is taken out of service, the excitation will be that corresponding to the base adjuster setting. The base adjuster is analogous to a main field rheostat in the DC exciter of Figure 7.2.

In this system, the generator terminal voltage is sensed in the input circuit or element supplied by VTs connected to the generator terminals. The intelligence element also has input from a current transformer in the generator leads. This current is used in compensation and limiting elements that will be described in the following paragraphs.

The voltage signal passes through a *reactive current compensator* before it reaches the voltage error detector. This reactive current compensator can be set to simulate the reactance drop to some point past the generator terminals. The reactive current compensator develops a voltage that is proportional to the reactive component of current times a reactance value set into the device. This voltage can be subtracted from the actual generator terminal voltage to simulate a point other than the generator terminals. When used in this manner, the device is usually set to compensate for approximately one-half the drop across the generator step-up transformer (GSU).

The compensator voltage can be reversed to add to the generator terminal voltage indication to cause the generator voltage to droop with increased reactive load. This feature is used when multiple generators are paralleled at their terminals without a transformer between each generator and a common tie bus.

The voltage level that the regulator is to maintain is set by the voltage adjuster, which supplies a reference signal to the voltage error detector. The voltage error detector output signal, by means of the signal mixer and firing circuits, controls the firing circuits of the thyristors in the power amplifier.

A minimum excitation limiter (MEL), an underexcited reactive ampere limiter (URAL), or any equivalent device (see Chapter 2) derives a signal from the generator terminal voltage and load current. It is typically set to simulate the steady-state stability limit of the generator operating on a power system with a given equivalent impedance. This device has no effect on the voltage regulator operation unless the preset characteristic is exceeded. In that case, the MEL output signal supersedes the voltage error signal and calls for a raise in excitation.

A *maximum excitation limiter* (MXL) is typically included in the AVR. In rotating exciters, this device responds to a signal from a transducer, which develops an output proportional to the millivolt drop across a shunt in the exciter field leads. The short-time thermal capabilities of the generator field are defined in ANSI C50.13–1989 and are given in Chapter 12. The MXL can be set to approximate, with some safety margin, those limits and act to reduce the excitation before the integrity of the generator field is in jeopardy. An *overexcitation protection* device is supplied by a transducer to reduce the excitation if the maximum excitation limiter fails or when the unit is on manual control of excitation. It can also be set to trip the unit and excitation as a last resort.

There are other optional elements that can be included in the AVR, depending on the user's special requirements or desires. One of these is a *volts-per-Hertz limiter,* which operates similarly to the maximum excitation limiter to limit the excitation based on the volts/Hertz. Volts-per-Hertz protection is discussed in Chapter 12.

Because the AVR regulates about a fixed level of excitation set by the base adjuster, it is important that the base adjuster setting be changed as the unit goes from no-load to full-load. If this is not done, removing the AVR from service could result in a significant change in generator voltage, which could jeopardize the reliability of the power system. To ensure that this does not occur, some users employ a *base adjuster follower circuit,* which causes the base adjuster setting to track the load on the unit.

The power for the exciter field and the voltage regulator in a rotating rectifier system can be supplied by a PMG. An excitation circuit breaker is used to connect the output of the PMG to the excitation control equipment. Excitation removal may be accomplished by opening the excitation circuit breaker and/or by phase control of the power amplifier as described earlier.

7.4 ABNORMAL EXCITATION SYSTEM CONDITIONS

Because of the many types of excitation systems in existence, as well as the different options that are included in their control systems, it is quite difficult to detail all possible problems. Additionally, there is considerable overlap between the excitation system problems and the field winding problems discussed in Chapter 5. Rather than exhaustively list all problems with all types of excitation systems, this chapter will present some of the most common as well as generic ones. The reader is encouraged to discuss the issue with a representative of the equipment manufacturer for more details.

7-4.1 Loss of Excitation

Complete or partial loss of excitation can be caused by a number of events. Among these are failure in the main or pilot exciter, shorted turns in the generator or exciter field, collector ring flashover, voltage regulator failure or malfunction, or operator error. Theoretically, the generator can operate as an induction generator, but it will not be able to deliver full power output, and the exciting current drawn from the power system will usually depress the system voltage and jeopardize the system reliability. Protection for this condition is discussed in Chapter 12.

7-4.2 Excessive Generator Field Current

Excessive generator field current may be the result of sustained generator overcurrent, depressed system voltage, a sustained fault on the transmission system, loss of sensing signal to the AVR, or AVR malfunction.

The possible effects are excessive heating of the generator field winding and possible excessive flux densities in the generator. As indicated previously, the short-time thermal capabilities for a generator field are defined in ANSI C50.13–1989. These limitations are listed in Chapter 12, but are shown in curve form in Figure 7.8. The AVR should include a maximum excitation limiter and overexcitation protection as described in Section 7.3. If the type of excitation system does not permit a direct measurement of the field current, as in the rotating rectifier system, the exciter field current should be used as an indication of the main field current. The exciter field current will not be exactly proportional to the main field current because of the exciter saturation curve, but the effect is to provide a greater margin of safety at the higher values of field current. This is shown by the lower curves of Figure 7.8.

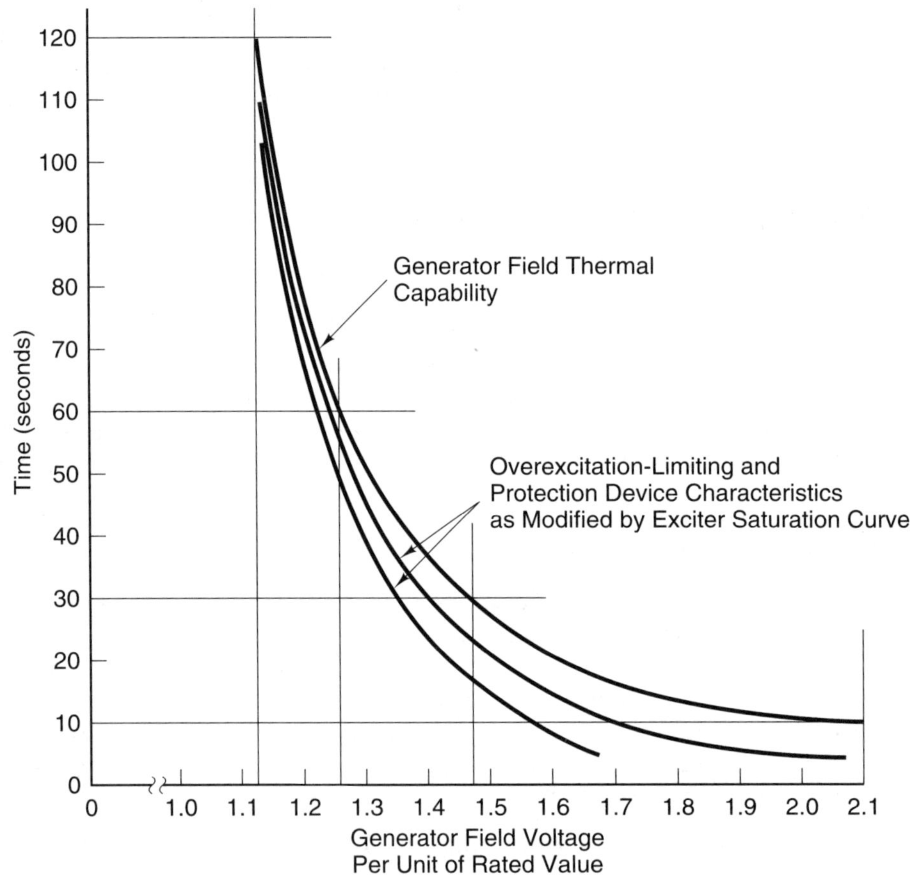

Figure 7.8 Generator field short-time capability.

7.4.3 Failure in Exciter

Failure of an exciter may be caused by foreign objects, contamination of insulation, overheating due to loss of cooling, failure of diodes or fuses in alternator-rectifier systems, or exciter field shorts or grounds. The effect of a failed exciter is a complete or partial loss of excitation on the main generator and possible damage to the exciter itself.

To minimize the possibility of exciter failure, it is desirable to monitor the exciter temperature, exciter ventilation, exciter field current, insulation to ground on the exciter armature and field circuits, and, in the case of brushless units, the condition of the diode fuses. If any of these conditions indicates a possible or impending failure, the condition should be corrected, if possible, or the unit should be removed from service. If an exciter failure occurs, the unit will probably be shut down by a loss-of-field relay in the generator protective relaying system.

7-4.4 Generator Field Failure

Generator field failure can be caused by collector ring flashover, open or shorted turns in the field winding, or failure of the field ground insulation. The effects of field circuit failure can be loss of excitation, thermal and/or magnetic unbalance, excessive vibration of the unit, and damage to the rotor forging. The field current and, where possible, field voltage should be monitored as well as the exciter field current and the field circuit insulation to ground. The manner in which these parameters are monitored will depend on the type of excitation.

7-4.5 Failure of Excitation Power Electronics Equipment

At least one manufacturer uses deionized water to cool the exciter rectifiers in a stationary excitation rectifier cabinet. If the water is not maintained within purity specifications, an impurity in the water can precipitate out of the rectifier and bridge the rectifier-to-ground clearing space, causing a ground indication in the rotor field winding. To clear the ground, that bank of rectifiers has to be removed and the cooling water path purged of deposits.

As long as the water quality is maintained within specification, no deposits should precipitate from the water. The trouble occurs when the deionizer is not properly maintained or if there is a defective batch of deionizing resin. In some cases, the utility has used a resin that is not recommended by the generator manufacturer. Other times, overheating of the deionizer can cause the water quality to deteriorate. Monitoring the conductivity of the water is the most common way of checking on the quality of the water and on the deionizer. See Chapter 10 for more information on the water system.

Proper cleaning of the affected components will be necessary if this situation has been allowed to develop. Removing rectifier modules makes it possible to do the work without disturbing the operation of the generator.

7-4.6 Failure of Voltage Regulator

Proper operation of the voltage regulator is critical to reliable operation of a modern turbine generator. A failure in some part of the voltage regulator can cause a number of effects, depending on the failure or malfunction mode.

One event, which is not a failure of the AVR, is loss of the voltage signal to the regulator. This may be the result of a blown or open VT fuse or a discontinuity in the VT secondary circuit. If this condition is not detected immediately and the regulator removed from service, the excitation of the unit will be forced to ceiling value rapidly. Devices or features in the AVR are available that compare the output voltages of the VTs supplying the AVR with the output of the metering or relaying VTs. If this function is not included in the AVR, it can be provided by a protective relay that is set to detect a loss of voltage intelligence to the AVR and to trip the AVR to manual operation.

A malfunction of the AVR may result in loss of excitation of the generator or high field current, depending on the nature of the malfunction. Some modern high-performance AVRs have the potential for increasing the excitation of a unit to a high level very rapidly. In these units there are, or should be, limiting devices with suitable backup devices to preclude sustained field forcing in the event of failure of some component or element. The monitoring of power supplies and other circuits in the AVR will depend on the design of the system and the probable mode of failure of the system.

7-4.7 Failure of Pilot Exciter

Failure of a pilot exciter may be caused by contamination, excessive vibration, foreign objects, or loss of magnetization of the permanent magnets. The result of a failure is, typically, reduced or interrupted voltage to the AVR and exciter field and partial or complete loss of excitation on the main field.

The pilot exciter output voltage should be monitored and suitable alarms provided to detect a failed unit immediately. If not detected, the unit could be tripped by a loss of field relay. In any event, it is likely that a failure of the pilot exciter will result in a unit shutdown.

7.5 EXCITATION SYSTEM PERFORMANCE

It is beyond the scope of this book to discuss excitation system performance in great detail. However, some comments about the topic are appropriate in this chapter.

The most common indicator of excitation system performance is the *excitation system voltage response ratio*, or simply, the *response ratio*, or *rr*. This parameter is defined in paragraph 3.2.6 of [IEEE, 421.2–1990] as

> The numerical value that is obtained when the excitation system voltage response, in volts per second measured over the first 1/2 sec. interval (unless otherwise specified), is divided by the rated load field voltage of the synchro-

nous machine [1, 2]. The basis of the concept underlying response ratio is developed in Appendix A.

Because the voltage of a rotating exciter does not increase linearly, but exponentially, the term is more easily understood from Figure 7.9. In this figure, the curve between points *a* and *d* is the actual exciter voltage response. A straight line is drawn from point *a* to point *b*, such that the area of the triangle *abc* is equal to that of triangle *adc*. The system response ratio is then defined as

$$\text{response ratio} = \frac{V_{bc}}{(V_{oa} * 0.5)} \tag{7.1}$$

This, however, is a simplified explanation of the concept of response ratio. The reader should refer to references [IEEE, 421.1–1986] and [IEEE, 421.2–1990] for a comprehensive treatment of excitation system performance.

The general principles of power transfer and power system stability concepts have been discussed in Chapter 2. The power transfer capability of a transmission line or system is proportional to the product of the voltage magnitudes at the two

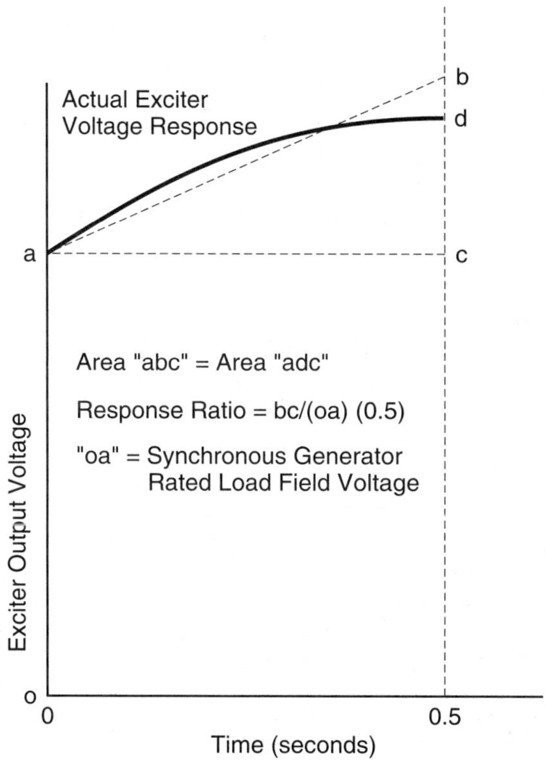

Figure 7.9 Excitation system voltage response.

ends of the line and inversely proportional to the reactance of the line. During steady-state or slowly changing conditions, it is not necessary for an excitation system to vary the excitation of the generator rapidly. However, during fault conditions and immediately thereafter, the ability of an excitation system to respond very rapidly may improve the transient stability of a power system.

When a severe fault occurs at a location electrically near a generator, the fault current in the armature (stator) induces a high value of current in the field (rotor) because the flux linkages in the generator cannot change instantaneously. The high armature current actually is trying to demagnetize the magnetic circuit, and induced field current is the result of the magnetic circuit trying to maintain a constant amount of flux linkages or net mmf. As the demagnetizing effect of the armature reduces the flux, the induced field current decays along with the magnetic flux. Therefore, when the fault is cleared by opening a transmission line circuit breaker, the generator voltage is below normal value, and the power transfer capability of the resulting system is reduced proportionally. Because the generator will have accelerated during the fault, the output power for the resulting phase angle difference will be less than with normal voltage. The net result is that the unit will continue to advance in phase until there is enough retarding torque to stop the advance. Then the unit will swing back to try to satisfy the power-angle diagram or equation. Ultimately, the unit will stabilize if the power limit has not been exceeded.

If, in the above situation, the excitation system could increase the field voltage to maintain the flux in the generator at the prefault level, the generator voltage at the time the fault is cleared would be the same as it was before the fault occurred, and the unit forward swing would be reduced. This is an ideal that some users attempt to approach with a high-performance excitation control system.

Without getting into a rigorous discussion of the definitions of excitation system performance, a general description should help a reader who is not familiar with them to have an understanding of their significance.

In any excitation system with standard performance, the exciter ceiling voltage will be at most 25% above the generator rated field voltage because ANSI C50.13–1989 requires that the generator field be capable of withstanding this amount of voltage for at least one minute. In a rotating exciter system with standard response, when the AVR calls for a sudden increase in excitation, the exciter voltage will get to the ceiling value in a little over 0.5 seconds. This is for an excitation system with a standard response ratio of 0.5 as defined in [IEEE, 421.1–1986]. Because modern protective relays and circuit breakers can detect and clear a transmission line fault in about three cycles, or 1/20 second, the standard excitation system will not have been very effective in approaching the ideal stated above.

A higher performance system designated as a *high initial response,* or *HIR,* system must attain 95% of ceiling voltage within 0.1 second. This performance can easily be obtained with a static exciter or with an alternator thyristor system described in Section 7.2. It can also be achieved with the alternator-rectifier system or the rotating rectifier system, but requires special considerations in the exciter and AVR design. The exciter must have a higher ceiling voltage, and means to reduce the exciter time constant to a low value must be employed. The AVR and PMG must also be capable of high field forcing of the exciter. The AVR will usually have additional compensating and stabilizing elements and safety features. If adjacent units in a generating station have different response ratios, those with the high-performance characteristics will tend to operate first to correct transient disturbances. If this is not desirable, an effort should be made to coordinate the performance of the excitation systems of the units.

It must be remembered that these high-performance excitation systems can represent a hazard to the generator because of their capability to apply high voltages to the generator field. Some means must be employed to prevent sustaining these high voltages on the generator field. These devices usually take the form of additional limiters in the AVR.

In the discussion of stability concepts in Chapter 2, the reference was primarily to steady-state stability, in which the excitation is assumed to be constant. During an actual disturbance as described above, the events occur very rapidly, and the excitation is not constant. In this case, transient stability of the system is the concern. Modern high-performance excitation systems can improve the transient stability of a power system, but caution must be taken.

A third type of instability is dynamic or oscillatory instability. This situation occurs when a unit or units swing or oscillate with a period of oscillation of one or two seconds. This type of instability can actually be caused by a high-performance excitation system because of time delays in the forward and feedback control loops. The condition is usually associated with generating units electrically remote from the load centers. A device known as a *power system stabilizer* (PSS) can be incorporated in the AVR to damp out the oscillations. The settings of these devices are rather complex and must be tailored to each application.

The IEEE has defined various models of excitation control systems (see [IEEE, 421.1–1990] and [IEEE, 421.5–1992]) for modeling the performance of the various types of excitation control systems in power system transient stability computer studies. Most excitation system types are represented by individual models. The advantage of these is that power system modellers can use standard representations for any type of excitation system being used in the study. The manufacturers will typically provide the values of the model parameters for their particular equipment.

SUMMARY

This chapter discussed the excitation system, the various types, their potential problems, as well as some of the performance characteristics and how they can significantly influence the stability and operation of the power system in which the generator resides. One significant element of the discussion is the control of excitation by AVRs.

Refer to {IEEE, 42.1–1986], [IEEE, 421.2–1990], [IEEE, 421.4–1990], and [IEEE, 421.5–1992] for information on excitation systems and performance.

REFERENCES

[ANSI, C50.13–1989] "Requirements for Cylindrical Rotor Synchronous Generators," American National Standard Institute Standard C50.13–1989, New York, 1989.

[IEEE, 421.1–1986] "IEEE Standard Definitions for Excitations Systems for Synchronous Machines," IEEE Standard 421.1–1986, New York, 1986.

[IEEE, 421.2–1990] "IEEE Guide for Identification, Testing, and Evaluation of the Dynamic Performance of Excitation Control Systems," IEEE Standard 421.2–1990, New York, 1990. (ISBN 1–55937–053–X).

[IEEE, 421.4–1990] "IEEE Guide for the Preparation of Excitation Specifications," IEEE Standard 421.4–1990, (ISBN 1–55937–054–8), New York, 1990.

[IEEE, 421.5–1992] "IEEE Recommended Practice for Excitation Systems for Power Stability Studies," IEEE Standard 421.5–1992, (ISBN 1–55937–218–4), New York, 1992.

CHAPTER 8

The Hydrogen Auxiliary System

The hydrogen auxiliary system exists to maintain the quality and quantity of the cooling gas required to ventilate the conducting and rotating parts of the generator. As seen in Chapter 4, hydrogen is not the only means of cooling the armature winding conductors. However, hydrogen cooling gas is utilized in the great majority of large generators. This chapter will describe the function and design of the hydrogen auxiliary system, as well as some possible malfunctions and their correction.

The hydrogen auxiliary systems of generators are basically quite reliable. Only a few conditions result in a shutdown of the generator. Nevertheless, proper maintenance of the hydrogen system is required for long-term benefits to the generator.

While some manufacturers employ hydrogen-cooled exciters, these will not be covered in this book, as most are air-cooled. Please consult the manufacturer if your unit is hydrogen-coded.

8.1 DESCRIPTION OF THE HYDROGEN AUXILIARY SYSTEM

It has already been established that all generators require some type of cooling mechanism in order to dissipate the heat generated by the I^2R and eddy current electrical winding losses and by the core losses, as well as losses generated by the mechanical friction and windage in the rotating parts of the generator. Chapters 4,

5, and 6 described how each of the major heat-generating components of the generator were either air- or hydrogen- and/or liquid-cooled. Air-cooling is more common for small and/or older generators. Therefore, they will not be covered in this book. This chapter will describe how the hydrogen is safely circulated and monitored to ensure its effectiveness as a coolant for the generator.

Hydrogen is the most effective gas available for ventilating a generator for several reasons:

- Hydrogen is a better insulator than air when it is pressurized.
- The density of hydrogen gas at a purity of 98% is approximately one-tenth that of air at comparable pressure. This reduces the fan and windage losses such that it is feasible to raise the pressure of the gas to as high as 75 lbf/in^2 gauge pressure. The higher pressures increase the quantity of hydrogen present within the generator, thus increasing its heat removal capability.
- The heat transfer characteristics of hydrogen are far superior to those of air. Its thermal conductivity is approximately ten times higher, its heat transfer coefficient is 35% more effective than air at the same pressure, and its specific heat is approximately 14 times higher.
- In a nearly pure state (98%), hydrogen does not support combustion.
- Hydrogen is not an oxidizing agent.

The major disadvantage to using hydrogen as the cooling medium is its flammability. Hydrogen is potentially explosive when its purity is between 4% and 75%. Fortunately, whereas hydrogen gas becomes flammable when mixed with air to a significant degree, pure or nearly pure hydrogen will not burn, even in the presence of sparks.

Therefore the objective of the hydrogen auxiliary system is *to supply and maintain a pure, dry supply of hydrogen gas within the generator at the proper pressure and temperature.*

To accomplish this objective, the following five requirements must be met:

- Safely charge and discharge the generator frame with hydrogen gas. Carbon dioxide is typically used as an intermediary gas.
- Maintain the desired gas pressure within the generator frame at all times.
- Maintain the purity of the gas within the specified range at all times.
- Maintain the hydrogen gas within the proper temperature range at all times.
- Maintain the humidity of the gas within a specified range at all times.

There are four basic subsystems in the hydrogen auxiliary system that will satisfy the above requirements: 1) the circulation subsystem, 2) the temperature control subsystem, 3) the gas supply subsystem, and 4) the humidity control subsystem.

8-1.1 The Hydrogen Circulation Subsystem

The hydrogen circulation is typically driven by a shaft-mounted blower. Powered by the main generator shaft, it rotates at constant speed, thus keeping a near-constant flow of hydrogen gas when it is needed. This is significant in that it makes monitoring of some other parameters somewhat easier knowing that the flow of coolant is (or at least should be) constant. Figure 8.1 describes a typical generator blower. This particular design is a single multi-stage axial flow compressor-type blower. The individual blades are dovetailed and locked in the blower hub for easy removal.

Figure 8.1 Description of a generator hydrogen blower. (Used by permission from Westinghouse Power Generation, Orlando, Florida.)

The paths followed by the gas differ widely depending on the design of the machine, which is affected by the following design issues:

- Conventional or conductor-cooled armature winding
- Hydrogen conductor-cooled or water conductor-cooled winding
- Conventional or conductor-cooling of rotor field windings
- Ventilation design of rotor forging
- Ventilation design of the main lead bushings and the main lead box
- Ventilation design of the stator core

Ventilation ducts with baffles in the frame are typically used to direct the cold gas into paths that are designed to cool specific areas. Figure 8.2 shows how baffles in the frame are used to direct hydrogen gas into radial ducts in the core, through the air gap, and into radial ducts in the rotor forging. If the stator bars are conductor-cooled, then other baffles direct cold hydrogen to the stator bar conductor vent tube stack. Baffling is also used to direct the hot gas into the hydrogen cooler inlet.

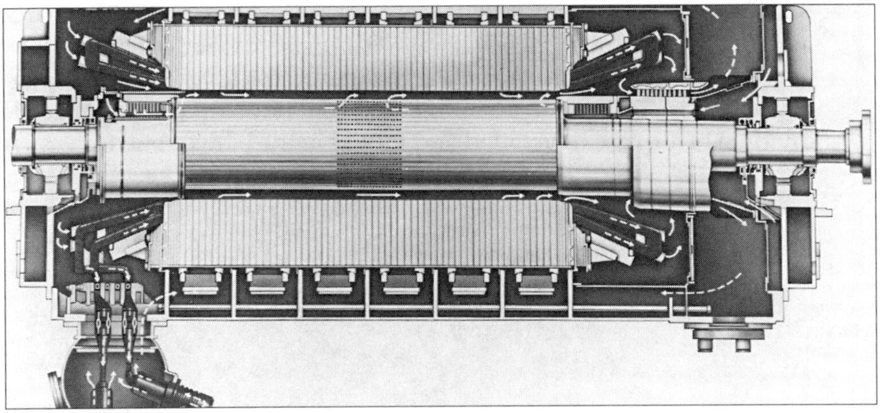

Figure 8.2 Ventilation circuits in a generator. (Used by permission from Westinghouse Power Generation, Orlando, Florida.)

8-1.2 The Hydrogen Temperature Control System

The hydrogen itself is maintained within a proper temperature range by hydrogen coolers located inside the frame, which use plant water as the secondary heat exchange medium. The hydrogen coolers are typically finned tube bundles through which service cooling water is pumped to remove the heat from the hydrogen cooling gas. The hydrogen gas flows around the outside of the finned tubes, making the tube walls the heat transfer surface. Because different manufacturers utilize different ventilation configurations, hydrogen coolers are mounted at various locations to suit the needs of the ventilation system of their generators. The coolers may have their long axes vertical or horizontal. They may be mounted horizontally, lengthwise or crosswise of the generator, on top. They may be vertical at one or both ends on the two sides. They may be in frame sections that are shipped separately and attached and sealed to the generator frame at destination. Figure 8.3 shows a photograph of a vertically mounted cooler for a generator.

Because the hydrogen coolers are located within the generator frame, the water pressure is typically designed to be lower than the gas pressure, so that small leaks in the hydrogen cooler will result in hydrogen leaking into the water system, rather than water leaking inside the generator. Water may leak into the generator in any case, if the leak is large enough. Circulation for the secondary water is supplied by the plant service water system.

In some designs, the hydrogen coolers are sectionalized, so that cooler sections operate independently of each other. This allows a generator to operate at reduced load when one of the cooler sections has to be taken out of service. This is called *sectionalizing the cooler.*

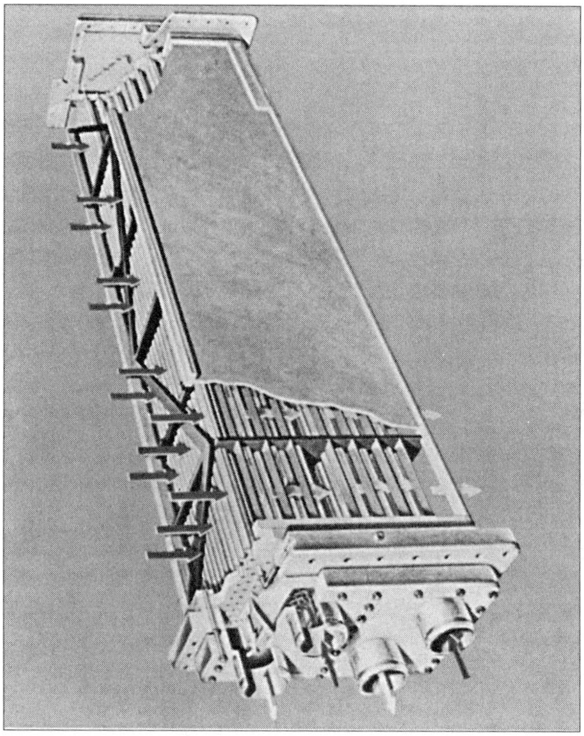

Figure 8.3 A typical hydrogen cooler. (Used by permission from Westinghouse Power Generation, Orlando, Florida.)

8-1.3 The Gas Supply System

The gas supply system provides the necessary pipes, valves, pressure sensors, pressure regulators, and other such devices necessary to introduce hydrogen gas into the generator frame during initial start-up or after an outage that required degasification of the unit.

Before hydrogen is piped into the generator, however, the latter must be purged of air by charging the generator casing with a heavy gas (typically carbon dioxide, CO_2) that pushes out the air. The heavy gas is then vented to the atmosphere, pushed out through vents located in the frame by the entering hydrogen gas. The gas supply system provides the equipment necessary to carry this out.

Likewise, when the machine is degassed after a shutdown, its hydrogen content must first be totally purged to avoid a potentially explosive mixture of hydrogen and air. The hydrogen is pushed out by the entering CO_2 and vented to the atmosphere. The fundamental rule to be observed when changing gas in the generator is that hydrogen and air should *never be* mixed within the frame.

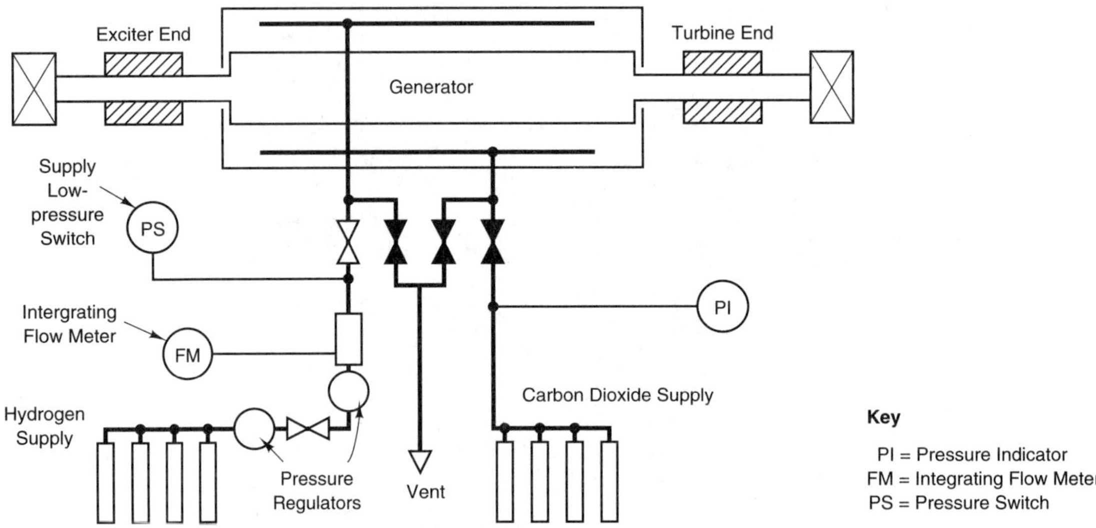

Figure 8.4 Schematic diagram of a generic hydrogen supply system.

The hydrogen supply system provides pure, fresh hydrogen to the generator from a bulk supply. The bulk sources of hydrogen are either hydrogen bottles, cylinders, or a trailer from a gas supplier. Pressure reducers are used to bring down the hydrogen pressure from the high supply pressures to a level more in line with the desired generator pressure.

Figure 8.4 shows a generic hydrogen supply system schematic diagram. During a degassing operation, CO_2 is introduced from a perforated pipe manifold in the lower parts of the generator, forcing the lighter hydrogen out of the generator through another such manifold located at the top of the enclosure. The purged hydrogen is then safely vented to the atmosphere. Upon complete purging of the hydrogen, the CO_2 is vented to the atmosphere until its pressure is equal to atmospheric pressure, at which point the generator *access covers* (also referred to as *man holes* or *hand holes*) can be removed to allow access to its internal elements. The CO_2, however, must be completely removed before entering the unit. This can be done either by applying clean pressurized instrument air through the purge lines or by blowing out the CO_2 with fans through the access holes.

Gassing the machine is done in a similar fashion. The air is purged out of the unit by introducing CO_2 into a closed generator through the lower manifold. Once the air is purged, the CO_2 is forced out of the lower manifold by the hydrogen gas. The hydrogen is introduced through the upper manifold and forces the CO_2 out through the lower one. The CO_2 is safely vented to atmosphere.

Although basically a closed system, hydrogen is difficult to contain, and the hydrogen system can have small, virtually undetectable leaks that result in the slight loss of gas. Such leaks can be in the coolers, in the bushing seals, or at the gasketed joints, as well as at damaged porcelain. Loss of hydrogen gas can also be the result of hydrogen absorbed by the oil at the gland seal interface. In rare instances, hydrogen may escape through the frame at welded joints, but in most of these cases, these leaks are detected and corrected during air tests at the factory. The hydrogen supply is used to make up for these slight losses and maintain the pressure at its proper value continuously during the operation of the generator. This can be done either manually or automatically.

A pressure regulator is typically supplied to automatically maintain the proper gas pressure in the generator at all times. See Figure 8.4. The regulator will sense the gap pressure inside the machine and introduce more hydrogen in response to low pressure levels, until the proper pressure is attained. Some operators, however, prefer to fill the generator manually and therefore eliminate automatic control of a volatile gas such as hydrogen. These procedures call for an operator to intentionally top off the generator when the latter's low hydrogen pressure alarm limit is reached. In such cases, a manual valve downstream of the pressure regulator (if one is supplied) is opened and closed as required.

The appropriate hydrogen pressure in the generator at any one time depends on the unit load and on the operator's philosophy. Because the heat removal capability of the hydrogen is directly related to its pressure, the generator capability curve typically specifies the operating limits of the generator at various pressure levels. Whenever the unit is being operated at full load, the hydrogen pressure should be kept at the unit's maximum rated value. At lower loads, however, the pressure can be reduced as long as the limits displayed in the capability curve are not exceeded. The advantages of running with reduced pressure in the event of reduced loads are reduction of hydrogen leakage losses and reduction of friction and windage losses. One disadvantage is that if the unit load is to be increased rapidly, the hydrogen pressure may have to be increased rather quickly in order to remain within the capability curve limits. It must be stated, however, that under normal circumstances, as long as the generator operates within the capability curve for reduced pressure, no harm will be done to the generator.

Occasionally, the introduction of contaminated sealing oil to the oil-hydrogen interface (to be explained in more detail in Chapter 9) will cause the hydrogen gas to lose its high purity level. When this is the case, two undesirable things can happen:

- Windage losses are increased because of the higher density of the gas.
- The danger of the gas reaching its flammable point are increased.

Some introduction of impurities is unavoidable in the operation of the generator. Excessive amounts of impurities, however, indicate a malfunction and will be examined later in this chapter. Regardless of the cause, the hydrogen supply system is used to refresh the supply of hydrogen in the generator in order to maintain proper purity levels. This is done by venting a certain amount of impure hydrogen and replacing it with fresh gas from the supply system.

Lastly, some manufacturers use the hydrogen supply system to provide a blanket of pure hydrogen to the water tank in the stator cooling water system (for units with water-cooled armature windings) in order to reduce the oxygen content of the water. (This link is not shown in Figure 8.4.) This blanket is intended to limit the oxygen content of the water to a very low level (20–60 ppb oxygen for most manufacturers) and therefore eliminate corrosion of the copper hollow conductors.

In contrast, other manufacturers recommend that a certain level of oxygen be maintained in the water (2–8 ppm oxygen). This level of oxygen is conducive to the formation of a tough and stable cupric oxide film in the hollow conductor surface that is in contact with the water. This film protects the copper conductor from corrosion and flaking and prevents consumption of the base metal.

8-1.4 The Humidity Control System

Excess moisture in the hydrogen gas can condense and collect on the various surfaces inside the generator frame. This can happen during periods of light load or, more likely, during off-line conditions at standstill with the unit still pressurized. Because relatively little or no heat is being generated by the unit, a hydrogen cold gas temperature that is lower than the dew point of the hydrogen may cause this condensation. Moisture in the hydrogen gas can lead to a number of undesirable things in the generator, including the following:

- Condensing moisture from the hydrogen can cause lead carbonate to form as a result of lead oxidation in the solder material of the fin-to-tube assembly, where the solder is used to attach the fins to the tubes. The circulation of the hydrogen can cause this highly conductive compound to collect on the conducting elements of the generator and result in tracking of the insulation and, consequently, in short circuits inside the generator. This will be discussed in Section 8.3.
- Condensation of moisture under the conditions described in the previous item can lead to stress corrosion cracks in critical elements such as the rotor end turn retaining rings, especially those made with austenitic steel. A cracked retaining ring can break under rotational load and lead to a catastrophic failure. The use of 18Mn–18Cr alloy steel for these parts has been shown to reduce the incidence of stress corrosion cracks, but not all retaining rings are made of such steel.
- Moisture and other contaminants reduce the purity of the hydrogen gas, which increases the windage losses in the machine and reduces its efficiency. A 6%-reduction in gas purity (for example, from 98% to 92%) will double the windage losses.

Some generator manufacturers presently recommend a dehumidification system for the hydrogen gas in their equipment. Others maintain that their hydrogen sealing system is capable of eliminating any ingress of humidity into the gas and thus do not require the additional equipment. This is partly related to the design of the hydrogen sealing system (described in Chapter 9), and its effectiveness in maintaining the interface between the sealing substance (oil) and the hydrogen gas. Furthermore, some designs of gland seal oil systems are also quite efficient at freeing the moisture from the oil that interfaces with the hydrogen. Nevertheless, other possible sources of humidity in the hydrogen gas are hydrogen cooler leaks, stator cooling water leaks, turbine gland seal leaks into the bearing lubrication oil, and a wet hydrogen supply from the manufacturer.

The hydrogen dryer consists of a chamber filled with activated alumina absorbent material, a desiccant through which a stream of hydrogen gas flows. By continuously circulating the gas stream through the dryer, the moisture content of the hydrogen within the generator can be controlled. If a dryer is used, most generator manufacturers connect it across the generator blower so that gas is circulated through it whenever the generator is running. However, some manufacturers locate it in the hydrogen supply line to the generator. Some dryer manufacturers utilize a fan to move small amounts of gas through the dryer when the generator is off-line.

The absorbent material can typically absorb approximately 2 lbs of water, at which point it becomes saturated and must be dried out. This is done by disconnecting the gas drying chamber from the gas stream and circulating a stream of hot, dry air through it, using an internal electric heater and a small blower. A thermostat protects the dryer material from overheating. When the moisture has been extracted, the flow of hydrogen is resumed to continue its drying action. Reactivation of the desiccant in a typical hydrogen dryer takes approximately four hours. Some dryers contain dual drying elements (called *towers*) so that hydrogen can be passed through one while the desiccant in the other tower is being dried.

When the unit is off-line and at either standstill or turning gear, but pressurized, the main blower will not be able to circulate the hydrogen through the dryer in the fashion described above. Thus, some unit owners will add external blowers that will force the hydrogen through the dryer. This is important, because as mentioned above, the effects of high humidity in the hydrogen gas are most critical during off-line conditions.

A dual tower dryer automatically switches back and forth between the two desiccant chambers (towers) after a number of operating hours, regardless of whether or not the desiccant is saturated. Thus, there is no interruption in the drying process, and a useful desiccant is assured at all times. Other systems require manual start of the reactivation cycle, as well as manual return to the absorbing mode. Figure 8.5 shows a schematic diagram of a single tower hydrogen dryer.

The dryness of the active material can be visually determined by the color of the desiccant material as seen through a window in the drying chamber. The material will be light blue when dry and will turn grayish pink when saturated. The desiccant will also lose its effectiveness because of the many cycles of drying hydrogen gas contaminates, particularly oil. Therefore, after a certain length of time, the desiccant should be replaced.

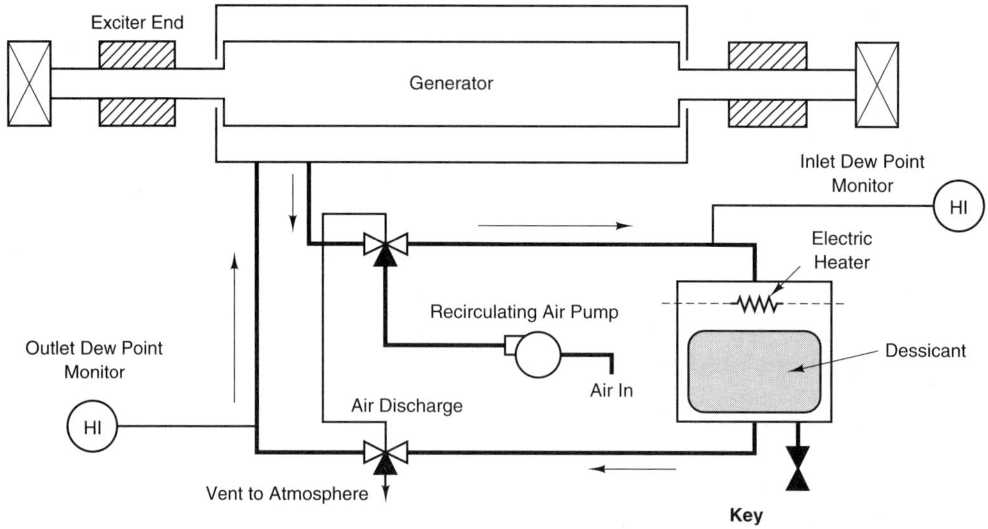

Figure 8.5 Single tower gas dryer.

8.2 PARAMETERS, SENSORS, AND MONITORS

The significant parameters of the hydrogen gas that need to be closely monitored are its

- pressure
- purity
- temperature
- humidity

The main objective of this section is to discuss how these parameters affect the performance of the generator and how and where they can be measured. This section will also describe the instruments which sense these, as well as other, less significant parameters in the hydrogen auxiliary system. Different generator manufacturers furnish different combinations of sensors to monitor a subset of the following parameters. The following sections represent most of the possible parameters that can be practically measured and used effectively for diagnosis of the gas supply system. It should be understood, however, that a particular unit is not likely to have the full complement of sensors described here.

8-2.1 Hydrogen Pressure

One of the most important parameters in the hydrogen system is its pressure because it directly affects the performance capability of a generator, as can be seen in the capability curve of Figure 2.19. The pressure is also used for a number of calculations as well as control features in the gland seal system. (These will be de-

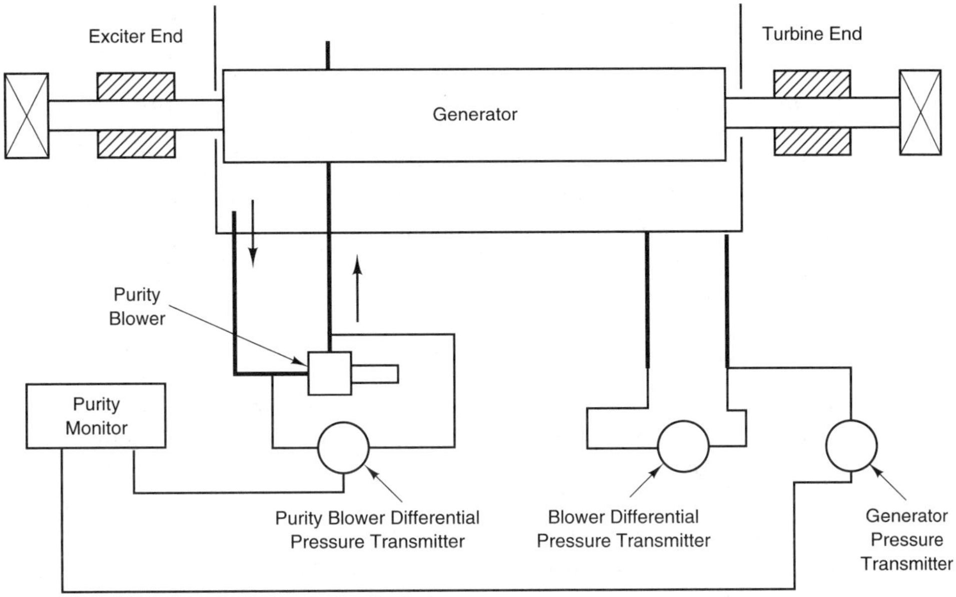

Figure 8.6 Hydrogen gas circulation monitoring and control.

scribed in Chapter 9.) The sensing point can be located anywhere, but it is typically found in the hydrogen supply system outside of the generator. The hydrogen pressure reading for U.S. applications is in pounds-per-square-inch (gauge), or lbf/in^2. It is typically an analog remote sensor, such as a pressure transmitter.

8-2.2 Generator Blower Differential Pressure

The purpose of the blower differential pressure measurement is to determine the integrity of the hydrogen blower and ventilation system. Improper blower or baffling assembly, or missing blades, will result in a decreased differential pressure rise across the blower when gas purity is constant. This sensor is important because this problem can affect the flow of hydrogen throughout the generator. The readings are given in inches of water pressure, and because it is a differential pressure, it is not related to atmospheric pressure.

The generator blower pressure is directly related to the gas density, which increases with lower hydrogen purity. Thus, the hydrogen blower pressure also serves as a backup for the hydrogen purity monitor described in Chapter 3.

Some generator manufacturers will normally supply local differential pressure gauges instead of transmitters. The plant owner should consider upgrading to transmitters in order to improve the generator diagnostic capability. Figure 8.6 shows a diagram of how the gas pressure monitor and the blower differential pressure monitor can be connected in the generator. The values of these two sensors are used to monitor the circulation of the hydrogen gas throughout the generator internal environment.

8-2.3 Hydrogen Supply Pressure

The hydrogen supply pressure is typically controlled by a simple pressure switch that alarms when the supply pressure gets low. The pressure setting depends on the recommendations of the generator manufacturer, but it is generally a few lbf/in^2 below the maximum rated hydrogen pressure in the generator (i.e., 30, 45, 60 or 75 lbf/in^2). Additionally, there are a number of pressure gauges in various locations throughout the supply system. Their number and location depend on the equipment manufacturer. The location of the blower differential pressure and the hydrogen supply pressure sensors can be seen in Figure 8.4.

8-2.4 Hydrogen Cold Gas Temperature

The temperature of the hydrogen is the next most important parameter measured because hydrogen is not useful unless it is capable of removing heat from the generator. Because the heat transfer capability of any coolant depends on the temperature difference between it and the surface to be cooled, the temperature of the hydrogen as it exits the cooler is very important. There is generally one cold gas thermocouple or resistance temperature detector (RTD) for every cooler section; their readings are in degrees Celsius. See Figure 8.7.

Some designs have an additional sensor that measures the average cold gas temperature for all the cooler sections. It is used to control the water flow to the cooler.

8-2.5 Hydrogen Hot Gas Temperature

The inlet temperature for the coolers is the hot gas temperature, also measured with either a thermocouple or an RTD. There is only one to serve all the sections, and it can be used to determine the effectiveness of the hydrogen coolers. Like the cold gas temperatures, its reading is in degrees Celsius. The temperature drop of the gas through the cooler can be measured in terms of the cold and hot gas temperatures:

$$\Delta T_{gas} = T_{hot} - T_{cold} \tag{8.1}$$

8-2.6 Hydrogen Cooler Service Water Temperature and Flow

It is often useful to monitor the service water flowing through the coolers. Many incidents in generators have occurred because of flow and/or temperature problems in the cooler service water system. Although these instruments are generally not part of the generator manufacturer's package, the owner should give serious consideration to retrofitting them.

The parameters of interest for the service water are flow and temperature. To be able to cover all contingencies, it is advisable to install temperature sensors at the inlet (T_{in}) and outlet (T_{out}) of the service water to the coolers. This will allow the detection of the inlet water temperature as well as the temperature difference of the water through the cooler. This temperature difference is

$$\Delta T = T_{out} - T_{in} \tag{8.2}$$

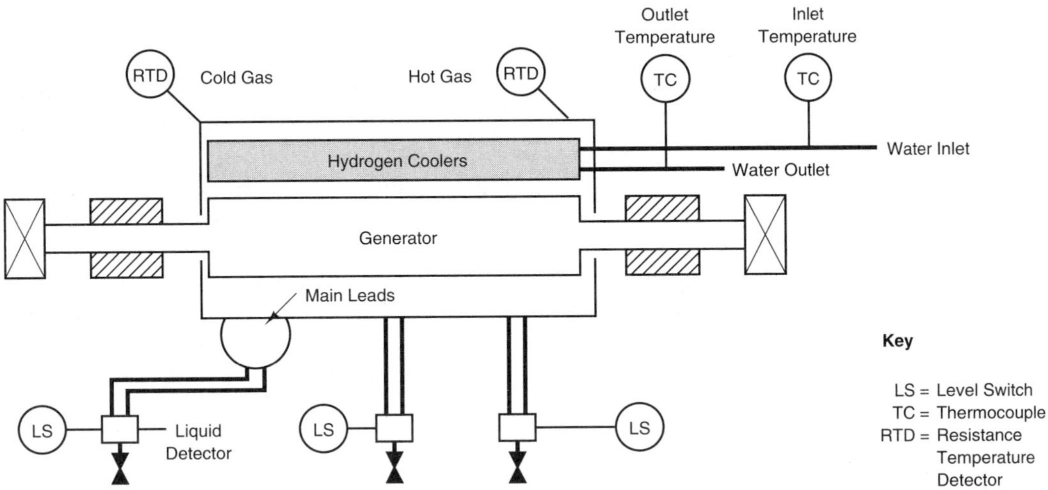

Figure 8.7 Typical hydrogen temperature monitoring.

Additionally, it would also be advisable to install a flow measurement device, either a direct in-line flow meter or a differential pressure monitor across an orifice plate. By knowing the water flow and its temperature difference, the efficiency of the coolers can be easily calculated.

While flow and temperature measurements can provide complete information as to the state of the coolers, often considerable information can be obtained from just one of the two types of sensors. Given that in-line flow meters are not easy to install and represent an obstruction in the water path, they are typically not the best choice. Differential pressure sensors are useful in determining flow, but provide little information as to the state of the cooler itself. Therefore, it is more practical to install thermocouples to measure the inlet and outlet temperatures of the service water. While not complete, they provide the most information for the investment. In this chapter, the existence of such thermocouples will be assumed.

8-2.7 Dew Point Temperature

The dew point is a measure of the relative humidity of the gas. It basically represents the temperature at which the relative humidity of the gas will be 100% and, therefore, will begin to condense. The higher the moisture content of the hydrogen gas, the higher the dew point temperature. It is typically measured in degrees Celsius or Fahrenheit. A hydrogen dew point of below 0° C (32° F) is generally acceptable [IEEE 1129–1992], although some manufacturers recommend an even lower dew point temperature. There is little consensus among the manufacturers and the utilities as to what constitutes a reasonable dew point temperature. The reader is encouraged to consult the manufacturer.

If a hydrogen dryer is supplied, it is highly likely that a dew point indicator is provided with it. Whether or not a dryer is installed, however, it is important that a dew point indicator be used to monitor the moisture content of the hydrogen gas.

Detecting the dew point is a complicated science, and, as a result, the sensors are typically not as reliable as other sensors in and around the generator. The probe requires frequent calibration in order to accurately reflect the real conditions. Moreover, the performance of the sensing element may be degraded by the presence of liquid water or oil in the gas stream. (See Section 3.8 for a detailed discussion of dew point monitors.)

If a single dew point monitor is used, it is typically placed at the inlet to the dryer. This provides a good indication of the general moisture content of the entire hydrogen volume in the generator. In some cases, however, this may not be the best location for the monitor because it is outside the generator and subject to different conditions. The manufacturer should be consulted to determine the best location.

Some manufacturers recommend that a second dew point monitor be installed in the outlet of the dryer to monitor the dryer's effectiveness. The difference in dew point temperature between the inlet and outlet hydrogen indicates the effectiveness of the dryer. It can also serve to determine the calibration of either probe by comparing their relative values. Common locations of the dew point probes can be seen in Figure 8.5.

8-2.8 Hydrogen Supply Integrating Flow (Consumption)

Some manufacturers include an instrument to measure hydrogen consumption in the control and instrumentation package to keep track of the volume of hydrogen introduced into the generator. It can be useful, but not necessary, for detecting significant hydrogen leaks in the generator.

8-2.9 Liquid Detectors

Liquid detectors are liquid level switches located in various places along the bottom of the generator, which collect any liquid (typically water and/or oil) that is leaking or condensing from the various sources in the generator. Although not directly related to the hydrogen auxiliary system, they can be useful in many situations.

Different manufacturers offer various locations for the detectors. However, common locations are 1) the exciter end of the generator, 2) the lead box, 3) the turbine end of the generator, and 4) underneath the hydrogen coolers. They can be wired either separately or in parallel, so the same alarm will register when one of the detectors trips. Some manufacturers design a piping system that taps various locations (10–12) along the bottom of the generator frame and drains all collected liquids into a manifold that leads into a single liquid level detector.

The first thing to do when the liquid detector(s) alarm is to determine which of the detectors is causing the alarm and whether the liquid collected is oil or water. The latter can be done by draining the contents of the detector through a drain valve located below it. The contents will be blown out because of the hydrogen pressure in the machine and should be collected in a container. The sensors are shown in Figure 8.6.

8-2.10 Hydrogen Purity Monitor

The hydrogen purity monitor was described in detail in Chapter 3.

8.3 MONITORING AND DIAGNOSTICS

This section will concentrate on the use of the sensors and monitors described in the previous section for the identification of problems in the hydrogen system. It will also discuss the corrective action to take in such circumstances.

Three problem areas will be discussed:

- Reduced cooling capability of hydrogen gas, which can result in improper ventilation of critical elements in the generator
- High moisture levels in the gas, which can produce stress cracking and general oxidation and degrade the integrity of the electrical insulation
- Hydrogen at a (relatively) low purity, which increases the rotational losses in the generator and can become flammable when in contact with a spark

As with all diagnostic situations, the validity of sensors should not be unquestionably assumed. Uncalibrated or malfunctioning sensors can cause a misdiagnosis and its attendant costs. The reader is advised to ensure to the greatest degree possible that all readings are accurate before any diagnosis is made.

8-3.1 Problem Area #1: High Humidity

As previously stated, one major consequence of high humidity levels in the hydrogen gas is that condensation can collect on soldered fin-to-tube joints in the coolers. If a lead-based solder was used in their construction, this condensation can form lead carbonate, a substance that can be blown and deposited on the electrically live parts of the generator, which, with the moisture, would cause surface discharge tracking across insulated parts. High humidity levels can also result in serious corrosion cracking of various critical components, such as the retaining rings.

Humidity can be easily detected by the dew point monitor (see Section 3.8). Generally, the dew point is to be maintained at 32° F for maximum safety. Different manufacturers, however, may have different recommendations. Nevertheless, caution is recommended when the dew point temperature begins to approach 45° F. While a dew point value of 45° F may not be a problem while the unit is under load and the internals of the generator are significantly warmer than this temperature, it may present a problem when the unit is out of service, pressurized with hydrogen, and cool (or cold) water is running through the hydrogen coolers.

To counteract the effects of high humidity, the cold gas temperature should be raised to a level above that of the measured dew point, but below the limit recommended by the manufacturer. If the increase is detected soon enough, this should not be a problem because the humidity will not have built up to a critical level. This approach, however, only addresses the symptoms, and not the causes, which must still be identified. Figure 8.8 contains a causality diagram showing the various potential sources of humidity in the hydrogen.

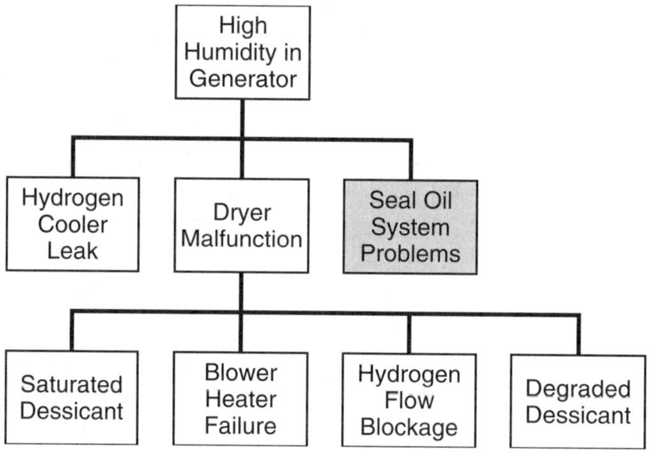

Figure 8.8 Causality diagram for high humidity problems in a generator.

Gland Seal System Problems

One common source of humidity in hydrogen gas is misoperation of the gland seal oil system. The exact mechanisms through which this takes place are many as well as complex. The block on Figure 8.8 that corresponds to this source of humidity is shaded with a dark background to indicate that the causes will be discussed in full detail in Chapter 9. Thus, there will be no diagnostic box included for this problem, and the reader is referred to Chapter 9.

Hydrogen Dryer Ineffectiveness

There are various possible causes for the dryer to malfunction. These will be considered separately and a diagnostic box will be included for each.

Saturated Desiccant. Desiccant saturation is likely to happen in dryers with a single drying element that must be manually reactivated. Reactivation may be neglected and result in an ineffective drying process.

■ **DIAGNOSTIC BOX 8.1** ■

PROBLEM: SATURATED DESICCANT

SENSOR INDICATION:

■ **The inlet dew point indicator** may give a high reading, which would be expected because the dryer is being ineffective.

- **The outlet dew point indication** (if it is available) may indicate little or no decrease in the dew point as the gas is streamed through the dryer. This will eliminate the possibility of the problem being inadequate flow of hydrogen.
- **The color of the desiccant crystals** may be grayish-pink instead of blue.

CORRECTIVE ACTION: If only one of the above indications is received, the validity of the sensor should be verified. If the sensor is found to be valid, place the dryer in the reactivating mode immediately. If humidity levels inside the unit are beyond the manufacturer's specifications, consult the manufacturer immediately for recommendations regarding shutting down the unit.

Reactivating Blower and/or Heater Failure. If the reactivation of the desiccant as described in Diagnostic Box 8.1 does not solve the problem, or if it is known for a fact that the desiccant has been reactivated recently, then the reason for the saturated desiccant could be improper operation of the blower and/or the heater.

DIAGNOSTIC BOX 8.2

PROBLEM: REACTIVATING HEATER AND/OR BLOWER FAILURE

SENSOR INDICATION:

- **The blower and/or heater failure alarm** will give direct and unequivocal indication of the problem. Some dryers monitor the operation of the heater and/or blower and alarm upon failure of either component.
- **The inlet dew point indicator** may give a high reading, which would be expected because the dryer is being ineffective.
- **The outlet dew point indication** (if it is available) may indicate little or no decrease in the dew point as the gas is streamed through the dryer.
- **The color of the desiccant crystals** is grayish-pink rather than blue.

CORRECTIVE ACTION: If only one of the above indications is received, the validity of the sensor should be verified. If monitored and in alarm, check to see if power to the heater and/or the blower was lost. If not, replace or repair the failed component. If not wired for alarms, manually test these components to determine their condition and replace or repair as necessary. Reactivate the desiccant immediately. If humidity levels inside the unit are beyond the manufacturer's specifications, consult the manufacturer immediately for recommendations regarding shutting down the unit.

Inadequate Gas Flow to Dryer. Inadequate flow of gas to the dryer can render the dryer ineffective. This can happen for various reasons, some obvious ones being a closed valve into the dryer, foreign material blocking the gas line to the dryer, corroded pipe lines (stainless steel is recommended by some manufacturers), or the size of the line leading into and out of the dryer being too small. Although the sizes of the hydrogen line to the dryer are standard, they are often undersized. The pressure drop because of the small size of the line may be too much to overcome for the relatively small differential pressure driving the gas through the dryer.

■ DIAGNOSTIC BOX 8.3 ■

PROBLEM: INSUFFICIENT HYDROGEN FLOW TO THE DRYER

SENSOR INDICATION:

- **The inlet dew point indicator** may give a high reading because the dryer's operation may not be having a significant effect on the humidity of the gas.
- **The outlet dew point indication** (if it is available) may indicate a normal decrease in the dew point of the gas as it is streamed through the dryer. This indicates that the dryer is operating properly.
- **The color of the desiccant crystals** is their normal blue color. One way to detect an inadequate flow of hydrogen is to observe the color and determine whether it turns gray in the normal length of time. If it takes significantly longer than normal to saturate the desiccant and the dew point meter continues to indicate high levels of humidity, this may mean a flow problem.

CORRECTIVE ACTION: If only one of the above indications is received, the validity of the sensor should be verified. If the indication is found to be valid, first determine whether any valves are closed that improperly shut off the gas flow to the dryer. If this is not the case, determine if a foreign object is blocking the line. Then check for corrosion in the pipes. Lastly, re-examine the design of the piping system to determine if the size of the line is adequate for the job. In many cases, the pipe should not be smaller than 1 in. However, this may vary in many cases and should be used only as a rule of thumb. Consult the manufacturer for specific recommendations. If humidity levels inside the unit are beyond the manufacturer's specifications, consult the manufacturer immediately for recommendations regarding shutting down the unit.

Desiccant Exhaustion. After many cycles of drying and being dried, the desiccant will degrade. When this happens, it loses its effectiveness and must be replaced.

■ **DIAGNOSTIC BOX 8.4** ■

PROBLEM: DESICCANT EXHAUSTION

SENSOR INDICATION:

■ **The inlet dew point indicator** may give a high reading, which would be expected because the dryer is ineffective.
■ **The outlet dew point indicator** (if it is available) may indicate no decrease in the dew point as the gas is streamed through the dryer.
■ **The color of the desiccant crystals** is grayish-pink rather than their normal blue.

CORRECTIVE ACTION: If the dryer remains ineffective after the desiccant has been in the regeneration mode without change in the indicator color and there is no other problem with the dryer as described above, replace the desiccant with a fresh lot. If this fails to solve the problem, consult with the generator or dryer manufacturer. If humidity levels inside the unit are beyond the manufacturer's specifications, consult the manufacturer immediately for recommendations regarding shutting down the unit.

Hydrogen Cooler Leak

Leaks in the hydrogen coolers are fairly common sources of moisture ingress into the hydrogen. If there is a liquid level alarm and if the liquid found is water, there is probably a leak in the hydrogen cooler. Additionally, for vertically mounted coolers, the first indicator to alarm will be the leak detector nearest the cooler.

Assuming that the relative pressures of the gas and water are appropriate (i.e., the gas is at higher pressure), then small leaks will not cause an inflow of water into the gas. Occasionally, however, either the pressures will be improperly set, or the leak may be large enough to introduce water into the gas, but small enough to be totally absorbed by the hydrogen. In these cases, the leak adds to the moisture content of the gas, but water may not collect in the liquid level detector(s).

■ **DIAGNOSTIC BOX 8.5** ■

PROBLEM: HYDROGEN COOLER LEAK

SENSOR INDICATION:

■ **Liquid detector(s)** may indicate the presence of water in the lower parts of the frame.
■ **The inlet dew point indicator** may give a high reading.
■ **The outlet dew point indicator** (if it is available) may indicate a normal decrease in the dew point as the gas is streamed through the dryer. This shows that the dryer is operating normally.

- **The color of the desiccant crystals** is blue (their normal color). This would be expected because the dryer is performing adequately. However, more frequent regenerations may be noticed if it is a single tower dryer.

CORRECTIVE ACTION: If only one of the above indications is received, the validity of the sensor should be verified. If the indication is found to be valid, because of the implications of high moisture content in a generator, some manufacturers recommend immediate shutdown upon discovery of the leak. That is also the recommendation made here. However, the manufacturer should be consulted as soon as possible to determine his specific recommendation, which should override any outside opinions. Temporary relief from the leak can be achieved by sectionalizing the cooler, thus isolating and discontinuing service water flow from the cooler section that leaks. This, however, results in some abnormal operating conditions that should not be allowed to persist for a long period of time. Upon coming off-line, sectionalizing the leaking cooler section, or determining after consultation with the manufacturer that the unit should remain on-line, the leak should be found and repaired by plugging the leaking tube. Consult the manufacturer for recommendations on how to find the leaking elements of the cooler.

8-3.2 Problem Area #2: Low Hydrogen Purity

As mentioned before, hydrogen gas will become flammable when its purity is between 4% and 75%. The purity of the hydrogen should be kept above 92% for safety reasons. However, for efficiency purposes, the purity should be kept at 98% or higher. This will provide an acceptable hydrogen density, at the same time remaining within a safety band as far as flammability. Alarms are generally set to activate when the purity decreases below 90%, but this depends on the owner's operating philosophy and the manufacturer's recommendation. It should be noted again that operation at 90% purity, although safe from a flammability standpoint, can cause the windage losses to double. In some large generators, this could be as high as 3 MW.

The main objective of this section is not to discuss the causes of impure hydrogen (they either have already been covered or will be covered in other sections), but rather to re-emphasize its negative effect on the operation of a generator, as well as its potential danger. However, one rare and trivial, but nevertheless occasionally valid, cause of low hydrogen purity will be presented. Figure 8.9 shows the causality diagram for this type of problem.

Gland Seal System Problems

One significant source of impurities in the hydrogen gas is a variety of problems in the gland seal oil system. This source of impurities will be covered in de-

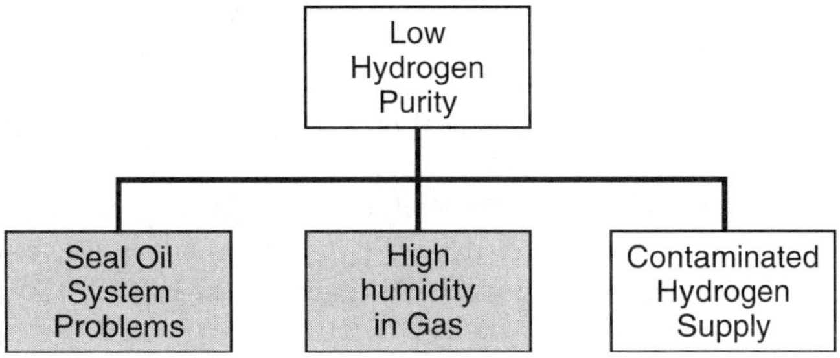

Figure 8.9 Causality diagram for low hydrogen purity conditions.

tail in Chapter 9. Note that the box in Figure 8.9 corresponding to this problem is shaded to indicate that this topic will not be discussed here.

High Humidity in Gas

High humidity has already been discussed in this chapter and will not be repeated here except to emphasize the point that the purity of the hydrogen can be reduced by the introduction of moisture into the gas. The box in Figure 8.9 corresponding to this problem is shaded to indicate that this topic will not be discussed here.

Contaminated Hydrogen Supply

While U.S. gas suppliers are generally quite reliable regarding the condition of the gas supplied, such may not be the case elsewhere. The gas can be delivered at a reduced impurity level or with a high moisture content.

■ DIAGNOSTIC BOX 8.6 ■

PROBLEM: CONTAMINATED HYDROGEN GAS SUPPLY

SENSOR INDICATION:

■ **The hydrogen purity meter** may read low even when the gas inside the machine is refreshed for the purpose of increasing its purity.

■ **The dew point monitor** may read high when the new gas is introduced in the generator.

CORRECTIVE ACTION: Take a sample of the gas from the main supply and analyze it for purity. If this confirms the hypothesis, obtain an alternative supply of pure hydrogen and refresh the hydrogen in the generator until the purity returns to a normal level. If humidity levels inside the unit are beyond the manufacturer's specifications, consult the manufacturer immediately for recommendations regarding shutting down the unit.

8-3.3 Problem Area #3: Loss of Cooling Capability of Hydrogen

The hydrogen gas can lose its ability to cool. There are basically three major reasons for this: 1) its pressure can be reduced through abnormal means, 2) its temperature can be above the upper design limit, or 3) low purity of the hydrogen can degrade its heat absorbing characteristics. The last item, however, will be discussed separately in its own section. Figure 8.10 depicts the causality diagram for this problem.

Loss of Seal

Loss of seal is also a gland seal oil system problem and will be discussed in Chapter 9. That is why it is shaded in Figure 8.10. Furthermore, if there is a loss-of-seal incident, the danger to the safety of plant personnel emerges as a more important issue than a loss of cooling capacity. For these reasons, no diagnostic box will be included for the loss of seal in this chapter.

Exhausted Hydrogen Supply

Occasionally, although not often, the supply pressure of the hydrogen will be lower than required to force the gas into the machine at the proper pressure. This can be due to exhaustion of the gas supply.

■ **DIAGNOSTIC BOX 8.7** ■

PROBLEM: EXHAUSTED HYDROGEN SUPPLY

SENSOR INDICATION:

- **The hydrogen pressure** reading may be low, and/or the alarm may be active.
- **The gas supply pressure switch** may alarm with a low reading.
- **The time taken to fill the generator** may be longer than usual for the same starting pressure.
- **The pressure indicators** in the supply system may all read low, even those on the supply side of the pressure reducer/regulator.

CORRECTIVE ACTION: If only one of the above indications is received, then the validity of the sensor should be verified. If the indication is found to be valid, obtain a new supply of hydrogen gas. In the meantime, reduce the generator load if capability curve limits are being exceeded.

Faulty Supply Pressure Reducer

In some cases, the reduced supply pressure is not from exhaustion of the supply, but rather from the misoperation of a pressure reducer in the supply system.

The hydrogen gas supply pressure is typically much higher than that required by the generator, and it has to be reduced to a more compatible pressure be-

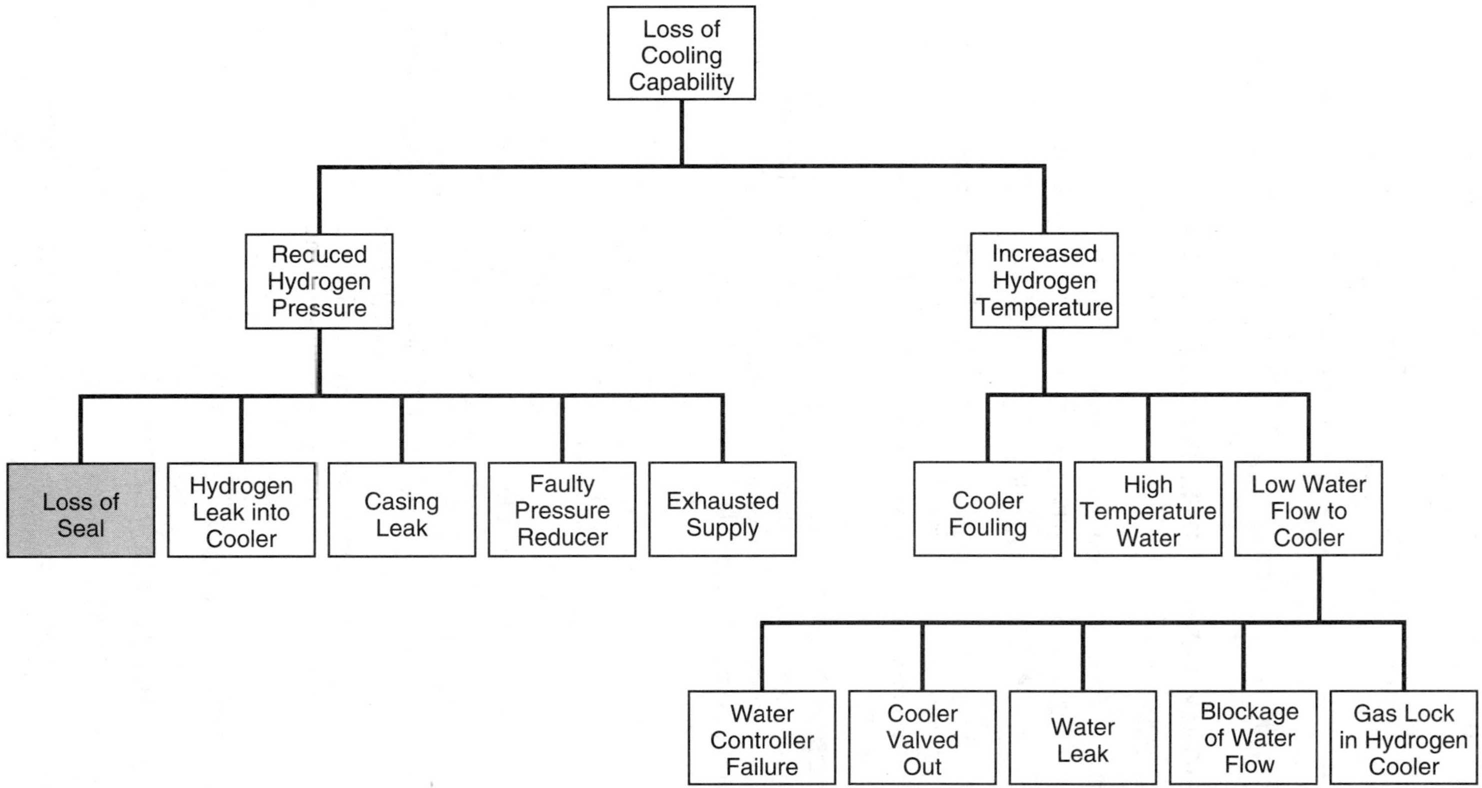

Figure 8.10 Causality diagram for loss of cooling capability of hydrogen gas.

fore it is introduced into the generator. This is typically done through a pressure reducer or regulator. A malfunction of this regulator could result in a decreased pressure level in the hydrogen gas supply manifold to the generator.

■ DIAGNOSTIC BOX 8.8 ■

PROBLEM: **FAULTY SUPPLY PRESSURE REDUCER**

SENSOR INDICATION:

- **Low hydrogen pressure** alarm may be active.
- **The gas supply pressure switch** may alarm with a low reading.
- **The time taken to fill the generator** may be longer than usual for the same starting pressure.
- **The pressure indicators** in the generator side of the reducer/regulator in the gas supply system may read low.
- **The pressure indicators** in the source side of the pressure reducer/regulator may read normal.

CORRECTIVE ACTION: If only one of the above indications is received, the validity of the sensor should be verified. If the indication is found to be valid, replace or repair the reducer/regulator. Reduce the load on the unit to ensure operation within the capability curve of the generator for the present hydrogen pressure.

Gas Leak into the Hydrogen Cooler

Because the service water pressure is typically designed to be lower than the gas pressure, a gas leak into the cooler is a more likely scenario than water leaking into the generator when there is a small leak in the cooler tubes. This will result in a loss of gas from the generator frame and a reduction in gas pressure. If the leak is large enough, it will also block the flow of service water through the cooler secondary hydraulic circuit. This is covered in Section 8.3.3, "Blockage of Water Flow in Cooler Tubes." The focus in this section is only on the loss of hydrogen. For this purpose, a small leak is assumed, which does not cause a blockage of service water flow.

■ DIAGNOSTIC BOX 8.9 ■

PROBLEM: **GAS LEAK INTO HYDROGEN COOLER**

SENSOR INDICATION:

- **The hydrogen pressure monitor** may identify a leak of hydrogen. This will be characterized by a relatively rapid drop-off of pressure after each top-off.

■ **Temperature rise** of the service water through the cooler secondary may be higher than normal because of the reduction in water flow.
■ **The hydrogen consumption meter** (if one is used) will indicate increased levels of consumption.

CORRECTIVE ACTION: If only one of the above indications is received, the validity of the sensor should be verified. If the indication is found to be valid, sectionalize the cooler section believed responsible for the leak, and repair the leaking tube if possible. Adjust the operating conditions of the machine to reflect the reduced cooling capability. Monitor the unit for increased levels of humidity in the gas. Consult the manufacturer for specific recommendations.

Frame Leak

Frame cracks are generally the result of long-term operation with high frame mechanical vibrations, possibly because of improper generator shaft alignment or poor support on the station foundation. Such cracks in the frame can cause hydrogen leaks to take place. Furthermore, gas will vent to the outside, creating a potentially explosive situation. Hydrogen leaks exceeding the maximum level recommended by the manufacturer should be investigated for the cause. If the consumption has an increasing trend, this indicates that a new problem is likely to be developing, which should be studied and corrected. Consult with the generator instruction manual or with the manufacturer to determine the unit's maximum allowable leakage rate.

Frame leaks were covered in Section 6.3.2. However, for the convenience of the reader, the diagnostics box is reprinted below:

■ DIAGNOSTIC BOX 8.10 ■

PROBLEM: FRAME LEAKS

SENSOR INDICATION:

■ **The hydrogen pressure monitor** may indicate a hydrogen pressure inside the unit that is dropping off faster than normal, indicating a continuous leak. This may be especially characterized by a relatively rapid drop-off of pressure after each top-off.
■ **A hydrogen detector** placed outside the frame may indicate the presence of hydrogen that has escaped from a crack in the frame.
■ **A tracer-gas detector** can be used for leaks that are difficult to detect. A small quantity of a tracer gas that is easier to detect than hydrogen is introduced into the frame of the generator. Then a detector designed for the tracer gas is used

outside of the generator to detect the presence of that gas and, therefore, the location of the leak.

- **A hydrogen gas consumption meter** may indicate larger-than-normal usage of hydrogen gas from the supply.
- **A frame vibration monitor** may register a high vibratory level, which will probably be at the running frequency of the rotor (60 Hz for a two-pole machine and 30 Hz for a four-pole unit).
- **The seal oil system** should be operating normally in the face of these leaks. This will eliminate the possibility of a hydrogen leak through the seals.
- **Hydrogen cold gas temperature** should be normal. This tends to decrease the possibility of a gas leak through the hydrogen coolers because of the presence of hydrogen in the water side of the cooler. This evidence, however, is not absolutely necessary for the presence of a gas leak into the cooler water.

CORRECTIVE ACTION: If only one of the above indications is received, the validity of the sensor should be verified. If the indication is found to be valid, and if the leak is not severe, continue monitoring the hydrogen pressure to ensure that an appropriate amount of cooling gas is present. Also, monitor the level of hydrogen outside the generator. This can accumulate in pockets under the machine. If the leak is significant and the amount of escaping gas can cause an explosion and/or fire, the unit should be shut down as quickly as possible and the crack repaired. In any case, the level of hydrogen consumption should be compared with the manufacturer's recommendations, and, if excessive, an early shut-down should be planned so the cause of the problem can be found and repaired. Consult with the manufacturer for recommendations on the action to be taken and the repairs to be made.

Hydrogen Cooler Fouling

The transfer of heat through the tube walls in the cooler between the gas and the water depends on the heat transfer properties of the transferring materials. Fouling of the inside of the cooler tubes will inhibit the transfer of heat through their walls by changing the effective nature of the materials in the walls to one that is less heat conductive. This will result in less efficient coolers and, consequently, in higher cold gas temperatures. In many installations, raw river water, with some mud and other debris, is used in the service water system. Moreover, in some instances, small clams enter the cooler and grow to foul the cooler tubes. The assumption made here is that the fouling does not cause any restrictions in the flow of water through the tubes, although this is often a consequence of the fouling process. Whereas some cooler fouling may always be present, the situation discussed here is fouling that has become severe enough to affect the capability of the cooler to operate within its rating.

■ **DIAGNOSTIC BOX 8.11** ■

PROBLEM: **HYDROGEN COOLER FOULING**

SENSOR INDICATION:

■ **Hydrogen cold gas temperature** may be higher than normal.
■ **The inlet temperature** of the cooling service water will be normal.
■ **Temperature rise (ΔT_{water})** of the service water through the cooler may be lower than normal for the same flow. This indicates that the amount of heat transferred to the water from the gas is reduced because of the fouling. Note that the flow of service water has to be the same for this observation to hold true.
■ **Flow of service water** (if measured) is constant.

CORRECTIVE ACTION: If only one of the above indications is received, the validity of the sensor should be verified. If the indication is found to be valid, the water flow through the cooler should be increased if it is not already at maximum allowed flow. If flow is already at maximum and the cold gas temperatures are at or above the maximum allowable point, reduce the unit load until the temperature is within normal range. Clean out the cooler at the next opportunity. Consult the manufacturer for proper cold gas temperature recommendations.

High Water Temperature

The rate of heat transfer through the tube walls depends on the temperature difference between the two mediums. The larger the difference, the more effective the transfer will be. Occasionally, the source of the cooling water warms up during the summer months such that the temperature difference is lower than that assumed in the design of the cooler. Consequently, the water cannot remove sufficient amounts of heat from the gas, and the gas temperature will rise. This condition does not reflect negatively on the cooler, but rather on the source of the service water.

■ **DIAGNOSTIC BOX 8.12** ■

PROBLEM: **HIGH WATER TEMPERATURE**

SENSOR INDICATION:

■ **The inlet temperature** of the cooling service water may be higher than normal. This is the key indication.
■ **Hydrogen cold and warm gas temperature** may be higher than normal.
■ **Temperature rise (ΔT_{water})** of the water through the cooler may be lower than normal for the same flow. This indicates that the amount of heat transferred

to the water from the gas is somewhat reduced because of the lower temperature difference. Note that the service water flow has to be the same for this observation to hold.

- **Service water flow** (if monitored) is constant.

CORRECTIVE ACTION: If only one of the above indications is received, the validity of the sensor should be verified. If the indication is found to be valid, increase the water flow through the cooler if it is not already at maximum. If flow is already at maximum and the cold gas temperatures are at or above the maximum allowable point, reduce the unit load until the cold gas temperature is within acceptable limits. Consult the manufacturer for specific recommendations.

Reduced Service Water Flow to the Cooler

Decreased service water flow to the cooler will reduce its ability to remove heat from the hydrogen gas. The volume of water available to carry away the heat is lower, thus the amount of heat carried away is also lower. The result is higher cold gas temperatures. There are various reasons for the decreased flow of water. They will be discussed individually in the following sections.

Cooled Valved Out. The closing of the cooler valve is the most common cause of decreased water flow. It happens relatively often when a unit is being brought back into operation after an extended outage. However, it can be a serious problem because the heat removal capability of the cooler is virtually eliminated.

DIAGNOSTIC BOX 8.13

PROBLEM: COOLER VALVED OUT

SENSOR INDICATION:

- **An unbalance** in the coolers because of a valve being closed could cause one side of the generator to be warmer than the other. Of course, this could happen if only one cooler is valved out. If water is interrupted to all coolers, this observation would be inapplicable.
- **Hydrogen cold gas temperature** may be much higher than normal and may be rising at a significant rate.
- **Hot gas temperature** may indicate a higher-than-normal value for the gas entering the cooler and may alarm.
- **The coil hot gas RTDs** (in gas conductor-cooled units only) may indicate higher-than-normal temperatures. This increase should be rather uniform across all coils.
- **Temperature drop** of the hydrogen across the cooler (ΔT_{gas}) may be nearly zero, but the absolute temperatures should be rising because there is no place

for the generated heat to go. The individual T_{hot} should track the increase in T_{cold}.

■ **The inlet temperature** of the cooling water may be normal.

■ **Temperature rise** (ΔT_{water}) of the service water through the cooler may be low or nearly zero. This indicates that the amount of heat transferred to the water from the gas is nearly zero.

■ **Service water flow meter** (if used) may show no water flow in the cooler secondary.

■ **Bearing vibrations** may steadily increase as the rotor body warps when under load.

■ **Hydrogen pressure** may be higher than normal because of the high temperatures within the frame.

CORRECTIVE ACTION: If only one of the above indications is received, the validity of the sensor should be verified. If the indication is found to be valid, the manufacturer should be immediately consulted. If recommended by the manufacturer, the water valves to the gas coolers should be immediately inspected, and if found closed, should be opened slowly so as not to cause thermal shock to the unit with rapid changes in temperature. If the situation has progressed very far, the unit should be shut down and the damage assessed before operation is resumed.

Water Flow Controller Failure. The service water circuit is typically controlled through a controller that attempts to maintain constant cold hydrogen gas temperature. A malfunction in this controller can result in decreased water flow through the coolers.

■ DIAGNOSTIC BOX 8.14 ■

PROBLEM: WATER FLOW CONTROLLER FAILURE

SENSOR INDICATION:

■ **Hydrogen cold gas temperature** may be higher than normal and into high alarm.

■ **Temperature drop** of the hydrogen across the cooler (ΔT_{gas}) may be lower than normal.

■ **The inlet temperature** of the cooling water should be normal.

■ **Temperature rise** (ΔT_{water}) of the water through the cooler may be higher than normal. The reduction of water flow will cause the temperature rise of the water to increase as well as cause a decrease in the amount of heat that is removed from the hydrogen gas because of a lower flow rate of service water.

■ **Service water flow meter** shows reduced water flow in the cooler secondary.

CORRECTIVE ACTION: If only one of the above indications is received, the validity of the sensor should be verified. If the indication is found to be valid, bypass the controller to resume normal water flow to the coolers. Inspect the controller, repair it, and return it into operation as quickly as possible.

Service Water Leak Outside of the Cooler. A leak of water outside and upstream of the generator cooler will result in reduced water flow into the cooler tubes. It is assumed that the leak is large enough to reduce the amount of water reaching the cooler. The symptoms seen here are similar to those seen in the previous section, except that a large leak will become obvious to the naked eye if the piping run is followed through the plant.

■ DIAGNOSTIC BOX 8.15 ■

PROBLEM: **WATER LEAK OUTSIDE OF THE COOLER**

SENSOR INDICATION:

- **Hydrogen cold gas temperature** may be higher than normal and into alarm.
- **Temperature drop** of the hydrogen across the cooler (ΔT_{gas}) may be lower than normal.
- **The inlet temperature** of the cooling water should be normal.
- **Temperature rise** (ΔT_{water}) of the water through the cooler may be higher than normal. The reduction of water flow will cause the temperature rise of the water to increase, but decrease the amount of heat that is removed from the hydrogen gas because of a lower flow rate in the service water.
- **An unbalance** in the coolers because of a leak on one of the coolers could cause one side of the generator to be warmer than the other. If the leak affects all the coolers equally, then this observation would not be applicable.
- **Service water flow meter** may indicate reduced water flow in the cooler secondary.
- **Significant water leak** is seen in the service water system.

CORRECTIVE ACTION: If only one of the above indications is received, the validity of the sensor should be verified. If the indication is found to be valid, inspect the service water system and locate the leak. Repair it as soon as possible. If the cold gas temperature is above the safe limit, reduce the load until temperature is within the acceptable range.

Blockage of Water Flow in Cooler Tubes. Occasionally, water flow to the cooler will become restricted, causing decreased water flow to the cooler. One possible cause is that air can become trapped in the cooler tubes, locking out the flow of water. This tends to happen more often during cold weather. Other causes of blockage could simply be foreign substances. The way to diagnose this problem is through elimination of all other problems linked with decreased water flow.

■ DIAGNOSTIC BOX 8.16 ■

PROBLEM: BLOCKAGE OF WATER FLOW IN COOLER TUBES

SENSOR INDICATION:

- **Hydrogen cold gas temperature** may be higher than normal.
- **Temperature drop** of the hydrogen across the cooler (ΔT_{gas}) may be lower than normal.
- **The inlet temperature** of the cooling water should be normal.
- **Temperature rise** (ΔT_{water}) of the service water through the cooler may be higher than normal. The reduction of water flow will cause the temperature rise of the water to increase, but decrease the amount of heat that is removed from the hydrogen gas because of the lower flow rate of the service water.
- **An unbalance** in the coolers because of a blockage on one of the coolers could cause one side of the generator to be warmer than the other. If the blockage affects all the coolers equally, this observation would not be applicable.
- **The service water flow meter** shows reduced water flow in the cooler secondary.
- **Other causes** for reduced water flow are not identified.

CORRECTIVE ACTION: If only one of the above indications is received, the validity of the sensor should be verified. If the indication is found to be valid, inspect the service water system and locate the blockage. Repair it as soon as possible. If the problem is an air lock, vent the cooler as soon as possible. If cold gas temperature is above the safe limit, reduce the load until the temperature is within the acceptable range.

Ventilation System Deficiency. A deficiency in the hydrogen ventilation system can be the result of improper assembly of the main blower and/or the baffling inside the generator. Improper assembly of the blower can be due to missing blades (if an axial flow blower is used) or reverse installation of the blower. This last condition has led to some serious consequences. Baffling can be either missing or improperly installed (i.e., gasketed) during installation or service. Either condition will permit leakage or bypass flows within the generator. These abnormal conditions will directly affect the effectiveness of the cooling system.

Although unlikely, blower blades can also be lost during operation of the generator as a result of a foreign object going through the blower. Although this can affect the pressure developed by the blower, the problem of a flying missile inside the generator becomes more critical at that point.

■ DIAGNOSTIC BOX 8.17 ■

PROBLEM: **BLOWER BLADE DAMAGE**

SENSOR INDICATION:

- **Blower differential pressure** may be lower than normal. This is the main indication of this problem.
- **Hydrogen cold gas temperature** may be higher than normal and into alarm.
- **Temperature drop** of the hydrogen across the cooler (ΔT_{gas}) may be somewhat higher than normal.
- **The inlet temperature** of the cooling water should be normal.
- **Temperature rise** (ΔT_{water}) of the water through the cooler may be slightly lower than normal.
- **Shaft vibrations** may increase because of the unbalance in the generator rotor.

CORRECTIVE ACTION: If only one of the above indications is received, the validity of the sensor should be verified. If the indication is found to be valid, inspect the fan during the next outage unless the problem becomes severe enough to warrant a forced outage. Consult the manufacturer immediately for recommendations.

SUMMARY

The purpose of this chapter was to describe the hydrogen auxiliary system and discuss how abnormal conditions in it can be identified and corrected. Four basic subsystems were described: 1) the hydrogen circulation subsystem, 2) the hydrogen temperature control subsystem, 3) the gas supply subsystem, and 4) the humidity control subsystem. The importance of each of these was discussed as well as how subsystem problems can affect the operation of the generator. The chapter included an introduction to some of the more commonly used sensors and monitors in the hydrogen system.

REFERENCE

[IEEE 1129–1992] "IEEE Recommended Practice for Monitoring and Instrumentation of Turbine Generators," IEEE Standard 1129–1992, New York City, 1992.

CHAPTER 9

Gland Seal System

The hydrogen cooling gas discussed in the previous chapter must be tightly sealed within the generator frame. Because the generator shaft is driven by an external source of power, the shaft must extend through the gas-tight frame. The shaft also extends through the other end of the generator opposite the drive to accommodate the generator field winding power input. The generator frame must be able to prevent the hydrogen gas from escaping at the points where the rotating shaft exits the frame. The gland seal system is designed to perform this function.

9.1 THE HYDROGEN SEALING SYSTEM

The two major components of the hydrogen sealing system are the gland seal rings used to seal the frame and the seal oil supply system used to provide the oil flow to the rings allowing the rings to perform their function.

9-1.1 The Gland Seal

Gland seals, supplied with oil under pressure, are used to prevent the leakage of hydrogen through the clearance between the shaft and the frame. Any such leaks will result in the hydrogen gas having to be continually replaced, and, much more importantly, the escaping hydrogen gas could also combine with air to create a volatile and potentially dangerous mixture that could result in equipment damage and in bodily harm.

A gland seal consists of a *gland seal assembly* that contains annular *sealing rings* supplied with oil under pressure. A gland seal assembly consists of housings that are supported from the outer end-shield at each end of the generator. The seal rings are typically made of bronze, with a tin-based babbitted interface at the shaft. There will be from one to three sealing surfaces at each end of the generator, depending on the type of system design. The clearances between the inner diameter of the ring and the shaft are typically 0.005–0.010 in, diametrical. The smaller the clearance, the smaller the volume of oil required to maintain a seal.

Figure 9.1 depicts one design of a gland seal ring. The oil is supplied to one or more annular grooves in the ring, from which oil flows axially along the shaft in opposite directions, through the clearance space between the shaft and the inner diameter of the ring. The pressure of this thin sheet of oil is typically set a few lbf/in^2 above the hydrogen gas pressure. The portion of the oil that flows toward the generator side of the seal comes into contact with the hydrogen gas, but the higher pressure of the oil keeps the gas from escaping through the clearance. The function of the gland ring is to restrict the flow of oil through the seal, as well as to provide the means for channeling the oil to the seal. The ring is typically allowed

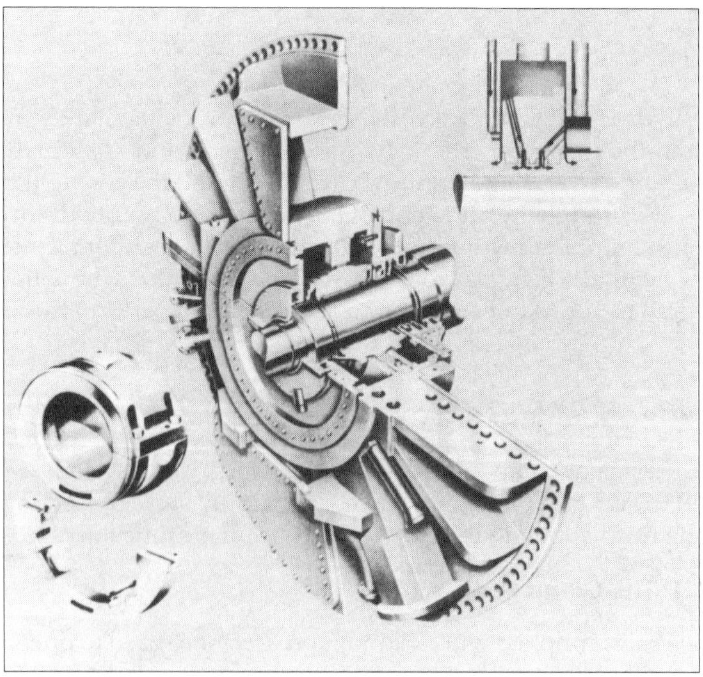

Figure 9.1 Typical gland seal ring. (Used by permission from Westinghouse Power Generation, Orlando, Florida.)

to float (move radially) on the shaft, but is prevented from rotating by a pin attached to the supporting structure. This helps maintain a constant and uniform clearance around the shaft in the presence of small changes in shaft elevation during operation. The ring is located along the shaft journal, inboard of the bearings.

The oil exiting the gland seal ring is collected in chambers on each side of the seal and is drained back into the supply system. With the small clearances presented by the rings, however, foreign particles in the seal oil, loss of seal oil, expansion and contraction of the shaft, or electrical currents flowing in the shaft can result in damage to the seals [Sheets, 1987]. The symptoms caused by this problem will be discussed later in this chapter.

9-1.2 The Seal Oil Supply System

The *seal oil supply system* is designed to furnish the oil required by the gland seal ring(s) in order to maintain a gas seal. It is a self-contained system that typically uses oil from the turbine generator shaft journal bearing lubrication system. Figure 9.2 is a photograph of the typical seal oil supply system equipment. Schematic diagrams of the two main types of systems will be presented in this chapter, and their

Figure 9.2 Seal oil supply equipment. (Used by permission from Westinghouse Power Generation, Orlando, Florida.)

operation will be described in the sections below. The general objective of the seal oil supply system, therefore, is to *maintain a seal at the generator shaft in order to safely contain the hydrogen gas within the frame.*

More specifically, it is the function of the seal oil supply system to lubricate the seals, prevent hydrogen from escaping through the seals, and to do so without introducing excessive amounts of contaminants into the hydrogen. There are various issues involved in maintaining a properly operating seal oil supply system that will accomplish these functions; they are as follows:

- The seal oil pressure must be kept above the hydrogen pressure in order to effectively seal the generator (6–12 lbf/in^2, depending on manufacturer recommendations). Lower pressures can result in a loss of seal, whereas higher pressures can cause inordinately high oil flow rates that can potentially cause oil to spill into the generator.
- The viscosity of the oil (and thus its temperature) must be kept within a relatively narrow band. Low viscosity oil, associated with high oil temperature, will result in a large volume of oil being pumped, which may exceed the drain capacity of the seal oil system. At the worst case, oil may spill over inside the generator frame.
- Because of the critical nature of the seal oil system, backup sources of oil must be maintained in case the main supply of seal oil fails to provide the necessary oil.
- The oil brought into contact with the hydrogen gas in the generator must be kept clean and free of contaminants.

The last of the above functions is significant because the oil, which is typically also used in the bearing lubrication system, can absorb an appreciable amount of any contaminant with which it comes into physical contact. If the gas is air and it contains water vapor, the oil will also absorb that moisture. As the oil passes through the annular shaft space to the generator side of the seal ring, the shearing action of the oil elevates its temperature and may cause it to foam. This foam may spill over through the labyrinth seals into the generator frame. The interface between the gas and the oil, even in the absence of excessive foaming, may also allow some dissolved gas/water to escape into the generator and mix with the hydrogen and contaminate it, lowering the latter's purity and increasing its moisture content. Thus, it is important to ensure that the oil that comes into direct contact with the hydrogen gas be free of contaminants.

This last requirement has led to various design philosophies on how to maintain hydrogen purity. All of them perform well in maintaining the other parameters mentioned above within specified ranges and are roughly equivalent in effectiveness. Nevertheless, the designs are rather different in how they accomplish their mission. In this chapter, two major designs will be discussed in detail:

- Double-flow system
- Single-flow system

The concept behind the double-flow system is that two nearly independent oil delivery systems are employed to ensure that the oil that interfaces with the hydrogen gas is maintained at a high level of purity. The gland seal ring for this type of system has two feed grooves, one for each oil supply. The different oil supplies, called the *air-side* (AS) and *hydrogen-side* (HS) systems, are kept separate until they interface at the rings. The HS oil is fed through the inner groove (closer to the generator) and flows inward along the shaft toward the generator; the AS oil is fed through the outer groove and will flow toward the bearing. The HS system supplies oil with dissolved hydrogen; the AS oil is obtained from the bearing oil system and is not particularly clean. Maintaining equal pressures between the two oil flows at the rings will minimize oil interchange in the clearance space between the two grooves where they actually come into contact. By minimizing the mixing between the two oil supplies, the purity of the HS oil is maintained. The combination of both systems at the ring provides the necessary flows to maintain a seal at the generator shaft.

The advantage of the double-flow system is that no oil processing equipment is necessary. Unfortunately, it does require an additional pump and cooler, plus all the piping, sensors, and controls required to operate a second system. Additionally, it can be difficult to balance the oil pressure and flows in the hydrogen side and the air side, which invariably results in some mixing between the two systems at the rings. This can result in a compromise of the purity of the HS oil.

Single-flow systems, on the other hand, require no additional pump, cooler, piping, or controls for the HS system, because there isn't one. Their disadvantage, however, is that oil purification equipment is necessary to maintain a clean oil supply. Furthermore, some designs require an additional pump to recirculate the oil through sprayers in the holding tank.

9-1.3 Double-Flow Seal Oil Systems

As mentioned above, the double-flow seal oil system is composed of the HS system and the AS system. Although the oil supplies are basically independent, there is some interaction between them at the ring. The two oil supplies are regulated to ensure that their respective pressures at the rings are equalized. This equalization is necessary to minimize mixing of the clean hydrogen-saturated HS oil and the contaminated AS oil. If the pressures at the rings between the HS and AS sides are not balanced, the oils will mix, and the purity of the oil that interfaces with the hydrogen degrades.

Hydrogen-Side Seal Oil System

The HS seal oil system interfaces directly with the hydrogen gas. The HS oil, therefore, will absorb hydrogen, but should be otherwise clean. Figure 9.3 shows a simplified version of an HS seal oil system. The gray blocks above the shaft journals represent the hydrogen seal rings. Also shown in the figure are the following components:

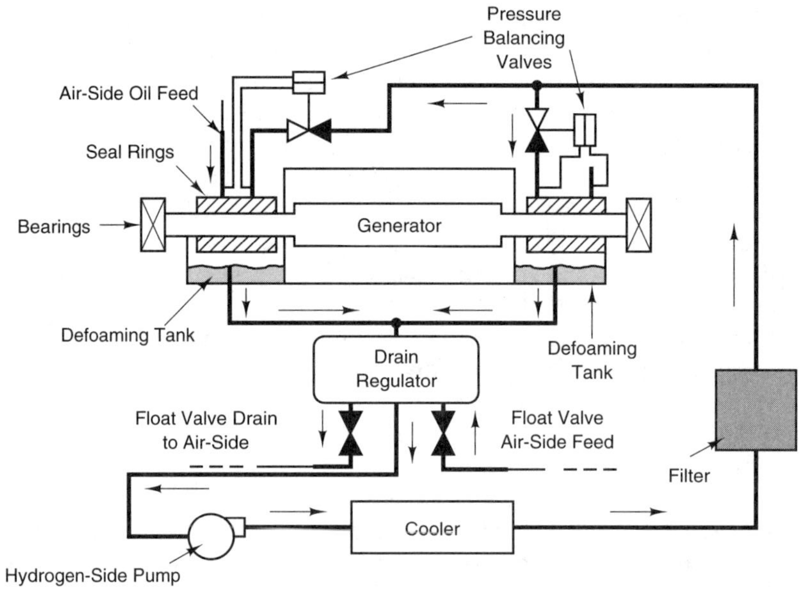

Figure 9.3 Simplified schematic diagram of hydrogen-side system.

Defoaming tanks: The defoaming tanks collect the draining HS oil. The defoaming tanks are located inside the frame, and thus are in the pressurized hydrogen environment. They are used to allow the dissolved hydrogen to defoam (detrain) before it is drained back to the HS system, thus permitting the excess hydrogen gas to remain within the pressurized generator environment. From the defoaming tanks, the seal oil is drained back into the closed loop HS system. There are two defoaming tanks, one on the turbine end and the other in the exciter (collector) end of the generator. When they fill to capacity, excess oil flow may overflow and result in an oil spill into the generator frame. The defoaming tanks are located in the bearing brackets of the generator. The oil level in the tanks is maintained by a standpipe overflow connection.

Drain regulator: Unequal pressures between the HS and AS oil supplies at the ring can cause oil from one system to flow into the other. Even when the pressures are kept constantly equal, small variations in pressure differences over a long period of time can introduce a gradual increase or decrease of the oil level being drained by the HS system through the defoaming tanks. The purpose of the drain regulator is to add or remove oil from the HS system in order to maintain the level constant within a range. Higher HS oil pressure (than AS) will force some HS oil to be drained along with the AS oil, resulting in a deficiency of HS oil in the drain regulator tank. Alternatively, lower HS pressures will allow some of the AS oil to enter the defoaming tanks and be drained into the drain regulator.

Both of these situations are corrected by two float valve-operated taps into the drain regulator tank from the AS system. These are shown in Figure 9.3.

One float valve operates on low oil level and allows AS oil to enter the drain regulator tank and make up the HS oil deficiency until the proper level is attained. The other float valve will operate on a high level and will open a drain into the AS system to remove excess oil collected in the drain regulator. The drain regulator will thus ensure a regulated supply of oil to the HS pump. The drain regulator also serves as a gas trap.

Hydrogen side seal oil pump: This pump is the prime mover behind the HS oil flow. It receives its supply from the drain regulator tank, and it pumps the HS oil to the rings.

Hydrogen side seal oil cooler: This cooler removes the heat from the HS seal oil. It uses plant service water in its secondary loop. Its proper operation is critical to ensure the desired viscosity of the oil.

Hydrogen side seal oil filter: As mentioned in the previous section, dirt or foreign particles in the oil can cause damage to the seal rings. The purpose of the filter, therefore, is to capture small foreign particles in the HS oil and thus maintain a dirt-free supply to the rings. The filter will accumulate dirt over a period of time and therefore has to be cleaned periodically. Otherwise, it will provide an impediment to the flow of oil that can reduce pressure and flow at the seal rings.

Pressure equalizing control valves: A pressure equalizing valve is provided in the HS feed line to each gland ring. These pressure regulators try to maintain the HS seal oil pressure as close as possible to that of the AS. Proper operation of these regulators is critical in avoiding the mixing of the HS and AS seal oils.

Other than any mixing done at the seal ring interface with the AS system, or any action by the drain regulator to make up a shortage of or to drain excess HS oil, the HS seal oil system is a closed-loop system. The generator can be operated at full rating even when the HS system is off-line, as long as the AS system is operating normally. The latter will provide additional oil flow to compensate for the absence of HS seal oil. The consequence is that the contaminated oil interfacing with the hydrogen imparts its impurities to the gas and results in decreased hydrogen purity and increased humidity.

Air-Side Seal Oil System

The AS seal oil interfaces with outside air over the oil in the bearing drain system, thus picking up all the impurities therein. During normal system operation, the AS oil is not intended to interface with the clean HS oil, but it sometimes does. The generator cannot be operated without the AS oil system in service, as the HS system cannot provide sufficient oil flow to maintain a seal. The AS oil system contains several backup sources of oil supply in case the primary AS pump fails.

Figure 9.4 shows a simplified schematic diagram of a typical AS seal oil system. The gray boxes located above the shaft journals represent the hydrogen seal rings. Note that the AS system's feed to the rings is located on the outboard sides

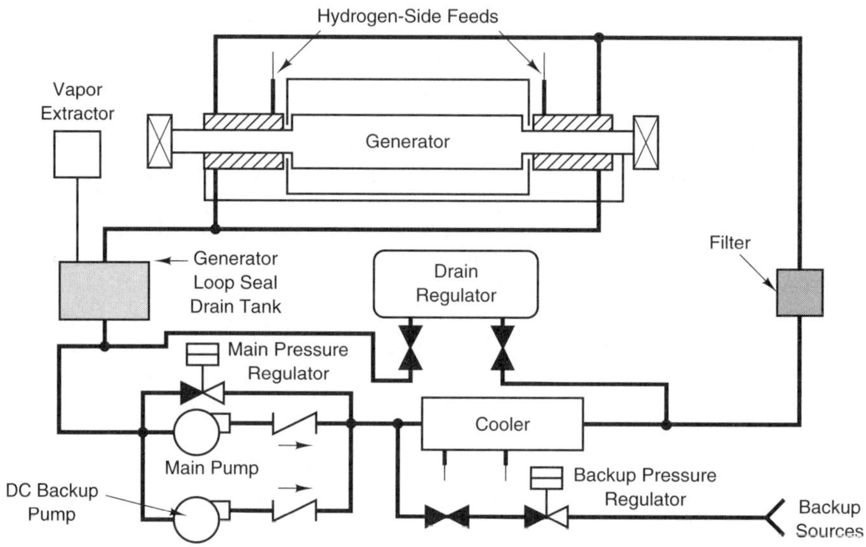

Figure 9.4 Simplified schematic diagram of air-side system.

of the rings as compared to those of the HS system shown on Figure 9.3. Also shown in this figure are the following components:

Main air-side seal oil pump: This pump is prime mover of the seal oil in the AS system. Its source is the combined bearing and AS seal oil drains.

Backup air-side seal oil pump: A DC backup pump is used as a second backup to the main seal oil pump. (The main backup comes from the bearing oil system through the backup regulator.) It is powered from the DC emergency bus in the plant so it can operate through a power blackout. It is isolated from the main pump through check valves. Some manufacturers recommend that when the DC backup pump is in operation, the generator hydrogen gas pressure be reduced to 3 lbf/in^2. This is because the last source of backup oil beyond the DC pump is low-pressure bearing oil. It is also important to ensure that the power source for the DC backup pump be at the proper voltage level. Because this source is typically a battery, it should always be charged to maintain the proper voltage.

Main air-side seal oil pressure regulator: This valve regulates the pressure and flow being produced by the main (as well as the backup) pump to maintain the seal oil pressure at approximately 12 lbf/in^2 above the gas pressure for some designs. It takes the hydrogen pressure reading from below the defoaming tank and the oil pressure from the AS seal oil line going to the ring. The regulator is connected as a return valve around the main and the backup pumps. A relief valve can be placed in parallel with it (not shown in Figure

9.4). The HS pressure follows the action of this valve through its pressure-balancing valves.

Backup pressure regulator: The AS system has a number of backup systems to ensure the supply of oil to the seals under any circumstances. The first backup is the high-pressure bearing oil. The backup pressure regulator will open to allow oil from the turbine sources when the pressure difference between the AS seal oil and the hydrogen gas is less than a preset value.

Air-side seal oil cooler: This cooler removes the heat from the AS seal oil, using plant service water in its secondary loop. Its proper operation is critical to ensure proper viscosity of the oil.

Air-side seal oil filter: As mentioned in the previous section, dirt or foreign particles in the oil can cause damage to the seal rings. The purpose of the filter, therefore, is to capture small foreign particles in the AS oil and thus maintain an oil supply to the rings that is free of debris. The filter will accumulate debris over a period of time and has to be cleaned periodically. Otherwise, it will impede the flow of oil and reduce pressure and flow at the seal rings.

Connections into hydrogen-side drain regulator through float valves: These aforementioned taps help regulate the drain of HS oil. These taps are controlled by float valves. One float valve opens on low drain regulator tank level to make up a HS oil deficiency from the AS supply. The other valve operates on high level to drain excess HS oil from the drain regulator to maintain the oil level and avoid overfilling the tank.

Generator loop seal tank: This tank collects the oil drained from the AS seal oil system as well as from the bearings. A vapor extractor is used to extract trapped hydrogen gas from the oil in the loop seal tank. The purpose of this loop seal is to prevent the hydrogen in the generator from escaping into the main oil reservoir in the event of a loss of seal. Otherwise, the result would be a sudden surge of hydrogen through the drain line.

Vapor extractor: The vapor extractor is designed to remove the minute amounts of hydrogen entering the bearing cavity either through the seal or through gasket leakage. On major seal failure, it can help to vent the large hydrogen leakage and avoid passing the hydrogen gas to the turbine lube oil reservoir. The vapor extractor is sometimes considered part of the bearing oil lubrication system and will also be covered in Chapter 11.

Seal Oil System Backups

Another important issue in the operation of the seal oil system is the emergency backups to the main seal oil pump. As mentioned before, the HS system need not be backed up because the generator can be operated normally with it being out of service, as long as the AS system is operating properly. The AS system, however, is indispensable in most designs, even if the HS system is in service. In normal operation, the main AS seal oil pump provides the pressure. The main pressure regulator is set to provide a preset pressure differential over the gas pressure, typically 6–12 lbf/in^2. There are, however, a number of backups for it, and these are as follows for one particular design:

1. The first backup is the high-pressure turbine oil sources. There are, however, more than one source of turbine oil. The shaft-driven main oil pump will be the first source of backup oil when the turbine is operating between two-thirds and full speed. During standstill, acceleration, or deceleration of the unit, the sources are either the turbine governor auxiliary oil pump or the auxiliary bearing oil pump. The backup pressure regulator will open when the pressure differential dips below a set value above gas pressure. It will maintain a specific reduced, but adequate, level of pressure differential.
2. The second backup is the DC backup pump. This pump is activated when the pressure differential drops below a second reduced level of pressure differential. Upon activation, it will operate to maintain a specific pressure differential.
3. The last backup is a low-pressure turbine oil. This is not considered to be an adequate backup system for running the generator, but it will serve to maintain a low gas pressure during standstill. Some manufacturers will recommend generator shutdown when the DC backup pump is operating because of the lack of available backups. One manufacturer's philosophy is that the amount of gas pressure to be maintained in the generator depends not on the seal oil source in operation at the time, but rather, on the pressure capability of the source of seal oil available from the next backup source. Generator gas pressure should be reduced in this case to a level compatible with the next source of backup oil.

Note that in some designs, if the AC pump is out of service and the seal oil is circulated by a shaft-driven pump, the generator can be operated without any adverse effects with the DC pump being inadvertently inoperable until the machine coasts down below a specific speed (typically near two-thirds speed), when the shaft-driven pumps can no longer provide sufficient pressure to maintain the seals. This points to the need to periodically test to ensure that the DC backup pumps are operable at all times. A scheme to automatically monitor the charge and/or voltage level of the battery that supplied the DC backup pump may be advisable in some cases.

9-1.4 Single-Flow Seal Oil Systems

There are a variety of single-flow system designs. The variations, however, are rather minor. Therefore, we will concentrate on only one particular design as being representative of single-flow systems. Figure 9.5 is a schematic diagram of the general system being represented. It makes use of bearing oil as the source of seal oil. Although a cooler is not shown in the diagram, the oil is cooled in the bearing oil system.

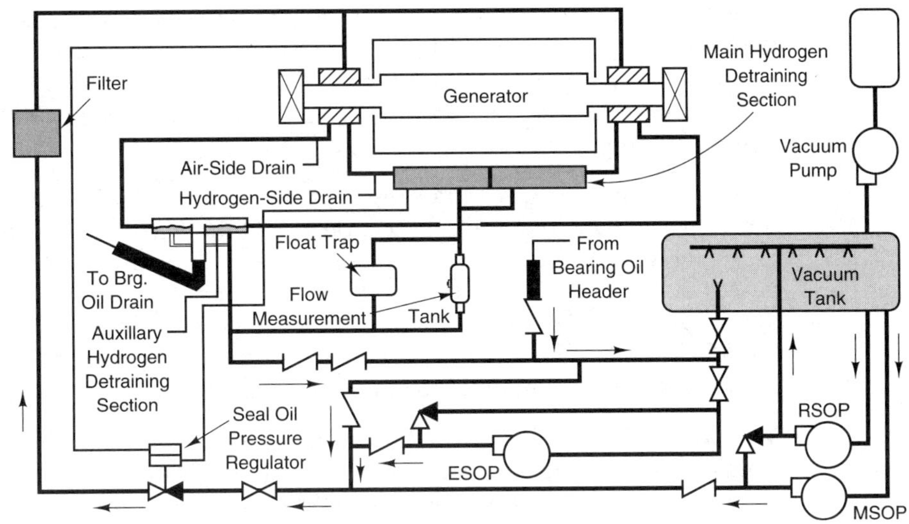

Figure 9.5 Simplified schematic diagram of single-flow system.

In a single-flow system, the bearing oil must interface with the hydrogen gas. Therefore, it must be cleaned through a vacuum process. This is shown as a vacuum tank and a vacuum pump. The oil is initially sprayed inside the tank through inlet nozzles to release the entrained vapors and is thereafter continuously recirculated through a different set of nozzles to multiply the spraying action. After flowing past the sealing rings, the oil is drained from the housing. The portion of the oil that flows toward the generator passes into the seal oil drain, while the oil that flows toward the bearing mixes with the oil discharged from the generator bearings. The following components make up the single-flow system discussed:

Main hydrogen gas detraining section: The detraining section can be an enlarged horizontal section of pipe that collects the oil flowing from the hydrogen side of the seal rings. It provides a large surface area that removes the larger hydrogen bubbles from the oil. This is equivalent to the defoaming tanks in the double-flow system previously discussed.

Drain trap and float valve: The outflow from the main detraining section passes through a trap containing a float valve that allows the detrained hydrogen to drain further but prohibits the hydrogen from progressing any further along the drain lines. During normal operation, the outflow from the float trap returns to the lubricating oil system through an auxiliary detraining section, which will be described below. During standstill, when the lubricating oil from the bearing header is not flowing, the outflow from the float trap recirculates directly to the vacuum tank through the two check valves instead of returning to the bearing oil header. This bypasses the cooling, but that is not necessary during standstill.

Drain trap and float valve bypass: In some circumstances, especially when the generator gas pressure is lower than 5 lbf/in^2, there may not be sufficient gas pressure to ensure the quick drainage of the oil from the trap through the float valve. In such cases, it may be desirable to open a bypass valve to assist in the draining process.

Auxiliary hydrogen gas detraining section: This section performs the same function as the main detraining section except that it collects the oil draining to the air side of the seal ring. It helps to detrain air from the oil before it passes to the main bearing oil drain. Partial dams allow the oil to recirculate to the seal oil system through check valves. This is important when the bearing oil system is out of service (when work is being done in the turbine), but gas pressure is to be maintained in the generator.

Vacuum tank: During normal operation, the vacuum tank collects lube oil from the bearing lube oil header and puts it through an inlet spray nozzle that is directed upward to separate some of the air and moisture that are entrained in the oil. The tank also contains another set of spray nozzles through which oil is recirculated to further remove the impurities. These are the recirculating set of nozzles, and they are pointed downward.

Vacuum pump: The air and moisture separated from the oil by the spraying are removed from the tank by a vacuum pump. The vacuum level maintained by this pump is typically 0.5–1.0″ Hg absolute pressure.

Main seal oil pump: This pump is the prime mover for the seal oil to the rings. It takes the oil from below the tank and pumps it to the rings through a filter and a pressure regulator. A relief valve is placed around it to return oil to the tank in case of overpressure.

Recirculating seal oil pump: This pump forces the recirculation of the oil through the recirculating spray nozzles described above. It takes oil from the bottom of the tank and returns it to the same place through the sprayer.

Emergency seal oil pump: This is the first backup to the main seal oil pump. It is driven by a DC motor and will automatically start in case of failure of the main pump, failure of the motor, or loss of AC power. Oil enters this pump from the bearing oil header, if available, or, more likely, from the HS drain lines through the two series check valves. The vacuum tank becomes inoperative when the emergency backup pump is operational. As with the double-flow system, a reliable source of DC power is required to maintain the integrity of this backup system.

Seal oil differential pressure regulator: This valve regulates the flow and pressure of the seal oil to the rings. The regulator maintains the oil pressure at a certain value over the gas pressure. It can be bypassed in case of its failure and the oil pressure maintained manually on a temporary basis.

Seal oil filter: As mentioned in the previous section, dirt or foreign particles in the oil can cause damage to the seal rings. The purpose of the filter, therefore, is to capture small foreign particles in the oil and thus maintain a clean supply to the rings. The filter will accumulate dirt over a period of time and therefore has to be cleaned periodically. Otherwise, it will provide an impediment to the flow of oil that can reduce pressure and flow at the seal rings.

Seal Oil System Backups

The primary source of seal oil in a single-flow system is, of course, the main seal oil pump. However, because of the critical nature of the hydrogen seals, backup sources must be available in order to reliably and safely operate the generator. Described below is the hierarchy of backup sources used in a single-flow system of the type described above. Please note that there are other types of single-flow system designs that use different backup sources, but the system described here represents a typical single flow design.

1. The first backup, or the second source of seal oil, is the emergency DC seal oil pump described above. It will activate whenever certain indications point to the failure of the main pump to deliver seal oil at the appropriate pressure.
2. The next source of backup oil is high pressure oil from the shaft-driven main turbine pump. Oil can be introduced into the seal oil system and delivered to the seals through the main seal oil differential pressure regulator. Normal operation can be continued under this mode of operation.
3. The last backup is not designed to allow normal operation of the generator, but can allow for sealing at lower hydrogen pressures. The oil used in this mode comes from the bearing oil header or from the HS drain system, bypassing the vacuum tank as well as the emergency seal oil pump.

9.2 PARAMETERS, SENSORS, AND MONITORS

The seal oil system has traditionally been under-instrumented with respect to the number of possible problems that can occur with it. Furthermore, most of the sensing is done through either local indicators or switches, which are generally inadequate for any diagnostic function beyond the bare minimum. Consequently, seal oil system problems tend to be neglected until they interfere with the operation of the generator. Thus, reactive troubleshooting, rather than incipient fault detection and preventive maintenance, is the norm in the seal oil auxiliary system. This section will describe the typical sensors used in monitoring seal oil supply systems and discuss some improvements that can be made to the diagnosis of problems in the seal oil system through the installation of additional sensors.

Sensors and monitors for the seal oil system will be divided into those for the double-flow system and those for the single-flow system. Although some of the sensors are similar in nature, others are different because of system differences. When the sensors are similar, they will not be described for a second time.

9-2.1 Double-Flow Seal Oil System Sensors and Monitors

As before, this type of system will be further divided into the HS and AS sensors and monitors.

Hydrogen-Side System Sensors and Monitors

Figure 9.6 shows the schematic diagram of Figure 9.3 with the sensors indicated. These are discussed in the following sections.

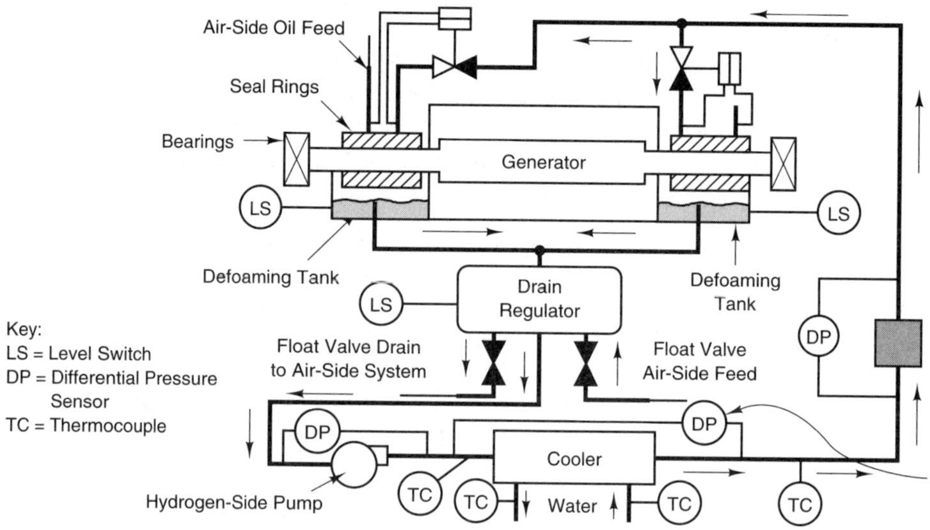

Figure 9.6 Hydrogen-side seal oil system schematic diagram including the sensors.

Defoaming Tank Level. When a defoaming tank overflows, excess oil can enter the generator frame. This is not a desirable situation, as will be discussed in Section 9.3. A level switch is typically used in each tank to provide a warning when a defoaming tank is in danger of overflowing. A local level indicator is also typically supplied in order to locally determine the oil level. These also verify the operation of the switch.

Because a switch operates only when the alarm level is reached, it would be more useful, albeit more expensive, to replace it with a level transmitter to continuously indicate the oil level in the tank.

Drain Regulator Tank Level. The drain regulator tank is typically monitored only for low level, because a specific oil level is required to provide a gas trap. However, a high level alarm is also important, and consideration should be given to furnishing this capability.

A high level of oil in the drain regulator tank can result from a plugged drain line, from misoperation of the drain float valve, or from misoperation of the makeup float valve. A full drain regulator can cause the oil to back up into the pipe leading it from the defoaming tanks, ultimately flooding the latter. This is generally more common during off-line conditions when the machine is degassed and the hydrogen pressure is not present to assist the draining process. Thus, a high level alarm in the drain regulator tank can serve as an advance warning of flooding the defoaming tanks. During on-line conditions, it can also help to identify misoperation of the float valves or a plugged drain line. A local level indicator is typically included for local readings.

The same argument that applies to using level transmitters in the defoaming tanks would apply here as well. However, because this represents only an advanced warning, the financial justifications for the added costs are not as clear as for the defoaming tanks.

Pump Differential Pressure. A differential pressure switch is placed across the main HS oil pump to monitor the differential pressure developed by the pump as a result of its operation. Its main objective is to determine whether the pump is working properly. A low, or zero, pressure differential across the pump means that the pump is not operating properly or that it has failed while in service. This switch is an important sensor.

Filter Differential Pressure. As the filter progressively accumulates more debris from the oil, it further impedes to the flow of oil, causing the pressure drop across the filter to increase commensurately. By monitoring the differential pressure drop across the filter, the state of the filter can be monitored. A differential pressure switch is typically used to alarm when the filter needs to be cleaned. However, by replacing the switch with a differential pressure transmitter, it can additionally be used as a low accuracy oil flow measurement.

Cooler Outlet Oil Temperature. A cooler outlet oil thermocouple and the accompanying local thermometer monitor the oil temperature at the outlet of the coolers. As mentioned before in this chapter, it is quite important to maintain the oil temperature within a certain range to regulate its viscosity in order to, among other things, keep the ring dimension (ID) fixed to avoid having it seize the shaft. A thermocouple is typically used to read the temperature in degrees Fahrenheit. It may also be advantageous, although not necessary, to install an inlet oil thermocouple, in order to be able to reason about the heat removal capability of the cooler.

Cooler Plant Service Water Inlet and Outlet Temperatures. Placing temperature sensors (most likely thermocouples) in the secondary water inlets and outlets of the oil coolers can be useful in diagnosing various malfunctions or abnormal conditions in the cooler. These are not typically supplied, but their addition can significantly increase the diagnostic capability as will be seen later.

Air-Side System Sensors and Monitors

The AS system is generally better instrumented than the HS, partly because of its greater complexity and partly because of its greater importance. Figure 9.7 shows a schematic diagram of the AS system with the sensors included.

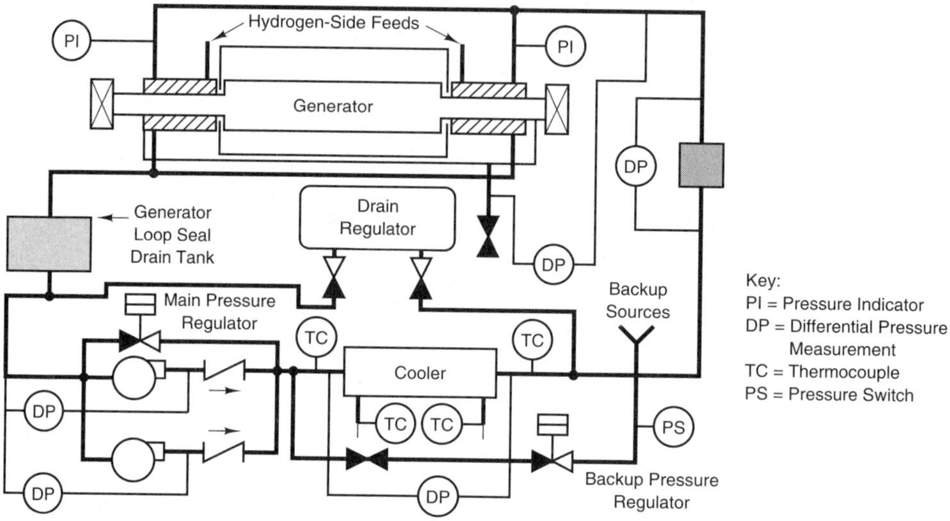

Figure 9.7 Air-side seal oil system schematic diagram including the sensors.

Pump Differential Pressure. As in the HS system, a differential pressure switch is generally placed around the main AS pump, as well as around the DC backup pump. Their objective is to detect failures of their respective pumps.

Filter Differential Pressure. This instrument is similar to that of the HS system. The same argument applies to the use of a transmitter in lieu of the switch.

Cooler Outlet Oil Temperature. The AS cooler outlet thermocouple description is the same as in the HS system. There is generally an accompanying local thermometer. It may also be advantageous, although not necessary, to install an inlet oil thermocouple.

Ring Oil Pressure. The AS seal oil system has one local pressure indicator on each side to measure the oil pressure at the point where it is most important, the ring. A case could be made to add a remote-reading pressure transmitter to allow monitoring from the control room or through the plant computer. In any case, a local indicator should be maintained.

Air-Side to Hydrogen-Side Oil Differential Pressure. The pressure difference between the HS and AS systems at the rings is the parameter that determines how much mixing there is between the oils from the two systems. Because this mixing has to be minimized, it is important to monitor this parameter. Although local gauges are typically used, a similar case can be made for using differential pressure transmitters for this function instead of local indicators. This parameter applies only to double-flow systems.

Turbine Backup Oil Pressure. The purpose of this device is to alarm whenever the turbine backup source oil pressure is low. It indicates the unavailability of this backup source of seal oil. Although this is typically supplied in double-flow systems, there is no reason why it cannot be used in a single-flow system that has such a backup. A pressure switch is generally used for this purpose.

Air-Side Oil to Hydrogen Gas Differential Pressure. The differential pressure between the seal oil and the hydrogen gas is important because it determines the ability of the seal oil system to maintain a seal in the generator. As its name implies, this instrument monitors the pressure difference between the AS seal oil and the hydrogen gas. Thus, it is probably the most important monitor in the seal oil system. Because the HS oil pressure follows the AS pressure, it is not generally necessary to monitor the HS oil/gas differential pressure. The specified normal pressure differential varies among the manufacturers, but is typically in the order of 3–12 lbf/in^2. Consult the generator instruction manual or the manufacturer for the appropriate value.

A switch is often supplied by some manufacturers, either in lieu of or in addition to, the transmitter described above. It monitors the same parameter as the transmitter, but it alarms when a critically low level of differential pressure is reached. In addition to alarming, it can also serve to activate the backup sources as required. This instrument is valuable if used in addition to the transmitter, rather than in lieu of it.

Cooler Service Water Inlet and Outlet Temperatures. As discussed in the section addressing hydrogen cooler plant service water temperature (Section 9.2.1), placing thermocouples in the service water inlet and outlet to the cooler can be useful in diagnosing various malfunctions or abnormal conditions in the cooler.

9-2.2 Single-Flow Seal Oil System Parameters, Sensors, and Monitors

Although some of the sensors in the single-flow system perform the same functions as in the double-flow system, there are others that are strictly used in a single-flow system (see Figure 9.8). These will be discussed in the following sections.

Main Detraining Section Level. A switch is typically used to measure the oil level in the main detraining section. It alarms when the oil level in the detraining section is too high. As in the defoaming tanks, excess oil in the detraining section can overflow into the generator casing and may cause damage. A level transmitter would be quite helpful in this function because it could provide advance warning of the flooding, as well as determine if the instrument is operational.

Vacuum Tank Oil Level. If the level of the oil in the vacuum tank is too high, it can render useless the inlet spray nozzles. A float switch can be used to alarm when the oil level in the tank is too high and threatens to affect the opera-

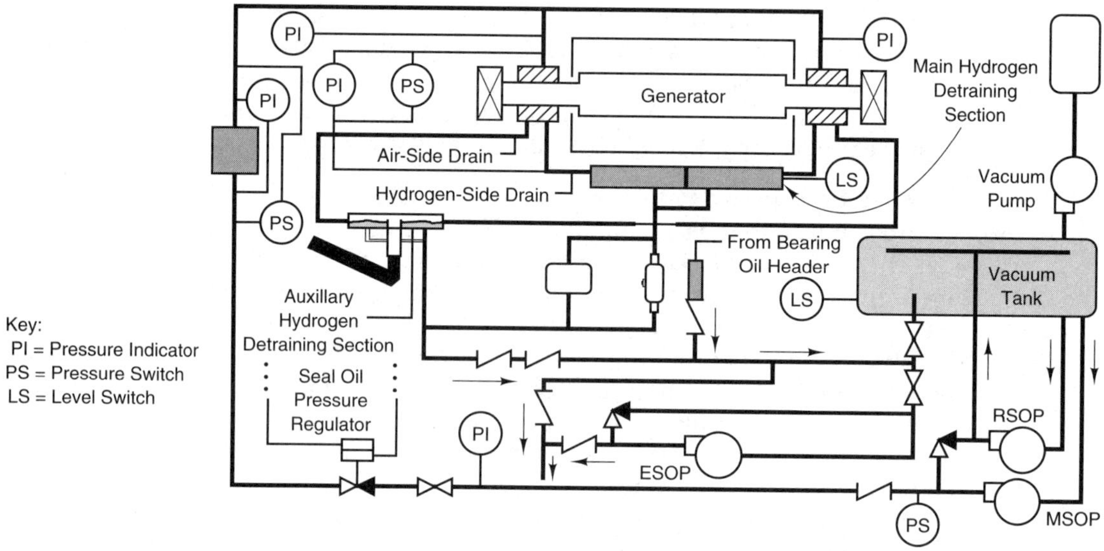

Figure 9.8 Single-flow seal oil system schematic diagram including the sensors.

tion of the spray nozzles. Likewise, a low oil level presents the potential unavailability of an oil supply to the main and recirculating oil pumps. Therefore, a second contact will alarm when the level is too low.

Main Pump Discharge Pressure. This instrument measures the pump discharge pressure and thus monitors the operation of the main seal oil pump. It performs the same function as the differential pressure switches used in the double-flow system. Similar instruments can be located to monitor the recirculating oil pump as well as the emergency backup pump. Manufacturer preference dictates whether a differential pressure element or a discharge pressure element is used.

Emergency Backup Pump Discharge Pressure. This instrument is the same as for the main pump except for the emergency backup pump.

Filter Differential Pressure. This instrument is the same as for the double-flow system. The same argument applies for the use of a transmitter to perform this function.

Ring Oil Pressure. The same comments apply to this instrument as for the double-flow system. The same argument applies for the use of pressure transmitters for this function.

Vacuum Gauge. This instrument could monitor the vacuum level of the vacuum tank. A remote sensing device such as a switch or transmitter would be preferable to a local gauge in this case.

Vacuum Pump Pressure. Although not shown in the schematic diagram, some implementations of the single-flow system include a vacuum pump failure switch that alarms when the pump fails. That failure is identified by measuring the discharge pressure (or vacuum) of the pump.

Hydrogen Gas to Seal Oil Differential Pressure. This instrument is the same as in the AS part of the double-flow system. A pressure switch as well as a transmitter is often used.

Turbine Backup Supply Failure. Some designs include an alarm when the pressure in the turbine backup system is low, indicating the inability of that system to serve as a backup source of oil.

9.3 SEAL OIL SYSTEM MONITORING AND DIAGNOSTICS

There are three basic problem areas in the seal oil system, plus a miscellaneous area to cover all remaining problems that cannot be classified under the first three. These four are as follows:

- Oil ingress into the generator
- Loss of hydrogen seal
- Dirty oil interface with the hydrogen gas
- Miscellaneous problems

As with all diagnostic situations, the validity of sensors should not be unquestionably assumed. Uncalibrated or malfunctioning sensors can be the cause of a misdiagnosis and its attendant costs. The reader is advised to ensure to the greatest degree possible that all readings are accurate before any diagnosis is made.

9-3.1 Problem Area #1: Oil Ingress to the Generator

Oil can be introduced into the generator because of a misoperation of the seal oil system. Oil in the generator may deposit in the windings or other electrically live parts, creating a potential for surface tracking discharge to ground and lubricating end-winding surfaces as well as contents of the slot (ripple springs), thus lowering friction and allowing the winding to vibrate. Figure 9.9 depicts a causality diagram for this problem area.

There are two basic causes for this oil ingress into the generator: degassed conditions in the generator and high seal oil flow. The former takes place during

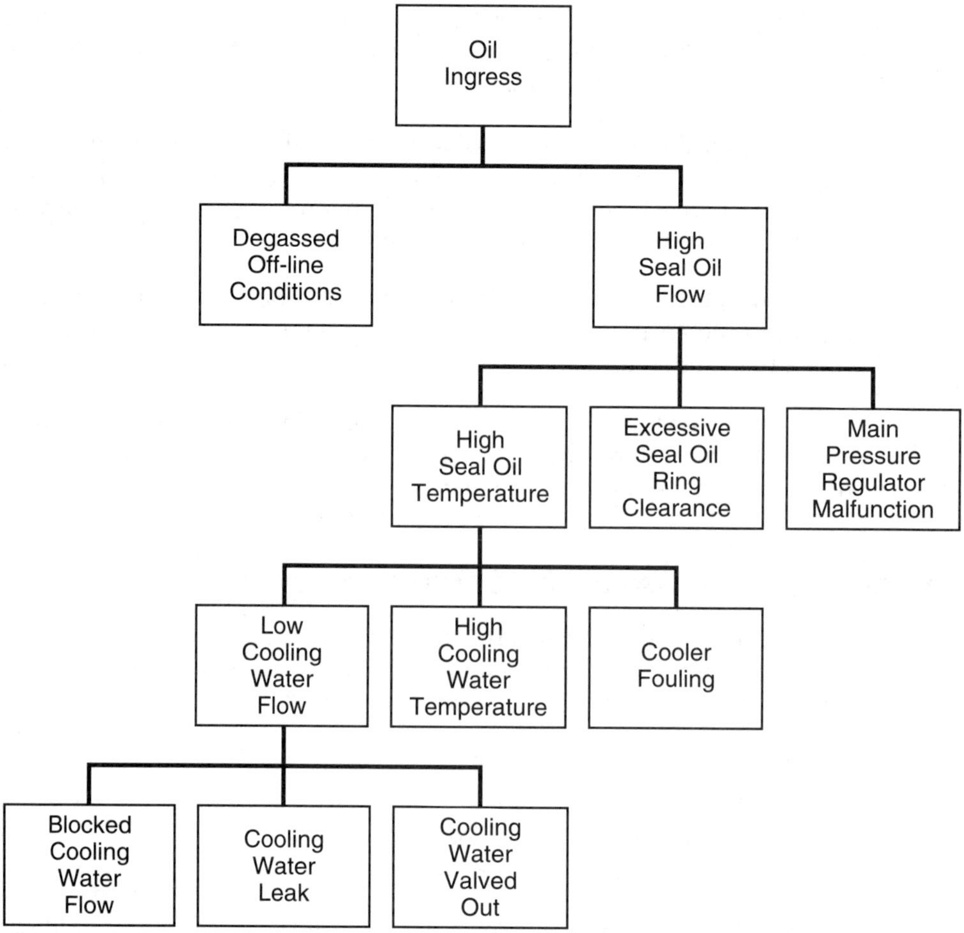

Figure 9.9 Causality diagram for ingress of oil into the generator.

off-line conditions only, whereas the latter can happen during on-line as well as off-line. High oil flow can be attributed to a number of different causes. The following subsections will be subdivided according to root causes.

Degassed Condition, with the Generator Off-Line

Oil ingress into the generator often takes place during off-line conditions when the generator is degassed. This is because the oil drained from the defoaming tanks (or the detraining sections) normally depends on the gas pressure to assist in draining the oil. Absence of this pressure makes the draining process much slower and thus far more susceptible to backing up under the slightest imbalance in the oil flow. The drain regulator tank (or the float trap) may fill to capacity and begin to back up through its inlet pipes to the defoaming tanks or detraining sections. Once these fill up, the excess oil may overflow into the generator.

There are generally no exact causes for this overflow. Some possibilities are slight imbalances between HS and AS systems, which, under fully gassed conditions, go unnoticed because of their rather insignificant nature. Another possibility would be that the oil cooler because the generator is not at speed, thus making the oil significantly more viscous. Lastly, small obstructions in the drain piping could represent obstacles to flow when the hydrogen pressure no longer helps in overcoming the effect of the obstruction. Diagnostic Box 9.1 discusses an imminent oil ingress problem.

■ DIAGNOSTIC BOX 9.1 ■

PROBLEM: OIL INGRESS BECAUSE OF DEGASSED CONDITIONS

SENSOR INDICATION:

■ **Liquid leak detectors** may alarm, the first direct indication of the presence of oil in the generator. Before the presence of oil can be confirmed, however, a sample of the contents of the leak detector needs to be examined to ensure that the liquid is not water, caused by non-seal oil related problems, such as a cooler leak.

■ **Defoaming/detraining tank level** alarm(s) may indicate a high level of oil. This may be true in either one or both sides (collector and turbine ends). This alarm should operate when the overflowing is imminent, but has not yet taken place.

■ **Drain regulator tank level** alarm will provide an early warning of the potential for flooding of the defoaming tanks.

CORRECTIVE ACTION: If only one of the above indications is received, the validity of the sensor(s) should be verified. If the situation is found to be real, the cause for the overflow must be investigated and corrected before any oil enters the machine. Because the machine is off-line, there is no immediate danger of failure. However, the oil ingress must be avoided or stopped, and the oil from the machine, if any has already entered, must be removed before the unit is restarted. Consult the manufacturer for specific recommendations.

High Seal Oil Flow

The other direct cause for ingress of oil into the generator is the high flow of seal oil. This mode of oil ingress occurs during on-line, fully gassed conditions. There are three main potential causes:

- Main pressure regulator malfunction
- Excessive ring clearance
- High seal oil temperature

The high seal oil temperature is generally caused by the malfunctioning of the seal oil coolers. That in itself could be due to a variety of causes, all which will be discussed in the following sections.

Excessive Ring Clearance. An oversized clearance of the seal oil ring will present reduced resistance to the oil flow, and result in increased flow. The flow of oil varies in relation to the cube of the clearance, so the process is quite sensitive to increases in clearance. The symptoms of this problem are similar to those of the high seal oil flow, but there is a significant difference: The pressure difference between the seal oil and the gas is normal. Because of the increased clearance, a normal pressure differential will drive more oil into the rings.

■ DIAGNOSTIC BOX 9.2 ■

PROBLEM: EXCESSIVE RING CLEARANCE

SENSOR INDICATION:

■ **One of the seal oil pressure indicators (transmitters)** at the rings may show a normal pressure while the other, a reduced pressure. Excessive clearance on the rings at both ends simultaneously is rather unlikely. This can be verified by inspecting the oil level in the defoaming tanks and noting that one of them may have a higher level than the other one, in addition to possibly more foam.

■ **The level switch(es)** alarm(s) in the defoaming tank(s) or in the detraining section may alarm because high flow can cause the oil to foam excessively and thus exceed the drain capacity of the tanks.

■ **Liquid leak detectors** may alarm if the excessive flow exceeds the drain capability of the defoaming or detraining tanks, and oil enters the generator. But this represents a worst-case situation. Before the presence of oil can be confirmed, a sample of the contents of the leak detector needs to be examined to ensure that the liquid is not water, caused by malfunctions not related to the seal oil system, such as a cooler leak. However, it would be desirable to detect the problem before any oil enters the generator. Thus, this indication is not a requirement for this problem to be diagnosed. It only expresses the urgency of the problem at hand.

■ **The level switch** alarm in the drain regulator tank will possibly *not* have alarmed.

■ **The seal oil-hydrogen differential pressure transmitter** will probably indicate a *normal* reading of seal oil pressure over the gas pressure.

■ **The seal oil-hydrogen differential pressure switch** will probably *not* alarm on a high reading (if it is set to do so.)

■ **The filter differential pressure switch (transmitter)** may alarm on a high reading, or the transmitter will indicate a high pressure drop across the filter, caused by the higher flow of seal oil. Because the state of the filters at the time of the incident may not be accurately known, this cannot be considered a reliable indication. However, it can serve to support the diagnosis.

- **The seal oil thermocouple** will probably read *normal* temperature.

CORRECTIVE ACTION: If only one of the above indications is received, the validity of the sensor(s) should be verified. If the situation is found to be real, inspect and correct during the next outage. If oil ingress is unacceptably high, then bring the unit down for the repair. Consult the manufacturer for more specific recommendations.

Main Oil Pressure Regulator Malfunction. Although malfunction of the main oil pressure regulator is not typically a problem, the regulator, being a mechani-cal device, can malfunction, allowing a greater seal oil flow to the rings than is required by the hydrogen gas pressure.

■ DIAGNOSTIC BOX 9.3 ■

PROBLEM: MAIN OIL PRESSURE REGULATOR MALFUNCTION

SENSOR INDICATION:

- **Liquid leak detectors** may alarm, the first direct indication of the presence of oil in the generator. Before the presence of oil can be confirmed, however, a sample of the contents of the leak detector needs to be examined to ensure that the liquid is not water, caused by problems unrelated to the seal oil system, such as a cooler leak. Nevertheless, it is desirable to detect the subject problem before any oil enters the generator. Thus, this indication is not a requirement for this problem to be diagnosed. It only expresses the urgency of the problem at hand.
- **Defoaming/detraining tank high level** alarm(s) may *not* have alarmed because the oil ingress is likely to be the result of excessive flow along the shaft, rather than an overflow of the tanks.
- **High level** alarm in the drain regulator tank may *not* have activated because the oil ingress is likely to be the result of excessive flow along the shaft, rather than an overflow of the tanks.
- **Seal oil differential-hydrogen pressure** transmitter may indicate a high reading of seal oil pressure over the gas pressure because of the higher-than-normal flow.
- **Seal oil-hydrogen differential pressure** switch may alarm on a high reading (if it is set to do so.)
- **Seal oil pressure indicators (transmitters)** at the rings may also indicate a high pressure over the gas pressure.

- **Filter differential pressure switch (transmitter)** may alarm on a high reading, or the transmitter may indicate a high pressure drop across the filter. This is caused by the higher flow of seal oil. Because of the unknown state of the filters at the time of the incident, this cannot be considered a reliable indication, but it can serve to support the diagnosis.
- **The seal oil thermocouple** will probably read *normal* temperature. This eliminates high seal oil flow caused by high oil temperatures.

CORRECTIVE ACTION: If only one of the above indications is received, the validity of the sensor(s) should be verified. If the situation is found to be real, the pressure regulator must be inspected and repaired. If the ingress is unacceptably large, then the unit should be brought off-line. Consult the manufacturer for recommendations.

Cooler Fouling. Cooler fouling is the first of the problems that cause high temperature seal oil. The problem generally is that high temperature oil decreases the viscosity of the oil and increases its flow. If the excess flow exceeds the oil drain capacity of the system, oil may spill into the generator. Moreover, high temperature oil tends to foam, which further impairs the capacity of the system to drain the oil. Much of the oil entering the unit enters as foam along the shaft through the labyrinth seals. If the cooler is not operating properly, the flow of oil at the same differential pressure (over the gas pressure) will be higher because of the decreased viscosity caused by the higher temperatures. Cooler fouling is one of the possible causes of a malfunctioning cooler, albeit a rare one if coolers are properly serviced during maintenance outages.

DIAGNOSTIC BOX 9.4

PROBLEM: COOLER FOULING

SENSOR INDICATION:

- **The seal oil cooler outlet temperature** (thermocouple or thermometer) may read *high*. This is the key symptom for identifying this problem.
- **Liquid leak detectors** may alarm. This is the first direct indication of the presence of oil in the generator. Before the presence of oil can be confirmed, however, a sample of the contents of the leak detector needs to be examined to ensure that the liquid is not water, resulting from leaks caused by malfunctions that are unrelated to the seal oil system, such as a cooler leak. Nevertheless, it would be desirable to detect the problem before any oil enters the generator. Thus, this indication is not a requirement for this problem to be diagnosed. It only expresses the urgency of the problem at hand.

- **The level switch(es)** alarm(s) in the defoaming tank(s) or in the detraining section may *not* alarm because it may be the foam above the liquid oil leaking into the machine that causes the liquid leak detectors to alarm. The foam in the defoaming/detraining tanks may not cause the level switches in the defoaming/detraining tanks to activate. Foam may be seen through the viewing glass of these tanks, if one is available.
- **The level switch** alarm in the drain regulator tank will *not* alarm.
- **The seal oil-hydrogen differential pressure transmitter** will probably indicate a *normal* reading of seal oil pressure over the gas pressure. However, in severe cases of very high temperature oil, the pressure differential between the oil and the gas may drop because the pump capacity is exceeded. This may cause a backup system (if one is available) to come on in spite of the main AC pump being in service.
- **The seal oil-hydrogen differential pressure switch** may *not* alarm on a high reading if it is set to do so.
- **The seal oil pressure indicators (transmitters)** at the rings will probably also indicate a *normal* pressure. However, in severe cases of very high temperature oil, the oil pressure at the rings may drop because the pump capacity is exceeded. This may cause a backup system (if one is available) to come on in spite of the main AC pump being in service.
- **The filter differential pressure switch (transmitter)** will alarm on a high reading, or the transmitter will indicate a high pressure drop across the filter caused by the higher flow of seal oil. Because the state of the filters at the time of the incident may not be accurately known, this cannot be considered a reliable indication. However, it can serve to support the diagnosis.
- **The temperature rise** of the cooling water, as calculated by the difference between the outlet (plant service) water thermocouple and the inlet water thermocouple, will be lower than normal because of the reduced amount of heat being removed from the oil across the fouled tube walls.
- **If a cooler inlet oil thermocouple is used, the temperature drop** of the oil should be lower than normal because of the reduced amount of heat being removed from the oil across the fouled tube walls.
- **The cooling service water inlet temperature** reading may be normal.

CORRECTIVE ACTION: If only one of the above indications is received, the validity of the sensor(s) should be verified. If the situation is found to be real, temporarily increase the plant service water flow to bring down the temperature of the oil (if possible). Clean out the cooler tubes during the next maintenance outage. If the oil ingress is unacceptably high, shut the unit down for immediate repair of the coolers. Consult the manufacturer for specific recommendations.

High Inlet Temperature Cooling Water. In many situations, the source of cooling water is shallow lakes, where the summertime temperature of the water will rise significantly over the design inlet water temperature rating of the cooler. It is likely that the coolers were underspecified because they were not designed to handle actual conditions. Little can be done about this problem in the short term. The long-term solution, of course, is to replace the coolers with adequate ones.

■ DIAGNOSTIC BOX 9.5 ■

PROBLEM: HIGH TEMPERATURE COOLING WATER

SENSOR INDICATION:

■ **The seal oil thermocouple (thermometer)** may read *high,* the key symptom for identifying this problem.

■ **The cooling water inlet temperature** reading may be high. This is another key indicator of this problem.

■ **The temperature rise** of the cooling (plant service) water, as calculated by the difference between the outlet water thermocouple and the inlet water thermocouple, may be near to or slightly less than normal.

■ **If cooler inlet oil temperature is measured, the temperature drop** of the oil should be near to or slightly less than normal.

■ **Liquid leak detectors** may alarm, the first direct indication of the presence of oil in the generator. Before the presence of oil can be confirmed, however, a sample of the contents of the leak detector needs to be examined to ensure that the liquid is not water, caused by malfunctions unrelated to the seal oil system, such as a cooler leak. Nevertheless, it would be desirable to detect the subject problem before any oil enters the generator. Thus, this indication is not a requirement for this problem to be diagnosed. It only expresses the urgency of the problem at hand.

■ **The level switch(es)** alarm(s) in the defoaming tank(s) or in the detraining section may *not* alarm because it may be the foam above the liquid oil leaking into the machine that causes the liquid leak detectors to alarm. The foam in the defoaming/detraining tanks may not cause the level switches in the defoaming/detraining tanks to activate. Foam may be seen through the viewing glass of these tanks, if one is available.

■ **The level switch** alarm in the drain regulator tank will probably *not* alarm.

■ **The seal oil-hydrogen differential pressure transmitter** may indicate a *normal* reading of seal oil pressure over the gas pressure. However, in severe cases of very high temperature oil, the pressure differential between the oil and the gas will drop because the pump capacity is exceeded. This may cause a backup system (if one is available) to come on in spite of the main AC pump being in service.

■ **The seal oil-hydrogen differential pressure switch** may *not* alarm on a high reading if it is set to do so.

■ **The seal oil pressure indicators (transmitters)** at the rings will probably also indicate a *normal* pressure.

■ **The filter differential pressure switch (transmitter)** will alarm on a high reading, or the transmitter will indicate a high pressure drop across the filter. This is caused by the higher flow of seal oil. Because the state of the filters at the time of the incident may not be accurately known, this cannot be considered a reliable indication, but it can serve to support the diagnosis.

CORRECTIVE ACTION: If only one of the above indications is received, the validity of the sensor(s) should be verified. If the situation is found to be real, temporarily increase the cooling water flow (if possible) to bring down the temperature of the oil. If the flow is maximized, then consider spraying water over the outside of the cooler shell as an emergency measure. Consider replacing the coolers with others that have a higher inlet water maximum temperature rating. Consult the manufacturer for specific recommendations.

Blocked or Leaking Service Water Flow. Occasionally, the flow of water to the cooler will become blocked by some foreign object. Alternatively, a leak of the service water will take place because of a broken pipe. In either case, the flow of water to the cooler will be reduced. This will cause a deficiency in the cooling capability of the cooler, which would likely lead to high temperature oil. Improper reassembly of the seal oil cooler after a maintenance outage can also result in reduced water flow. If an automatic water flow regulator is used, this may not surface as a problem until the flow is maximized and the cooling deficiency in the cooler remains.

■ **DIAGNOSTIC BOX 9.6** ■

PROBLEM: BLOCKED OR LEAKING COOLING WATER FLOW

SENSOR INDICATION:

■ **The seal oil outlet thermocouple** (thermometer) may read *high*, the key symptom for identifying this problem.

■ **The temperature rise** of the cooling water, as calculated by the difference between the outlet water temperature and the inlet water temperature, will be higher than normal because of the reduced cooling water flow.

■ **If cooler inlet oil temperature is measured, the temperature drop** of the oil should be lower than normal because of the reduced heat transfer.

■ **The cooling water inlet temperature** reading should be normal. This will eliminate the possibility of high inlet water temperature being the problem.

■ **Liquid leak detectors** may alarm, the first direct indication of the presence of oil in the generator. Before the presence of oil can be confirmed, however, a sample of the contents of the leak detector needs to be examined to ensure that the liquid is not water, caused by malfunctions unrelated to the seal oil system, such as a cooler leak. Nevertheless, it would be desirable to detect the subject problem before any oil enters the generator. Thus, this indication is not a requirement for this problem to be diagnosed. It only expresses the urgency of the problem at hand.

■ **The level switch(es)** alarm(s) in the defoaming tank(s) or in the detraining section may *not* alarm because it may be the foam above the liquid oil leaking into the machine that causes the liquid leak detectors to alarm. The foam in the defoaming/detraining tanks may not cause the level switches in the defoaming/detraining tanks to activate. Foam may be seen through the viewing glass of these tanks, if one is available.

■ **The level switch** alarm in the drain regulator tank may *not* alarm.

■ **The seal oil differential pressure transmitter** will probably indicate a *normal* reading of seal oil pressure over the gas pressure. However, in severe cases of very high temperature oil, the pressure differential between the oil and the gas will drop because the pump capacity is exceeded. This may cause a backup system (if one is available) to come on in spite of the main AC pump being in service.

■ **The seal oil-hydrogen differential pressure switch** may *not* alarm on a high reading (if it is set to do so.)

■ **The seal oil-hydrogen pressure indicators (transmitters)** at the rings may also indicate a *normal* pressure.

■ **The filter differential pressure switch (transmitter)** will either alarm on a high reading, or the transmitter will indicate a high pressure drop across the filter. This is caused by the higher flow of seal oil. Because the state of the filters at the time of the incident may not be accurately known, this cannot be considered a reliable indication. Hovever, it can serve to support the diagnosis.

CORRECTIVE ACTION: If only one of the above indications is received, the validity of the sensor(s) should be verified. If the situation is found to be real, temporarily increase the cooling water flow to bring down the temperature of the oil (if possible). Find the blockage or leak and correct the problem. If the ingress of oil is unacceptably high, then shut the unit down for cleanup and repair. Consult the manufacturer for specific recommendations.

Service Water to Cooler Valved Out. Operating with the service water flow to the cooler shut off is a fairly common problem, especially when bringing a unit on-line after a maintenance outage. It can be quite dangerous because of the potential for oil ingress caused by the resulting high temperature of the oil and the subsequent low viscosity and high oil flow. Fortunately, it is easy to correct once it has been detected.

■ DIAGNOSTIC BOX 9.7 ■

PROBLEM: **COOLING WATER VALVED OUT**

SENSOR INDICATION:

■ **The seal oil cooler outlet temperature** (thermocouple or thermometer) may read *high,* the key symptom for identifying this problem.

■ **The temperature rise** of the cooling (plant service) water, as calculated by the difference between the outlet water thermocouple and the inlet water thermocouple, may be near zero. This is another key symptom for this condition.

■ **If cooler inlet oil temperature is measured, the temperature drop** of the oil should be significantly lower than normal.

■ **The cooling water inlet temperature** reading should be normal. This would eliminate the possibility of high inlet water temperature being the problem.

■ **Liquid leak detectors** may alarm. This is the first direct indication of the presence of oil in the generator. Before the presence of oil can be confirmed, however, a sample of the contents of the leak detector needs to be examined to ensure that the liquid is not water, caused by malfunctions unrelated to the seal oil system, such as a cooler leak. Nevertheless, it would be desirable to detect the subject problem before any oil enters the generator. Thus, this indication is not a requirement for this problem to be diagnosed. It only expresses the urgency of the problem at hand.

■ **The level switch(es)** alarm(s) in the defoaming tank(s) or in the detraining section may *not* alarm because the foam above the liquid oil leaking into the machine may cause the liquid leak detectors to alarm. The foam in the defoaming/detraining tanks may not cause the level switches in the defoaming/detraining tanks to activate. Foam may be seen through the viewing glass of these tanks, if one is available.

■ **The level switch** alarm in the drain regulator tank may *not* be in alarm.

■ **The seal oil-hydrogen differential pressure transmitter** may indicate a *normal* reading of seal oil pressure over the gas pressure. However, in severe cases of very high temperature oil, the pressure differential between the oil and the gas will drop because the pump capacity is exceeded. This may cause a backup system (if one is available) to come on in spite of the main AC pump being in service.

■ **The seal oil-hydrogen differential pressure switch** may *not* alarm on a high reading (if it is set to do so.)

■ **The seal oil-hydrogen pressure indicators (transmitters)** at the rings may also indicate a *normal* pressure.

■ **A filter differential pressure switch (transmitter)** will alarm on a high reading, or the transmitter will indicate a high pressure drop across the filter. This is caused by the higher flow of seal oil. Because the state of the filters at the time of the incident may not be accurately known, this cannot be considered a reliable indication, but it can serve to support the diagnosis.

CORRECTIVE ACTION: If only one of the above indications is received, the validity of the sensor(s) should be verified. If the situation is found to be real, open the cooler water valves. Consult the manufacturer for specific recommendations in case the unit was damaged. Put this item in the start-up procedure checklist.

9-3.2 Problem Area #2: Loss of Hydrogen Seal

Loss of hydrogen seal is a serious problem because it represents a failure of the seal oil system to carry out its main mission: to contain the hydrogen gas inside the generator frame. The hydrogen gas can escape through the seals, creating a serious fire hazard as well as depleting the supply of gas inside the generator.

The problem is typically caused by insufficient flow of oil to the rings. Because many backups are available, this is a rare occurrence. For this reason, all of the causes to some degree depend on the failures of the main, as well as some of the backup, sources of seal oil. Nevertheless, the serious consequences of this problem make it an important one to cover. Figure 9.10 depicts the causality diagram of this problem area.

Clogged Seal Oil Filter

If any foreign substance buildup occurs in the seal oil path, it is most likely to occur in the filter, because its function is to collect foreign substances. The seal oil filters become clogged by the accumulation of foreign particles in the oil.

The seal oil filters are typically rotating plate type (commonly referred to as *cuno-type*), or *cartridge-type*. If cuno filters are used, the filter is cleaned by turning the handle through a complete revolution; if cartridge type, the cartridge is removed and disposed of and a new one is inserted into the filter. Generally, duplex filters are supplied in the last case, so that one cartridge can be replaced while the other one is in service.

■ **DIAGNOSTIC BOX 9.8** ■

PROBLEM: CLOGGED SEAL OIL FILTER

SENSOR INDICATION:

■ **The filter differential pressure switch/transmitter** may indicate a high value and/or alarm. This is the primary indication of this problem.

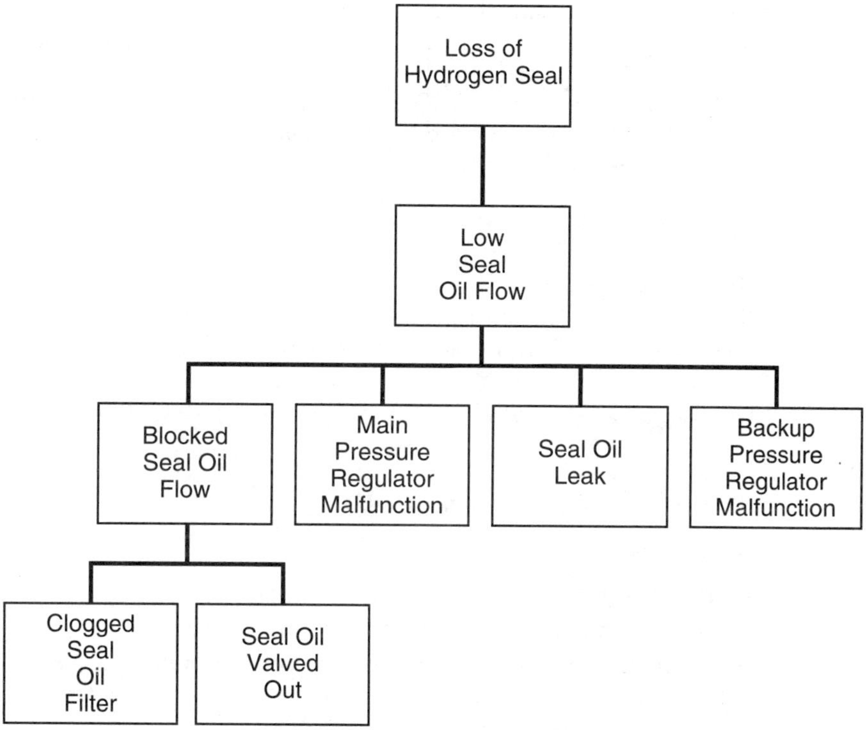

Figure 9.10 Causality diagram for loss of hydrogen seal problem area.

- **The seal oil-hydrogen differential pressure transmitter** may indicate either a normal or low reading of seal oil pressure over the gas pressure, depending on where the oil pressure measurement is taken. If taken upstream of the filter, then the pressure will probably read normal. If it is measured downstream of the filter, however, the differential pressure reading may be low.
- **The seal oil-hydrogen differential pressure switch** may alarm on a low reading (if it is set to do so.) This depends on where the oil pressure measurement is taken. If taken upstream of the filter, the pressure will not read low. If it is measured downstream of the filter, then the reading may be low enough to alarm.
- **The seal oil pressure indicators (transmitters)** at the rings may indicate a lower-than-normal pressure.
- **The seal oil cooler outlet temperature** should read normal. A normal reading will eliminate high flow caused by low viscosity as one of the possibilities. However, it is not absolutely necessary that it be this way.

CORRECTIVE ACTION: If only one of the above indications is received, the validity of the sensor(s) should be verified. If a cuno-type filter is used, simply turn the handle to clean it. Consider the installation of the motorized cleaning feature so that the filter is turned automatically after a certain period of time. If a cartridge type is used, switch the flow to the alternate side and replace the cartridge on the clogged side. Regular turning or cartridge replacement should be scheduled as recommended by the manufacturer. Consult the manufacturer for specific recommendations.

Seal Oil Valved Out

When the seal oil is valved out, seal oil flow to the ring is blocked. This is a very serious situation, but not very likely to happen under load.

■ **DIAGNOSTIC BOX 9.9** ■

PROBLEM: SEAL OIL VALVED OUT

SENSOR INDICATION:

■ **The filter differential pressure switch/transmitter** may indicate zero differential pressure (if a transmitter). This is a key indication.

■ **The seal oil-hydrogen differential pressure transmitter** will most likely indicate off the scale on the low side, if measured downstream of the closed valve. If measured upstream, it may read normal.

■ **The seal oil-hydrogen differential pressure switch** may alarm on a low reading if its seal oil sensor is located downstream of the closed valve. If measured upstream, it may not activate.

■ **The seal oil pressure indicators (transmitters)** at the rings will indicate a very low pressure, or zero.

■ **The emergency backup pump** will most likely be in operation if the seal oil differential pressure switch and/or transmitter reads or alarms low. This may, however, depend on the location of the incorrectly closed valve.

■ **The seal oil temperature differential** across the cooler will show little or no decrease because there is no oil flowing through the cooler.

CORRECTIVE ACTION: If only one of the above indications is received, then the validity of the sensor(s) should be verified. If the situation is found to be real, open the seal oil valves. Consult the manufacturer for specific recommendations. Add this item to the startup checklist.

Main Oil Pressure Regulator Malfunction

Main pressure regulators can fail by supplying too little oil as well as too much (Section 9.3.1). This section will concentrate on the deficient flow failure mode. In systems where there is a turbine backup regulator, this is not a significant issue, unless both regulators happen to be malfunctioning at the same time, which is not likely to happen. It is more serious in systems that have only one regulator or in systems in which the backup supply flows through the faulty regulator. When the regulator acts to reduce the flow of oil to the seals, the general indications are of low seal oil-to-hydrogen gas pressure differential. The effects of the malfunctioning regulator depend on whether it is a flow-through or a by-pass regulator.

■ DIAGNOSTIC BOX 9.10 ■

PROBLEM: MAIN OIL PRESSURE REGULATOR MALFUNCTION

SENSOR INDICATION:

■ **The filter differential pressure transmitter** may indicate a low value.

■ **The seal oil-hydrogen differential pressure transmitter** may indicate a low reading of seal oil pressure over the gas pressure.

■ **The seal oil differential pressure switch** may alarm on a low reading. If a backup exists, it may support the flow of oil as long as it doesn't go through the same faulty regulator, but it may be at a lower pressure.

■ **The seal oil pressure indicators (transmitters)** at the rings may indicate a low pressure.

CORRECTIVE ACTION: If only one of the above indications is received, the validity of the sensor(s) should be verified. If the situation is found to be real, open the bypass valves to resume flow, and control manually. If robust backups are available, then operate on backup temporarily. Inspect and repair the valve. Take the generator off-line if loss of seal is imminent. Consult the manufacturer for specific recommendations.

Seal Oil Leak

A leak of seal oil anywhere between the pump and the rings will result in a reduced amount of oil to the rings. A small leak would probably not be enough to impact the operation of the system because of compensation by the backup systems. However, a large leak could result in a loss of seal. Of course, the best way to find it is to look for a puddle of oil somewhere along the pipe. But the leak could also take place in the cooler (into the water) and not be visually obvious. If a large leak were to take place between the AC pump and the rings, where the pressure is measured, the backup supplies (if any) would activate and proceed to dump more oil through the leak with little effect on the seal oil-to-gas differential pressure. This problem, however, is rare and should not be considered the first suspect.

■ **DIAGNOSTIC BOX 9.11** ■

PROBLEM: SEAL OIL LEAK

SENSOR INDICATION:

■ This problem is highly dependent on the location and size of the leak as well as the arrangement of the seal oil piping and backup systems. Thus, no general sensor indications can be confidently provided. A puddle of oil may be noticed through periodic visual inspections of the seal system skid and piping.

CORRECTIVE ACTION: Find the leak and repair it. Periodic inspections of the entire system should provide an early warning. Consult the manufacturer for specific recommendations.

Backup Pressure Regulator Malfunction

Backup pressure regulator malfunction is not a likely occurrence in systems where the backup supply flows through a backup regulator, and it is not applicable to systems that do not have a backup pressure regulator. For this condition to be noticeable, the main pressure regulator must be out of service, because of either its own failure, or that of the main pump. This may also take place when the unit is down and the seal oil is supplied by the turbine oil backup system.

■ **DIAGNOSTIC BOX 9.12** ■

PROBLEM: BACKUP PRESSURE REGULATOR MALFUNCTION

SENSOR INDICATION:

■ **The emergency seal oil pump** will be running. This is annunciated in some system designs and is the first indication that there is something wrong with the turbine backup system. For this to happen, however, the main pressure regulator or the main pump must have already been out of service for other reasons.

■ **The seal oil-hydrogen differential pressure transmitter** may indicate a normal reading of seal oil differential pressure when the backup pump is in service. This reading, however, may not be equal to the differential pressure setting regulated by the backup regulator (depending on the design of the system).

■ **The seal oil pressure indicators (transmitters)** at the rings may indicate a normal pressure corresponding to the pressure when the emergency backup pump is operating. This may not coincide with the value normally maintained by the backup regulator.

- **Turbine backup oil pressure alarm** will be off. If it was active, the backup pump would be running because the pressure in the turbine backup system had failed for some other reason.

CORRECTIVE ACTION: If only one of the above indications is received, the validity of the sensor(s) should be verified. If the situation is found to be real, place the main oil system back on line, if possible. Otherwise, operate with the backup pump until the unit is shut down, the main system is brought back on line, or the backup regulator is repaired. If on DC pump backup, some manufacturers will recommend that load be reduced or the unit taken off line and gas pressure in the machine be reduced to near zero. Consult the manufacturer for specific recommendations.

9-3.3 Problem Area #3: Dirty Oil Interface with Hydrogen Gas

By allowing the hydrogen gas to interface with oil that has moisture, dirt, and other contaminants entrained in it, the purity of the gas will deteriorate as these impurities are imparted to the gas. As discussed before, reduction in gas purity is not desirable because it increases the windage losses in the generator blower and it reduces its heat transfer capabilities. Additionally, if the hydrogen purity is severely reduced, the potential for its flammability may rise. This section discusses the causes of contaminated oil interface. This problem area, in general, does not have the serious impact of the prior two. The unit can still be operated as long as the hydrogen gas is refreshed periodically to maintain its purity above the minimum recommended by the manufacturer (typically 90%–92%). Nevertheless, it can be expensive to operate in this fashion, and it should be avoided if possible. Figure 9.11 shows the causality diagram for this problem area.

In general, this problem can be detected through an increased humidity and decreased purity of the hydrogen gas. If the oil is very wet, some of the humidity may collect in the liquid level detectors at the bottom of the machine. Also, the water will separate from the oil in the defoaming/detraining tanks. This means that, generally, only the detectors close to the ends of the generator will collect the water.

Vacuum Pump Failure

Vacuum pump failure applies only to single-flow systems. If the vacuum pump fails, then the oil detraining system becomes greatly deficient. This can lead to the introduction of impurities into the hydrogen gas because the dirty oil is the interface to the hydrogen gas. This problem is rather easy to detect and correct, however. Because the seal oil system can operate without the vacuum pump running for an extended period of time, a failed pump poses no immediate safety hazard to either the generator or the personnel as long as the hydrogen gas is refreshed periodically.

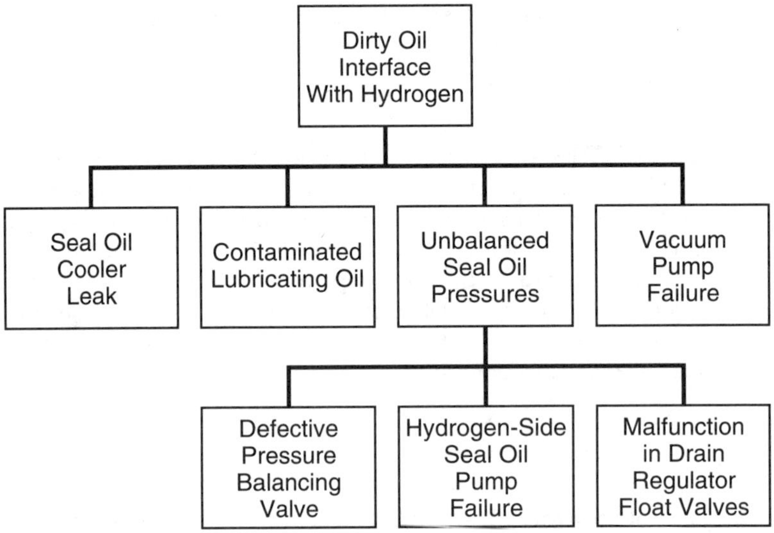

Figure 9.11 Causality diagram for the dirty oil interface problem area.

■ DIAGNOSTIC BOX 9.13 ■

PROBLEM: VACUUM PUMP FAILURE

SENSOR INDICATION:

■ **Vacuum pump failure** alarm may activate. This is the primary indication, and it is a very direct one.

■ **The vacuum sensor** may indicate no vacuum in the vacuum tank. This is another primary indication of this problem.

■ **The hydrogen purity meter** may indicate decreasing purity.

■ **The hydrogen dew point monitor** may indicate an increasing dew point in spite of proper operation by the dryer. This indication may not necessarily be present if the dryer is working properly and the leak is not sufficient to overcome the dryer's capacity to extract the moisture.

■ **The amount of water** taken out by the dryer has increased or is increasing. This is a way to detect the presence of increased water extraction from the hydrogen gas when the hydrogen dryer is functioning properly and within its capability.

■ **The liquid leak detectors** in the bottom of the frame may indicate water ingress into the unit. This, however, is also not necessary for the diagnosis because the amount of moisture may not be sufficient to collect in the detectors.

CORRECTIVE ACTION: If only one of the above indications is received, the validity of the sensor(s) should be verified. If the situation is found to be real, repair the pump and place it back in service at the earliest possible time. In the meantime, refresh the hydrogen gas periodically to maintain a high level of purity and low moisture. Consult the manufacturer for specific recommendations.

Defective Pressure-Balancing Valves Between Hydrogen-Side and Air-Side Systems

The problem of defective pressure-balancing valves applies only to the double-flow systems. If the regulating valves that keep the HS ring oil pressures equal to the AS pressures fail to regulate properly, the contaminated AS oil will mix with the clean HS oil. The consequence will be a degradation of the purity of the oil that interfaces with the hydrogen gas and the imparting of these impurities to the gas. It is obvious why this would happen when the AS pressures are significantly higher than the HS. Not so obvious is that when the converse is true, the drain regulator tank will require makeup oil from the air side in order to maintain an adequate supply to the HS pump. This mixing will ultimately lead to a deterioration of the HS oil quality.

DIAGNOSTIC BOX 9.14

PROBLEM: **DEFECTIVE PRESSURE BALANCING VALVES BETWEEN THE HYDROGEN-SIDE AND THE AIR-SIDE SYSTEMS**

SENSOR INDICATION:

- **Air-side to hydrogen-side oil pressure** differential may be higher than normal (in either direction), the main indication of this problem.
- **The hydrogen purity meter** may indicate decreasing purity.
- **The hydrogen dew point sensor** may indicate an increasing dew point, even though the dryer may be working and all dryer malfunctions have been ruled out. This indication is not necessary because the dryer may be operating properly and the amount of moisture ingress through the dirty oil may not be sufficient to overcome the capacity of the dryer to extract it.
- **The amount of water** taken out by the dryer may have increased or may be increasing. This is a way to detect increased water extraction from the hydrogen gas.
- **The liquid leak detectors** in the bottom of the frame may indicate water ingress into the unit. This, however, is not a necessary condition for the diagnosis because the amount of moisture may not be sufficient to collect in the detectors.

CORRECTIVE ACTION: If only one of the above indications is received, the validity of the sensor(s) should be verified. If the situation is found to be real, adjust or repair the regulator. In the meantime, refresh the hydrogen gas periodically to maintain a high level of purity and low moisture. Consult the manufacturer for specific recommendations.

Hydrogen-Side Seal Oil Pump Failure

Pump failure applies only to double-flow systems. It is an easy problem to detect because there is a direct sensor that indicates this problem.

■ **DIAGNOSTIC BOX 9.15** ■

PROBLEM: **HYDROGEN-SIDE SEAL OIL PUMP FAILURE**

SENSOR INDICATION:

■ **The hydrogen-side pump differential pressure switch** will indicate low alarm, or simply, "H2 pump is off." This is the main indication, and it is typically a reliable one.

■ **The hydrogen purity meter** will indicate decreasing purity.

■ **The hydrogen dew point sensor** will indicate an increasing dew point, even though the dryer may be working and all dryer malfunctions have been eliminated. This indication is not necessary because the dryer may be operating properly and the amount of moisture ingress through the dirty oil may not be sufficient to overcome the capacity of the dryer to extract it.

■ **The amount of water** taken out by the dryer has increased or is increasing. This is a way to detect increased water extraction from the hydrogen gas.

■ **The liquid leak detectors** in the bottom of the frame indicate water ingress into the unit. This, however, is not necessary for the diagnosis because the amount of moisture introduced through the contaminated oil interface may not be sufficient to collect in the detectors.

CORRECTIVE ACTION: If only one of the above indications is received, the validity of the sensor(s) should be verified. If the situation is found to be real, repair the HS pump and return it to service as soon as possible. In the meantime, refresh the hydrogen gas periodically to maintain a high level of purity and low moisture. Consult the manufacturer for specific recommendations.

Misadjusted or Malfunctioning Drain Regulator Float Valves

Problems with drain regulator float valves apply only to double-flow systems, but they are quite common. If there is a malfunction or a misadjustment of either the feed valve or the drain valve (or both) in the drain regulator tank, the AS oil will be mixed with the HS oil, contaminating the latter. Typically this manifests itself by having both valves open simultaneously. If the feed valve is prematurely open, the drain valve will have to open to drain some of the excess oil. Conversely, if the drain valve opens prematurely, the feed valve will have to open to make up the excess drainage. Furthermore, if neither one opens when it should, then a high or a low level in the tank will result and be detected by the appropriate sensors. Although a low level indication/alarm in the drain regulator tank is typically available, this is not so for high levels. Unless specifically added, the latter can only be visually indicated through the local level indicator on the tank. It can be difficult to detect this problem because there are no direct sensors to the simultaneous opening of the valves.

■ DIAGNOSTIC BOX 9.16 ■

PROBLEM: MALFUNCTIONING DRAIN REGULATOR FLOAT VALVES

SENSOR INDICATION:

■ **The temperature of the pipes** leading from the drain regulator tank through the float valves of interest may be high if there is oil flow within them. If both pipes are hot at the same time, both valves are likely to be open simultaneously. This is the main indication. While such a sensor is not normally supplied, it may be useful to add one to remotely monitor this temperature if the problem happens relatively frequently.

■ **The hydrogen purity meter** will indicate decreasing purity.

■ **The hydrogen dew point sensor** will indicate an increasing dew point, even though the dryer may be working and all dryer malfunctions have been eliminated. This indication is not necessary because the dryer may be operating properly and the amount of moisture ingress through the dirty oil may not be sufficient to overcome the capacity of the dryer to extract it.

■ **The amount of water** taken out by the dryer may have increased or is increasing. This is a way to detect the presence of increased water extraction from the hydrogen gas.

■ **The liquid leak detectors** in the bottom of the frame may indicate water ingress into the unit. This, however, is not necessary for the diagnosis because the amount of moisture may not be sufficient to collect in the detectors.

CORRECTIVE ACTION: If only one of the above indications is received, the validity of the sensor(s) should be verified. If the situation is found to be real, repair or adjust the misoperating valve(s). In some cases, this can only be done with the unit down and degassed. Consult the manufacturer for specific recommendations.

Seal Oil Cooler Leak

If there is a water-to-oil leak in the seal oil coolers, especially in the hydrogen side of a double-flow system, the oil may become wet and impart its water content to the seal oil through the interface at the rings. In most cases, however, the oil pressure is higher than the water pressure; thus any leaks will cause oil to leak into the water. However, it is not impossible for water to get into the oil because of incorrect operating pressures, or a large enough leak could cause mixing in both directions.

■ DIAGNOSTIC BOX 9.17 ■

PROBLEM: SEAL OIL COOLER LEAK

SENSOR INDICATION:

■ **Visual inspection** or testing of the coolers is the best way to identify this problem with certainty. However, this is difficult, if not impossible, to do with the unit in service.

■ **The hydrogen purity meter** may indicate decreasing purity.

■ **The hydrogen dew point sensor** may indicate an increasing dew point, even though the dryer may be working and all dryer malfunctions have been ruled out. This indication is not necessary because the dryer may be operating properly, and the leak may not be sufficient to overcome the capacity of the dryer to extract it.

■ **The amount of water** taken out by the dryer has increased or is increasing. This is a way to detect the presence of increased water extraction from the hydrogen gas.

■ **The liquid leak detectors** in the bottom of the frame may indicate water ingress into the unit. This, however, is not necessary for the diagnosis because the amount of moisture may not be sufficient to collect in the detectors.

CORRECTIVE ACTION: If only one of the above indications is received, the validity of the sensor(s) should be verified. If the situation is found to be real and if the leaking tube(s) can be identified, then plug it (them). In some designs, however, this can be done only with the unit out of service. If the dryer is keeping up with the ingress of moisture so that the dew point can be kept below the alarm level, the unit may be operable until the next maintenance outage. Consult with the manufacturer for specific recommendations. In the meantime, refresh the hydrogen gas periodically to maintain a high level of purity and low moisture content.

9-3.4 Problem Area #4: Miscellaneous Problems

There are other problems that do not neatly fall into the first three categories, yet are of sufficient importance to be discussed in this chapter. Figure 9.12 depicts the

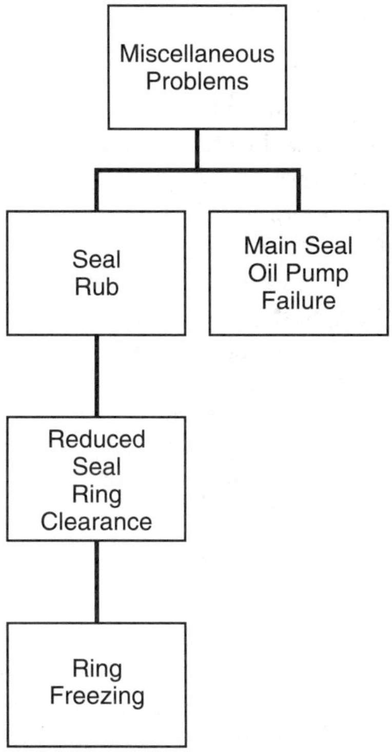

Figure 9.12 Causality diagram for miscellaneous problems.

■ DIAGNOSTIC BOX 9.18 ■

cause of the various problems in a graphical mode. Because all of these problems are not directly related, as were the previous problem areas, the diagram does not show causality except for the seal rub problem.

Seal Rub

Seal rub occurs when the ring freezes into one position rather than floating on the shaft. The consequence will be nonuniform ring clearances and, therefore, uneven heating around the ring related to the shearing action of the oil. This may cause bearing vibrations to take place. Unfortunately, it can be difficult to distinguish seal-related vibrations from the many other causes of bearing vibrations. Incorrect machining of the ring or incorrect assembly of the gland seal support brackets may reduce the ring's ability to maintain a uniform clearance to the shaft.

PROBLEM: SEAL RUB

SENSOR INDICATION:

- **Bearing vibrations** may be high.
- **High readings of seal metal temperatures** may help in detecting these rubs or misalignments. However, it is not common to monitor this parameter.
- **Seal oil temperature** may fluctuate. However, this may not be noticeable anywhere else other than at the ring.

CORRECTIVE ACTION: If only one of the above indications is received, the validity of the sensor(s) should be verified. If the situation is found to be real, temporarily increase the seal oil temperature in order to heat the ring and increase its clearance. This, however, is only a short-term solution. Beware of oil ingress caused by the increased temperatures. In double-flow systems, deactivate the HS pump to move the ring slightly. Observe the vibration levels. Consult the manufacturer for specific recommendations.

Reduced Ring Clearance

Reduced ring clearance is related to improper machining of the internal diameter of the gland seal ring. The reduced amount of oil flowing in the annular region cannot absorb the heat being generated by the oil shearing taking place, and the oil temperature at the ring increases rather drastically. As this happens, the ring expands somewhat, and the amount of oil flow increases. However, as the flow increases, the temperature decreases, causing a reduction in the clearance. That, in turn, causes the original problem of reduced clearance and the attendant local oil heating, characterized by wide fluctuation in the oil temperature at the ring. Because the temperature fluctuations involve only a small amount of oil, these may not be noticeable anywhere else other than the ring. The result is typically increased bearing vibrations.

■ DIAGNOSTIC BOX 9.19 ■

PROBLEM: REDUCED RING CLEARANCE

SENSOR INDICATION:

- **Bearing vibrations** will be high.
- **High readings of seal metal temperatures** may help in detecting this condition. However, it is not common to monitor this parameter.
- **Seal oil temperature** may fluctuate. This, however, may not be noticeable except at the ring itself.

CORRECTIVE ACTION: If only one of the above indications is received, the validity of the sensor(s) should be verified. If the situation is found to be real, consult the manufacturer for specific recommendations.

Main Seal Oil Pump Failure

It is rather simple to detect main oil pump failure, given that in most seal oil system designs there is a direct indication of this problem. When backups are available, as they are almost without exception, this is not a serious problem, and normal operation of the generator can continue. However, if the pump seal fails, the pump may provide less oil than required but still be running. If the decrease is significant enough, the backup sources will activate even though the main pump is in operation.

■ **DIAGNOSTIC BOX 20** ■

PROBLEM: MAIN SEAL OIL PUMP FAILURE

SENSOR INDICATION:

■ **Main seal oil pump differential pressure switch** may alarm on low pressure difference across the pump or on low discharge pressure. This is a direct sensor and is quite reliable. Although there are other indirect indications of this problem, they are probably not necessary in the presence of this indication. If the seal on the pump fails, then backup sources will become activated while the main pump remains in operation. This may cause the main pump "off" indication to not alarm.

CORRECTIVE ACTION: If only one of the above indications is received, the validity of the sensor(s) should be verified. If the situation is found to be real, repair the pump and return it to service at the earliest possible time. Oil leaks on the floor near the pump should be investigated if noticed during inspection. Consult the manufacturer for specific recommendations.

SUMMARY

The normal operation of the seal oil system is critical for the proper operation of the generator. Seal oil systems will provide long-term and reliable service when properly maintained and monitored. One problem is that the system is so reliable that little attention is typically paid to it until a major incident is experienced. In the worse case, faulty operation of the seal oil system may lead to a dangerous leakage of hydrogen gas from the generator. Other problems that may be caused by a malfunctioning seal oil system are the introduction of oil into the generator and the imparting of impurities to the hydrogen gas, thus reducing its purity and increasing its moisture content. Lastly, failure of the seal oil system that results in damage to the shaft and/or to the ring will cause expensive repairs and an outage of a few days.

REFERENCES

[Sheets, 1987] SHEETS, W. J., "Electric Generators," Electric Power Research Institute Power Plant Electrical Reference Series, Palo Alto, Calif.: Electric Power Research Institute, 1987.

CHAPTER 10

The Stator Coil Cooling Water Auxiliary System

As indicated in Chapter 4, some higher rated generators employ direct cooling of the stator winding by circulating deionized water through the coils (bars) themselves. In fact, there exist some older generators that circulate oil in a similar arrangement. The high thermal conductivity and specific heat of water permits a reduction in stator bar depth; consequently, the stator diameter and its weight can be reduced.

Therefore, in generators whose stator windings are water-cooled, an additional auxiliary system exists that is dedicated to supplying the cooling water to the stator bars and monitoring its flow, temperature, pressure, conductivity, and perhaps oxygen content (if the water cover is hydrogen). That system is the *stator coil cooling water auxiliary* (SCCW) system.

10.1 DESCRIPTION OF THE STATOR COIL COOLING WATER SYSTEM

Compared to the hydrogen auxiliary system and the gland seal oil system, the SCCW system is rather simple in nature. It needs only to maintain a steady water flow through the stator windings with the water at the proper conductivity and temperature. Yet, its task is critical, because failure to perform this function properly can result in unplanned outages and, in the worst case, a catastrophic failure of the generator caused by overheating of the stator windings. Like the gland seal oil system, the SCCW system is typically packaged on a skid and shipped to the site preassembled. Figure 10.1 shows a photograph of the physical equipment that embodies most of the components to be discussed in this chapter.

The objective of the SCCW system is to *provide the generator stator windings with cooling water at the appropriate temperature, pressure flow, and conductivity.*

Generally, water is pumped from a collecting tank in a closed system, and its temperature, pressure, and conductivity are monitored to ensure appropriate operating conditions. Make-up water is periodically introduced in the SCCW system in order to compensate for the small losses inherent in any "closed" process system.

Because high mineral content in the water increases its electrical conductivity, the cooling water purity must be maintained high to keep the insulation integrity of the electrical conductors. The cooling water is therefore demineralized to avoid undesirable paths for the electrical current. Failure to maintain the conductivity within the specified range can result in an electrical flashover across ground insulation. A typical conductivity value of the water is in the range of 0.1–1.0 micromhos/in [IEEE, 1129–1992].

There exist many variations of SCCW system design. In one typical design, water is driven through the system by a set of two AC pumps, one in operation, and the other on stand-by. Additionally, some systems also include one DC pump for emergency shutdown. The cooling water is passed through a set of coolers and filters before it reaches the stator windings inside the generator. A small percentage

Figure 10.1 Stator Coil Cooling Water System Equipment. (Used by permission from Westinghouse Power Generation, Orlando, Florida.)

of the water in the system is continuously diverted to a demineralizer plant in order to maintain a low level of conductivity. Once inside the generator, the water is delivered to the windings through headers, manifolds, hoses, waterboxes, or a combination of all of these.

The gaseous content of the cooling water is determined by the choice of gas covering the water reservoir. The water will become saturated by this gas. This is handled in different ways by the various manufacturers. Some manufacturers cover the water with hydrogen and take precautions that air is excluded. Others vent the water supply to atmosphere and maintain an air cover over the water.

Figure 10.2 shows a schematic diagram of a typical SCCW system. Following is a description of the various components that make up a typical stator coil cooling water auxiliary system.

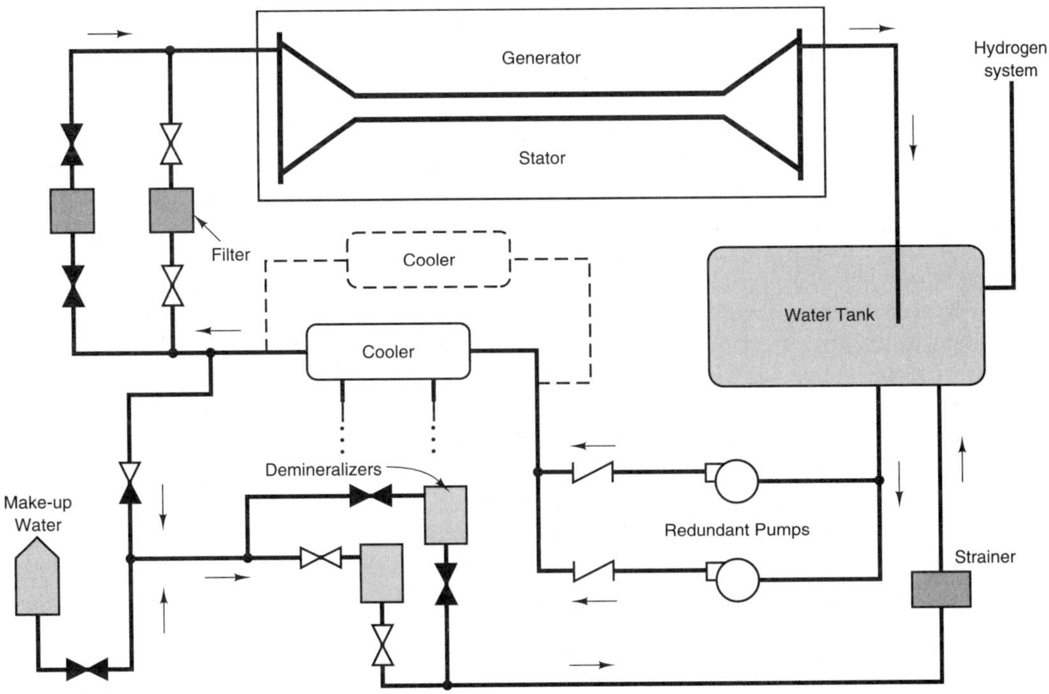

Figure 10.2 Schematic diagram of the SCCW system.

Dual pumps: Two identical AC pumps are typically used to drive the water through the system. One is in regular operation while the other is on stand-by. Upon failure of the pump designated as the main pump, the backup pump immediately starts and assumes the load. The roles of the pumps are interchanged periodically to allow for equal wear, as well to ensure that both pumps are operational at all times. Some SCCW designs include an additional DC pump for emergency shutdown.

Dual coolers: The coolers generally use low pressure service water as the secondary heat transfer medium. Heat exchangers are typically designed to cool the water to its specified stator water inlet temperature based on the anticipated service water temperature. In the system design shown in Figure 10.2, the heat exchangers are fully redundant so there is a backup available at all times. Another design builds in partial redundancy by supplying two half-size coolers in series. The advantage of this configuration is that if one cooler fails, the generator can still be operated at a significant fraction of full load if the defective cooler is taken out of service. Redundancy of any kind, however, is not always designed into the SCCW, as some system designs have only one full-size cooler in the loop.

Filters: A filter or a set of filters is typically used to remove all foreign particles, some as small as 5 μm, from the cooling water. A filter is typically cleaned by replacing its cartridge, which may last approximately one year under normal conditions. The cartridge may be replaced with the generator in service by bypassing the filter temporarily while the cartridge is being replaced. If a redundant filter exists, the flow is redirected to the redundant filter.

Water tank: A water tank is used to collect the water as it exits the stator and to act as a source of cooling water for the pumps. In some designs, the tank is covered by a blanket of clean hydrogen to control the oxygen content of the water at a proper level to minimize oxidation of the hollow strands in the coils. The water pressure is always maintained lower than the hydrogen pressure so that any perforations or cracks in the water-carrying hollow conductors will cause hydrogen gas to leak into the water system, and not vice versa. Therefore, any introduction of gas into the water system will register as an increase in pressure of the hydrogen blanket, if one is used. This is a key parameter to monitor when trying to detect the presence of such perforations or cracks. This will be discussed in greater detail later in this chapter.

Demineralizers: Maintaining the conductivity of the water within the specified range is a critical function of the SCCW system. To achieve this, it continually routes a small portion of its flow through a set of demineralizers or "polishers." These demineralizers pass the water through a bed of resin that extracts the minerals, and thus the conductive elements, from the water. The resin is typically effective for a finite period of time (typically one to two years, depending on the quality of the water), and the demineralizer must be refilled with new resin to restore its effectiveness.

Makeup water subsystem: As mentioned previously, some loss of water is inherent in any such closed system, and makeup water compensates for such losses. Before the water can be pumped into the tank to re-establish its normal level, it is passed through the demineralizers to ensure that its conductivity is consistent with the specifications. A strainer can be used to capture any larger foreign particles before the water is introduced into the water tank. The source of makeup water is usually the boiler feedwater system.

Although the SCCW system is comparatively simple, its function is so critical that it must be monitored extensively to detect any abnormal conditions within it. This next section will describe the monitoring and sensing requirements for the SCCW system.

10.2 SENSORS AND MONITORS

The main parameters that have to be monitored and controlled in a typical SCCW system are the following:

- Flow of water into the stator windings at the proper pressure
- Temperature of the water as it enters the stator winding
- Conductivity of the water

The sensors that will be discussed are used to support these monitoring functions. Some of these may be supplied by some generator manufacturers and others may not. Although they are not all necessary for the proper monitoring of the generator, it is highly recommended that they be installed by the station owner for maximum effectiveness in monitoring and diagnosing problems with the SCCW system. Figure 10.3 shows the schematic diagram of the SCCW system depicting the monitors and sensors that are discussed in the following sections.

Sections 10.2.1–10.2.5 deal with the issue of water flow and pressure, directly or indirectly. Sections 10.2.6–10.2.9 cover the problems associated with the temperature of the stator cooling water, and Sections 10.2.10 and 10.2.11 address water conductivity. Lastly, Sections 10.2.12 and 10.2.13 cover sensors used to identify and locate leaks of hydrogen into the water system.

10-2.1 Main Pump Differential Pressure

The differential pressure across the water pump can be used to determine if it is operational. Generally, a simple differential pressure switch is used, one for each pump. Alternatively, discharge pressure can be monitored. If a pump has failed, the switch will immediately act to energize the other pump through a secondary contact, with a minimum of interruption in the water flow. An alarm is also typically sounded.

10-2.2 Filter Differential Pressure

The filter differential pressure is mainly used to determine the internal state of the filtering element. If it is excessively clogged with filtered material, the element will resist the flow of water, resulting in an abnormally large pressure drop across it.

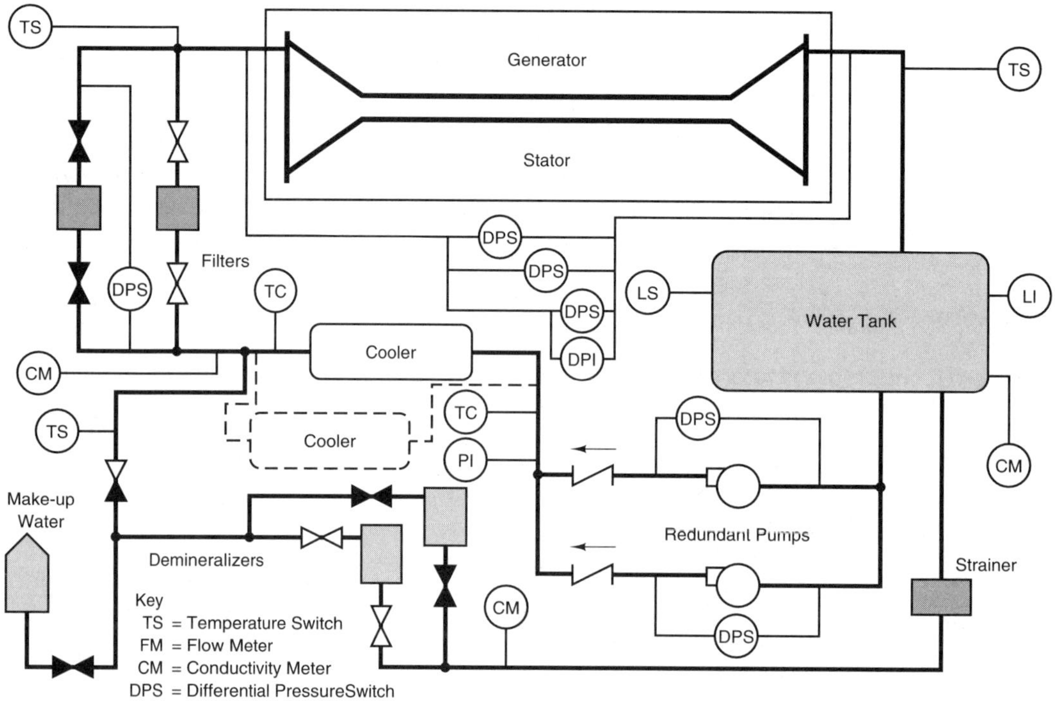

Figure 10.3 Schematic diagram of the SCCW system including sensors and monitors.

This is a good indication that the filter needs to be cleaned. Additionally, filter differential pressure can also serve as a gross measurement of water flow in the SCCW system. A differential pressure of zero, or near zero, through the filters indicates that there is no water flow through the system.

The most commonly used instrument for detecting the pressure drop across the filter is a differential pressure switch set to alarm at high values of pressure differential. Unlike the pump switches, these are typically not used to automatically route the water through the standby filter.

Nevertheless, from a monitoring and diagnostics standpoint, there is a definite advantage to using a remote analog monitor (i.e., a differential pressure transmitter) to perform this function. This allows for the determination of low (or zero) water flow through the system, as well as of a dirty filter element.

10-2.3 Generator Stator Water Flow

The stator water flow represents the main means of determining if water is flowing in the stator coils. Some designs use an inline water flow meter, whereas others use a differential pressure device (switch or transmitter) across the entire stator. The advantage of the former is that the measurement is direct and unambiguous.

However, depending on the instrument used, it may introduce a potential obstacle to the flow in the main water line to the generator in case of instrument failure. This is typically not desirable. The other method, although indirect, does not introduce this obstacle into the water line, and may thus be preferable.

Typical arrangements are either a transmitter to provide an analog measurement or a switch to alarm when the flow reaches a dangerous level. One scheme uses three differential pressure switches:

- High differential pressure, indicating either high water flow or blockage of flow inside the stator
- Low differential pressure, indicating low water flow
- Very low differential pressure, indicating very low or no water flow

10-2.4 Stator Inlet Water Pressure

The purpose of the inlet water pressure sensor is to ensure that the cooling water is at the proper pressure and flow. Typically, a local indicator is used, but it can be upgraded to a transmitter if desired. In general, if the flow and the temperature of the water are monitored, there is little need to convert this instrument to a transmitter.

10-2.5 Water Tank Level

The water tank level can indicate some problems with the system flow, such as leaks into, as well as out of, the system. More practically, it can be used to determine whether makeup water is required. A switch can be placed in the tank to alarm when the level is low, as well as when it is high. However, a transmitter can provide the trends that become significant as a way to predict incipient problems.

10-2.6 SCCW Cooler Inlet and Outlet Temperatures

Thermocouples typically measure the temperature of the stator cooling water as it enters and exits the coolers. Although the outlet measurement is quite important, the inlet measurement is only marginally so and can be omitted as long as there is some monitoring of inlet and outlet temperatures of the service water to the cooler. Thermometers are sometimes used for local indication, either in lieu of or in addition to the remote indication.

10-2.7 Stator Inlet and Outlet Temperatures

The temperature of the cooling water can be monitored at the inlet and outlet of the stator to detect when the temperature is too high. Temperature switches are generally used for this purpose, and they are set just below the maximum water temperatures specified by the manufacturer.

With the existence of the bulk coolant thermocouple (introduced in Chapter 4), the inlet switch may not be necessary. However, it could be used as backup evidence for detecting a malfunctioning cooler or as a means of detecting malfunctioning sensors.

The outlet temperature switch is more useful because it can detect cooling flow problems. However, it can be redundant with the presence of the coolant bulk temperature thermocouple that measures the water temperature in the outlet manifold. This sensor was also introduced and discussed in Chapter 4.

10-2.8 Cooler Service Water Inlet and Outlet Temperatures

As for the gland seal oil coolers and the hydrogen coolers, the measurement of inlet and outlet service water temperatures can be useful in determining the state of the SCCW coolers. The difference between the outlet and inlet readings can provide an indication of how much heat is being removed from the stator cooling water. Additionally, the value of the inlet temperature of the service water can also be used to determine whether or not it is within the specified range.

10-2.9 Cooler Differential Pressure

In some designs, the differential pressure of the stator cooling water across the cooler can be used to detect flow across the cooler. Although not accurate enough to detect the actual flow, it can be used to identify variations in flow, which can be used to detect plugged cooler tubes. However, this is not a commonly used instrument.

10-2.10 Water Conductivity

Conductivity cells are used throughout the SCCW system to monitor the conductivity of the stator cooling water and ensure that its electrical conductivity is within limits. Because conductivity is such a critical parameter, cells are placed at various locations in the system.

One such cell is typically placed at the outlet of the demineralizers to monitor their effectiveness in reducing the conductivity of the water routed through the demineralizer plant. A second conductivity cell is often placed in the water tank to measure the overall conductivity of the water that will be pumped into the stator coils. Some designs also include a third cell near the inlet to the generator to measure the water conductivity closer to the point where its value is most significant.

10-2.11 Flow Measurement of the Water Routed to the Demineralizer

If the stator cooling water is not properly routed to the demineralizer, high conductivity levels will result. The flow is measured by a flow measurement device. It can be a local indicator, or it can be a switch or a flow transmitter. Naturally, for best diagnosis, a transmitter is desirable; but, in this case, a switch may be adequate to detect low flow.

10-2.12 Water-Hydrogen Differential Pressure

As mentioned previously, the pressure of the water is typically designed to be lower than that of the hydrogen gas, so that if a puncture or crack occurs in the hollow conductor, the water will not leak into the generator frame. Instead, the hydrogen gas will leak into the water system. Given the sometimes wide variations of hydrogen pressure in the generator, coupled with small variations in the water system pressure, the proper differential pressure may not exist at all times. By monitoring this parameter, it can be easily determined if the desired differential pressure between the gas and the stator water is present. This sensor is typically a differential pressure switch set to alarm at a low value of pressure differences of gas to water.

10-2.13 Hydrogen Leakage Detector

By monitoring the hydrogen blanket pressure or the volume of excess hydrogen vented from the tank to the atmosphere, a gas leak into the water system can be detected. This can be a significant piece of evidence in detecting punctures and/or cracks in the hollow conductors or plastic hoses that carry the stator cooling water inside the generator. It should be noted, however, that a small amount of hydrogen gas may normally permeate through the Teflon water hoses and be vented through the tank. Thus, a certain amount of leakage should be overlooked in some generator designs. The manufacturer should be consulted to determine what amounts to a normal leakage. This detector applies only to those designs that maintain a hydrogen blanket over the water tank.

10-2.14 Makeup Water Activation Indication

A pressure switch is sometimes placed in the makeup water line to detect when makeup water is being introduced into the system.

10.3 MONITORING AND DIAGNOSTICS

As in any process control system, there are a number of things that can go wrong with its operation. There are basically three main areas of problems that we will discuss here:

- Cooling water flow deficiency
- High cooling water temperature
- High water conductivity

As with all diagnostic situations, the validity of sensors should not be unquestionably assumed. Uncalibrated or malfunctioning sensors can be the cause of a misdiagnosis and its attendant costs. The reader is advised to ensure to the greatest degree possible that all readings are accurate before any diagnosis is made.

10-3.1 Problem Area #1: Cooling Water Flow Deficiency

A deficiency in the cooling water to the stator bars can result in unplanned outages and, at worst, in overheated coils and subsequent electrical failure. It is a serious problem if the deficiency is significant. A small deficiency could be resolved by reducing load levels until the temperatures are normal. Figure 10.4 displays the causality diagram for this problem area.

If the water flow is interrupted or reduced to a very low value, the load-carrying capability of the armature winding is greatly reduced. Generator kVA load should be immediately reduced because the winding will overheat very rapidly under these circumstances. The manufacturer may have provided an automatic run-back of the turbine valves to a predetermined value of kW load following loss of cooling-water flow. However, this may not always be the case. This predetermined value will vary depending on the manufacturer. It is important that field current also be reduced to keep the power factor and kVA load at reasonable values. If the voltage regulator is in operation, it will accomplish this. Overfluxing of the core can result if excitation is on manual control, or equivalent, at the time of run-back.

Cooling Water Pump Failure

Failure of the cooling water pump is not a significant problem as long as a backup pump is operational and correctly comes on-line when required. Although the pump can experience a loss of AC power while in operation, in this discussion it will be assumed that the backup pump fails to come on-line, because that is the most common case.

■ DIAGNOSTIC BOX 10.1 ■

PROBLEM: COOLING WATER PUMP FAILURE

SENSOR INDICATION:

■ **Pump #1 differential pressure** (or discharge pressure) switch is in low alarm. This is a direct indication that pump number 1 has failed ("Pump #1 off").

■ **Pump #2 differential (discharge) pressure** switch is in low alarm. This is a direct indication that pump number 2 has failed ("Pump #2 off"). It must be noted that this case should not be considered trivial. Both this switch and the one above must be in (low) alarm.

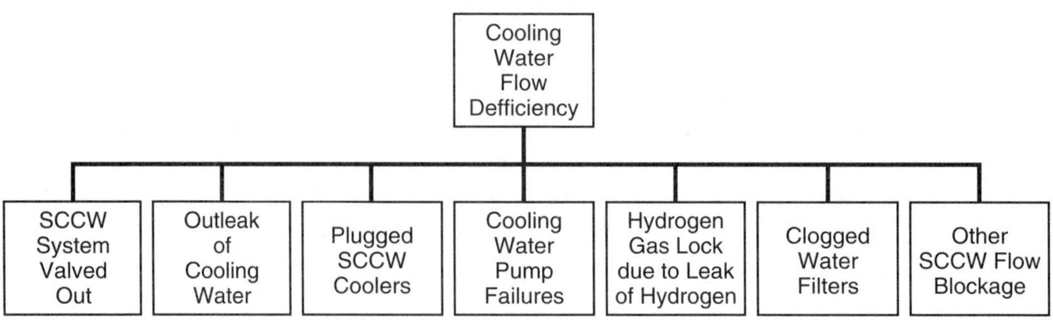

Figure 10.4 Causality diagram for the cooling water flow deficiency problem area.

■ **Stator differential pressure** or flow may alarm on low, as well as on very low.
■ **Filter differential pressure** (drop) may be zero if monitored by an analog sensor. If a switch is used instead, it may not be of any use because the switches are typically set to alarm only on high differential pressure.
■ **Cooler differential pressure** may indicate zero. The same comment as above applies if a differential pressure switch is used.

CORRECTIVE ACTION: If only one of the above indications is received, the validity of the sensor(s) should be verified. If the situation is found to be real and if a DC backup pump is available, place it in operation immediately. Otherwise, immediately shut down the generator, or, if so designed, reduce the load ("run back") to a minimum allowable level. Consult the manufacturer immediately for more specific recommendations.

Clogged Water Filters

The stator cooling water filters have a relatively long life. However, they might become clogged with the debris they gather, creating an obstacle to the flow of water and possibly reducing its flow to the stator coils. When the filtering element becomes clogged, it must be either cleaned or replaced. Some filters do not allow cleaning of the filtering element, but rather require that the cartridge be replaced. Care must be taken to ensure proper installation and replacement of the cartridge.

■ **DIAGNOSTIC BOX 10.2** ■

PROBLEM: CLOGGED WATER FILTERS

SENSOR INDICATION:

■ **Filter differential pressure** (drop) may alarm on high differential pressure. This is the primary indication of this problem.
■ **Stator differential pressure** or flow may alarm on low, depending on the severity of the clogging and its subsequent reduction of water flow. This may serve as an indication of how severe the flow deficiency is.
■ **Cooler differential pressure** may be low because of the reduced flow.
■ **Stator cooling water temperature rise** across the stator will be higher than normal because of the reduction in cooling water flow. It can be calculated by subtracting the bulk inlet water temperature from the outlet.
■ **Stator water outlet temperature** switch may alarm on high if the reduction is significant enough to cause the water temperature to rise to the alarm point.

CORRECTIVE ACTION: If only one of the above indications is received, the validity of the sensor(s) should be verified. If the situation is found to be real, immediately switch to the other filter if there is a backup. Otherwise, bypass the existing one temporarily and clean it or replace the cartridge as quickly as possible, then place it back in service. Send any debris captured in the filter for analysis. Consult the manufacturer for more specific instructions.

Outleak of Stator Cooling Water

A significant cooling water leak into either the service water system or to atmosphere (between the pump and the stator inlet), may reduce the flow of water to the stator coils. Its effect on the operation of the generator would depend on how severe the leak is, as well as where it is located. Because of the many variations of these two parameters, which can have a significant effect on the symptoms seen, this topic will not be discussed in detail. The best way to handle the problem is to inspect the SCCW skid and the connected piping periodically for leaks.

SCCW System Valved Out

Valving out the SCCW system can happen when bringing the unit on-line after a prolonged outage during which the SCCW system was deactivated. It is not likely to happen while the unit has been in normal service for a length of time.

■ DIAGNOSTIC BOX 10.3 ■

PROBLEM: SCCW SYSTEM VALVED OUT

SENSOR INDICATION:

■ **Filter differential pressure** may indicate zero if an analog measurement is used. If a switch is used instead, it may not be of any use because switches are typically set to alarm only on high differential pressures.

■ **Stator differential pressure** or flow may alarm on low and on very low, or it may read zero.

■ **Cooler differential pressure** may be zero (if an analog measurement is used).

■ **Neither pump differential pressure** switches will be in alarm. This may distinguish this problem from failed water pumps.

■ **SCCW outlet temperature** switch may alarm on high if the load is significant enough to cause the water temperature to rise to the alarm point.

CORRECTIVE ACTION: If only one of the above indications is received, the validity of the sensor(s) should be verified. If the situation is found to be real, no general recommendation can be given for this condition, other than to take the unit off-line for precautionary reasons. Immediately consult the manufacturer for a specific recommendation. If the water system valve is to be opened, care must be taken not to cause thermal shock in the stator bars by introducing cold water to a hot conductor. The manufacturer is in the best position to say what the correct course of action should be. Opening the SCCW valves should be added to the start-up checklist, if not already on it.

Plugged SCCW Coolers

Cooler tubes can be blocked by foreign substances, which can impede the flow of the stator cooling water. This is fairly easy to detect if an analog differential pressure indication across the cooler is available, but not so easy otherwise.

■ DIAGNOSTIC BOX 10.4 ■

PROBLEM: **PLUGGED COOLERS**

SENSOR INDICATION:

- **Cooler differential pressure** may be high. This is the primary indication of this problem.
- **Filter differential pressure** may indicate lower-than-normal reading if an analog measurement is employed. If a switch is used instead, it may not be of any use because the switches are typically set to alarm only on high differential pressures.
- **Stator differential pressure** or flow may alarm on low, depending on the severity of the problem. This reading serves as a good indication of the severity of the problem.
- **The stator cooling water temperature rise** across the stator may be higher than normal because of the reduced water flow.
- **SCCW stator outlet temperature** switch may alarm on high if the load is significant enough to cause the water temperature to rise to the alarm point (because of the reduced flow).
- **Stator inlet pressure** may also be lower than normal.

CORRECTIVE ACTION: If only one of the above indications is received, the validity of the sensor(s) should be verified. If the situation is found to be real, switch to the alternate cooler and inspect the clogged one. If only partial redundancy is available, reduce the unit load to the appropriate level for one cooler. If no redundancy exists, bypass it temporarily to allow for the cooler to be cleaned out. However, if the maximum coil temperatures are being exceeded without the cooler, then consider reducing the load or taking the unit off-line. Consult with the manufacturer for specific recommendations.

Gas Blocking of SCCW Caused by Hydrogen Leak into the Water

Gas blocking condition was covered in Section 4.3.2. However, described here is a somewhat different condition in which the leak does not take place within the coils, but rather in the water manifolds, connection rings, or hoses. This variation converts the problem into a more general one, which involves the generator as a whole, rather than one specific coil. The introduction of large amounts of hydrogen into the water system can cause a blockage in the piping that inhibits the flow of water to or from the coils.

Depending on the location and the magnitude of the blockage, this problem may not be detectable at all, other than through the hydrogen leak detector. Because hydrogen leakage into the water system can cause the blockage, the manufacturer's recommendation for allowable rates of hydrogen leakage into the water system should be followed. It may require a visual inspection after the unit is removed from service in order to detect this problem.

Because of the many variations of location and size of gas leaks, the diagnostic box for this problem will not be extensive. The manufacturer should be consulted as soon as the gas leakage rate exceeds their recommended alarm conditions.

■ DIAGNOSTIC BOX 10.5 ■

PROBLEM: **GAS BLOCK OF SCCW CAUSED BY HYDROGEN LEAK INTO THE WATER**

SENSOR INDICATION:

■ **Hydrogen leak detector** will indicate an abnormal amount of hydrogen leakage into the water system. This is the primary indicator.

CORRECTIVE ACTION: If the above indication is received, the validity of the sensor(s) should be verified. If the situation is found to be real, take the unit off-line to look for the leak. Consult the manufacturer for more specific recommendations.

Other Flow Blockage in the SCCW

Diagnostic Box 10.6 covers any blockages that are not in the cooler, in the filters, or in the stator itself. It represents the default case.

■ DIAGNOSTIC BOX 10.6 ■

PROBLEM: **OTHER FLOW BLOCKAGE IN THE SCCW**

SENSOR INDICATION:

■ **Filter differential pressure** may read lower than normal if an analog signal is used. This will distinguish this problem from a clogged water filter. If a differential pressure switch is used that is set for high alarm, this indication will be irrelevant.

■ **Stator differential pressure** or flow may alarm on low or very low, depending on the severity of the clogging and its subsequent reduction of water flow. This may serve as an indication of how severe the flow deficiency is.

■ **Cooler differential pressure** may be low if an analog signal is used. This indication can distinguish the problem from plugged coolers.

■ **Temperature drop** of the stator cooling water across the cooler may be lower than normal because of the reduced flow.

- **Stator water temperature rise** across the stator may read high. This can be calculated by subtracting the bulk inlet water temperature from the outlet.
- **SCCW stator outlet temperature** switch may alarm on high if the reduction is significant enough to cause the water temperature to rise to the alarm point.

CORRECTIVE ACTION: Reduce the load immediately until the temperatures are within range. If this is not possible, take the unit off-line as a precaution. Consult the manufacturer for specific recommendations.

10-3.2 Problem Area #2: High Cooling Water Temperatures

If the temperature of the stator cooling water at the stator inlet is higher than specified by the manufacturer, the water loses its cooling capacity, resulting in increased coil temperatures and, consequently, in reduced insulation life. This can lead to unplanned outages and, in the extreme case, melting of the copper bars. This situation typically points to a problem in the cooler operation. Because coolers have been covered in detail in the last two chapters, some of this material will not be new to the reader. However, in the interest of completeness, it will be covered in full detail here as well. Figure 10.5 shows the causality diagram for this problem area.

High Service Water Temperature

High service water temperature can be quite easy to detect if there are temperature measurements in the station water inlet to the cooler(s). It is assumed here that everything else is normal.

This problem may not be detected if the service water flow to the cooler is automatically controlled so that it is increased in order to maintain the SCCW within the specified temperature limits. As long as station water flow capacity is not exceeded, there is no consequence to this condition. This will only become a problem when the station water flow rate is at a maximum and can no longer compensate for the high service water temperature. Diagnostic Box 10.7 makes that assumption.

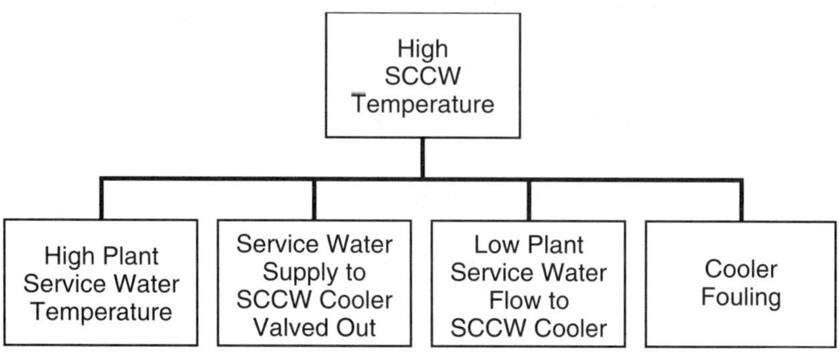

Figure 10.5 Causality diagram for then high cooling water temperature problem area.

■ DIAGNOSTIC BOX 10.7 ■

PROBLEM: HIGH SERVICE WATER TEMPERATURE

SENSOR INDICATION:

■ **Cooler station water inlet temperature** is high. This is the primary indication.

■ **Cooler station water temperature rise** may be normal or near-normal across the cooler.

■ **Stator water cooler outlet temperature** may read higher than normal because of the lower amount of heat being removed from the stator coil water.

■ **Stator cooling water temperature rise** across the stator may read normal or near normal.

■ **Stator cooling water inlet and outlet temperatures** may read high, depending on the severity of the condition. The switches may alarm on high if the stator cooling water temperature is high enough to cause an alarm.

CORRECTIVE ACTION: If only one of the above indications is received, the validity of the sensor(s) should be verified. If the situation is found to be real and if any of the stator water temperature readings exceed the limit, reduce the load immediately until the temperatures are within limits. Increase the flow of service cooling water to the cooler, if it is not already at maximum. Investigate the cause of the high inlet water temperature in the water system and correct it. If the problem is not immediately solvable, spray water on the outside of the cooler to enhance its cooling capability. Consult the manufacturer for specific recommendations.

Deficient Service Water Flow to Cooler

Deficient service water flow to the cooler does not reflect on the performance of the cooler itself, but will definitely have an effect on the amount of heat removed from the stator coil cooling water. Low service water flow to the cooler could be due to failure of the service water supply system. This could happen during normal operation of the generator because it is related to the service water system.

■ DIAGNOSTIC BOX 10.8 ■

PROBLEM: DEFICIENT SERVICE WATER FLOW TO COOLER

SENSOR INDICATION:

■ **Stator water cooler outlet temperature** may read higher than normal because of the lower amount of heat being removed from the stator coil water.

■ **Cooler service water** inlet temperature is normal.

■ **Service water temperature rise** is higher than normal across the cooler. This is caused by the decreased flow of service water.

■ **Stator water inlet and outlet temperatures** in the stator itself may read high. The temperature switches may alarm on high if the flow of water is sufficiently reduced.

■ **Stator water temperature drop** may be lower than normal across the cooler. This reflects less heat being removed by the service water.

■ **SCCW differential pressure** across the cooler is normal. This suggests normal flow in the SCCW system.

CORRECTIVE ACTION: If only one of the above indications is received, the validity of the sensor(s) should be verified. If the situation is found to be real and if the stator water temperature exceeds the limit, reduce the load until the temperatures are within limits. Increase the flow of service cooling water to the coolers. If this is not possible, spray water on the outside of the cooler in order to enhance its cooling capability. If it is still not possible to maintain the temperature within limits, take the unit off-line. Consult the manufacturer for specific recommendations.

Service Water Supply to SCCW Cooler Valved Out

Unlike the previous malfunction, which can take place during normal operation, valving out the service water to the cooler typically happens immediately after an outage and is usually attributed to human error. The consequences of this problem are very similar to the previous one, except this one can be solved by opening a valve.

■ DIAGNOSTIC BOX 10.9 ■

PROBLEM: **SERVICE WATER SUPPLY TO SCCW COOLER VALVED OUT**

SENSOR INDICATION:

■ **Stator water cooler outlet temperature** will read higher than normal because of the much reduced amount of heat being removed from the water.

■ **SCCW stator inlet and outlet temperatures** will read higher than normal. Temperature switches may alarm on high if there is no service water flow.

■ **Cooler service water temperature** rise across the cooler is zero or near zero because of the no-flow conditions.

■ **SCCW temperature rise** across the stator may read normal or slightly lower than normal.

■ **SCCW temperature drop** across the cooler is much lower than normal, or zero, across the cooler. This reflects little heat being removed by the service water.

■ **SCCW differential pressure** across the cooler is normal. This reflects normal stator cooling water flow.

CORRECTIVE ACTION: If only one of the above indications is received, the validity of the sensor(s) should be verified. If the situation is found to be real and if the limit of the temperature of the water to the coils is exceeded, reduce the load until the temperatures are within limits. If service water flow is not resumed quickly, the stator coil temperatures may rise beyond the safe limits, leading to potentially major damage to the stator coils. Valves should be opened slowly to avoid thermal shock to the cooler. If it is not possible to bring the SCCW water temperatures within limits, take the unit out of service. It is important that the manufacturer's recommendation for the specific unit in question be obtained. Opening the service water valve to the SCCW coolers should be added to the start-up checklist, if not already on it.

Cooler Fouling

Deposits or scaling in the cooler tubes can deteriorate their heat transfer capabilities. This can result in less heat being removed from the stator cooling water and, consequently, higher temperature in the stator water. Cooler fouling manifests itself in decreased effectiveness of the cooler in removing the heat from the stator water. This means a lower temperature rise of the service water across the cooler and possibly a lower temperature drop of the stator water across the same cooler. If service water flow can be increased to maintain appropriate temperatures, this problem may be hidden, only to be faced when the water flow has reached the maximum available.

■ DIAGNOSTIC BOX 10.10 ■

PROBLEM: COOLER FOULING

SENSOR INDICATION:

■ **Stator water cooler outlet temperature** may read higher than normal because of the much reduced amount of heat being removed from the stator water for the same amount of service water flow. This may manifest itself only when the flow rate of the service water has reached a maximum.

■ **Cooler service water temperature** rise is lower than normal for the same flow because of the decreased effectiveness of the cooler.

■ **Cooler service water** inlet temperature will read normal. This will eliminate high temperature service water as a possible problem.

■ **Stator cooling water** inlet and outlet temperatures may read higher than normal. Temperature switches may alarm on high if the situation is severe enough.

■ **Stator cooling water temperature** rise across the stator may read normal or slightly less than normal.

■ **Stator cooling water temperature** drop across the cooler is lower than normal. This reflects a reduced amount of heat being removed by the service water.

■ **Differential pressure** of the stator cooling water across the cooler is normal. This reflects normal flow in the stator cooling water.

CORRECTIVE ACTION: This is generally not a serious problem until the situation becomes severe enough to affect the stator coil temperatures. If only one of the above indications is received, the validity of the sensor(s) should be verified. If the situation is found to be real and if a backup cooler is available, switch over to the backup and clean out the first one. If partial redundancy is available, the fouled cooler can be removed from service while the load on the unit is reduced to an appropriate level. If a backup is not available and the stator water temperature exceeds the safe limit, reduce the load until temperatures are within limits. If this is not possible, take the unit off-line. Consult with the manufacturer for recommendations that are specific to the unit in question.

10-3.3 Problem Area #3: High SCCW Conductivity

As mentioned previously, maintaining the electrical conductivity of the stator water within the specified range is one of the primary tasks of the SCCW system. High conductivity can result in electrical tracking to ground through the hoses connecting the individual coils to the water inlet and outlet manifolds. Additionally, it can lead to scaling within the hollow strands in the coils, which may result in constraining the water flow. The conductivity of the water must generally be maintained in the 0.1–1.0 micromhos/in range. Check with the manufacturer for the range applicable to the specific unit in question.

Discussion of high SCCW conductivity after the cooling water deficiency and the high cooling water temperature problem areas by no means implies that it is not a significant problem. In fact, it is as significant as the other two. There are three abnormal conditions with which to be concerned here:

- Demineralizer failure
- Service water inleak into stator water system through the cooler
- Introduction of highly contaminated makeup water

The relationship in time between detection of high conductivity by the three conductivity cells can help determine the cause of the problem. Figure 10.6 shows the causality diagram for this problem area.

Demineralizer Resin Loss of Effectiveness

The most common cause of malfunction in a demineralizer is that the resin has lost its effectiveness and needs to be replaced. The resin bed should be replaced periodically according to the manufacturer's schedule. The effective life of the resin, however, could be reduced from the manufacturer's expectation if the condition of the makeup water is uncommonly poor.

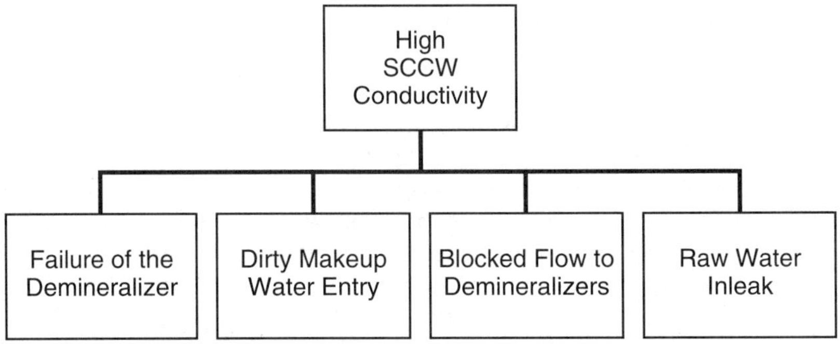

Figure 10.6 Causality diagram for the high SCCW conductivity problem area.

■ DIAGNOSTIC BOX 10.11 ■

PROBLEM: DEMINERALIZER FAILURE

SENSOR INDICATION:

- **The conductivity cell** at the outlet of the demineralizer may indicate high water conductivity. In the case of spent resin, the increasing trend in the conductivity may take place gradually over time.
- **The conductivity cell** at the water tank may not show any increase in conductivity until much later than the indication at the demineralizer outlet because the water through the demineralizer is only a small fraction of the entire flow.
- **The conductivity cell** at the stator inlet may behave similarly to the one at the water tank.
- **Demineralizer water flow indication** may indicate normal flow.
- **Water tank level** may be normal.
- **Makeup water introduction** rate may be normal.

CORRECTIVE ACTION: If only one of the above indications is received, the validity of the sensor(s) should be verified. If the situation is found to be real, the problem could be solved by simply replacing the demineralizer resin. Consult the manufacturer for recommendations applicable to the specific unit in question.

Service Water Inleak Through the Cooler

If there are leaks in the cooler, some of the "contaminated" service water may contaminate the "clean" stator water. This could be a very serious problem for units located near the seashore where seawater is used as the service water. Generally, the service water pressure is kept lower than the stator water to avoid this problem, but larger leaks can cause some interchange to take place. In this situation, the stator inlet conductivity cell, which is located downstream from the coolers, will be the first to indicate high conductivity. The tank conductivity cell will follow suit, but more slowly, whereas the demineralizer outlet cells will not show any increase at all, assuming that they are working properly.

■ DIAGNOSTIC BOX 10.12 ■

PROBLEM: SERVICE WATER INLEAK THROUGH THE COOLER

SENSOR INDICATION:

■ **The conductivity cell** at the stator inlet will indicate a high conductivity first because of its location immediately downstream of the coolers.

■ **The conductivity cell** at the water tank may not show any increase in conductivity until much later than the indication at the stator inlet. The volume of the tank will cause the dirty water to be diluted.

■ **The conductivity cell** at the outlet of the demineralizer will indicate normal water conductivity. It may not show any increase in conductivity as a result of this condition.

■ **Demineralizer water flow** indication will indicate normal flow.

■ **Water tank level** may be higher than normal, depending on how much water is leaking.

■ **Makeup water introduction** rate may be less than normal because of the introduction of external water through the cooler.

CORRECTIVE ACTION: If only one of the above indications is received, the validity of the sensor(s) should be verified. If the situation is found to be real, reduce the load to meet the manufacturer's recommendation based on the reduced cooling capacity when the leaking cooler is taken out of service. If redundancy exists, shift operation to the other cooler and find and repair the leaking tube(s). Consult the manufacturer for recommendations applicable to the specific unit in question.

Introduction of Contaminated Makeup Water

Although not a common occurrence, if the makeup water being introduced is excessively contaminated, it may overcome the efforts of the demineralizer to reduce its conductivity to the required level. The symptoms for this problem look very much like loss of effectiveness of the demineralizer resin. The difference is that the high conductivity water may come in "slugs" that correlate in time to the introduction of makeup water. This problem may be more noticeable when the demineralizers are small in capacity.

■ **DIAGNOSTIC BOX 10.13** ■

PROBLEM: **INTRODUCTION OF HIGHLY CONTAMINATED MAKEUP WATER**

SENSOR INDICATION:

■ **The conductivity cell** at the outlet of the demineralizer may periodically indicate high water conductivity, which may coincide with periods of introduction of makeup water.

■ **The conductivity cell** at the water tank may not show any increase in conductivity until much later than the indication at the demineralizer outlet because it takes more time for the conductivity of a larger volume of water to be increased.

■ **The conductivity cell** at the stator inlet may behave similarly to the one at the water tank.

■ **Demineralizer water flow** indication will indicate normal flow.

■ **Water tank level** may be normal, but not necessarily.

■ **Makeup water introduction** rate may be normal, but not necessarily.

■ **Replacing the resin** in the demineralizer does not improve the situation significantly.

CORRECTIVE ACTION: If only one of the above indications is received, an off-line measurement of the conductivity of the makeup water should be carried out to confirm the diagnosis. Cut off the makeup water source until the problem is corrected. In the meantime, keep a close watch on the tank level to ensure an appropriate supply to the SCCW pumps. Consult the manufacturer for recommendations applicable to the specific unit in question.

Reduced Flow to the Demineralizer

This condition can take place because of a blockage or a leak in the system. This means that less water will be routed through the demineralizer in order to control the overall conductivity of the water. It is easily detected with the water flow indicator. This may not be totally true if there is a leak between the indicator and the demineralizer, but such a possibility can be discounted, depending on the length of the pipe, or eliminated through simple inspection.

■ **DIAGNOSTIC BOX 10.14** ■

PROBLEM: **REDUCED FLOW TO THE DEMINERALIZER**

SENSOR INDICATION:

■ **Demineralizer water flow indication** will indicate lower-than-normal flow. This will be the primary indication of this problem.

- **The conductivity cell** at the outlet of the demineralizer will indicate normal water conductivity.
- **The conductivity cell** at the water tank will indicate a slow and steady increase in conductivity.
- **The conductivity cell** at the stator inlet will behave similarly to the one at the water tank.

CORRECTIVE ACTION: Investigate the blockage or the leak and rectify it. If the conductivity rises beyond the safe limits, the unit should be taken off-line as a precaution. Consult the manufacturer for recommendations applicable to the specific unit in question.

Oxygen Ingress Through Leak to Atmosphere

The oxygen content of the water may increase through excessive contact with oxygen. A leak to atmosphere in the water can cause this interface with oxygen. Although this may not increase the conductivity, it may increase the dissolved oxygen content beyond the manufacturer's limits. This problem is only applicable to systems that are isolated from atmosphere through a hydrogen blanket. This leak would most likely occur in the water tank. Water leaks in the piping may not cause a significant amount of interface to result in drastic increases in oxygen content unless the leak was severe, in which case other problems would take precedence.

DIAGNOSTIC BOX 10.15

PROBLEM: OXYGEN INGRESS THROUGH LEAK TO ATMOSPHERE

SENSOR INDICATION:

- **Hydrogen leak detector:** It is possible that the hydrogen leak detector (if one is available to measure outflow of hydrogen from the blanket over the water tank) may show less flow than normal. This is because the gas would also be escaping through the leak.

CORRECTIVE ACTION: Investigate the leak and rectify it. If the oxygen content rises beyond the safe limits, consult the manufacturer for recommendations applicable to the specific unit in question.

SUMMARY

The critical issues with stator water systems are the flow, temperature, and conductivity of the stator cooling water. The sensors and diagnostic techniques are geared to detecting problems that lead to any of these conditions.

REFERENCES

[IEEE 1129–1992] "IEEE Recommended Practice for Monitoring and Instrumentation of Turbine Generators," IEEE Standard 1129–1992.

CHAPTER 11

Bearing Lubrication Oil System

The turbine generator bearings must be lubricated whenever the rotor is turning at any speed. The *bearing lubrication oil system* (*lubrication oil system*) provides a clean and temperature-controlled supply of oil to serve this purpose. Its objectives are as follows:

1. Supply lubricating oil to the bearings while the unit is in operation, on startup, or on coast-down.
2. Supply lubricating oil to the bearings when the unit is on turning gear.
3. Supply lubricating oil to the turning gear itself.
4. Serve as a source of oil to the gland seal oil system.
5. Supply oil to other subsystems, depending on the design of the turbine generator and its auxiliaries.

The generator bearings are, in most cases, lubricated by the turbine bearing lubrication system. Thus, it is difficult to isolate discussion of the generator bearing lubrication system from that of the entire turbine generator shaft system. Therefore, the supply system to all the bearings in the turbine generator shaft system will be discussed in this chapter, although only problems that affect the generator bearings will be discussed.

The following sections will describe the operation of a typical bearing lubrication oil system, the sensors that are used to monitor its operation, and any problems that could result from abnormal operation of the lubrication oil system.

11.1 System Description

Before a discussion on the bearing lubrication oil system, it may be appropriate to describe the bearings in a turbine generator. This may be useful in understanding the discussion of the lubrication oil system afterwards.

11-1.1 Turbine Generator Bearings

The size and type of bearings used in turbine generators largely depends on the unit rating and its speed. Large turbine generators invariably use *sliding surface* (also called *hydrodynamic*) type bearings that are babbitted with tin or other soft metal that will act as sacrificial material to save the shaft journal from being damaged. Such bearings are used to support the generator rotor at the shaft extensions where the journals are located. The bearings are themselves typically held up by the *bearing brackets.* Separate pedestals external to the generator are also used. The babbitt lining of the bearing must be securely bonded to allow proper lubrication.

There are many designs of sliding surface bearings used in turbine generators. The most common is a simple split-casing two-groove cylindrical bearing. The housing of one such bearing is shown in Figure 11.1. A cross-sectional diagram in-

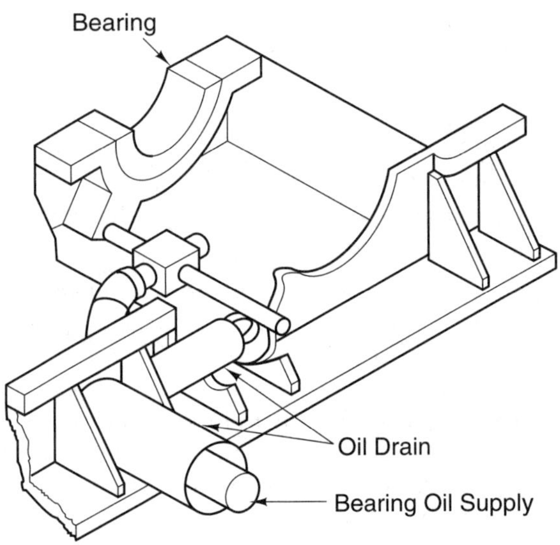

Figure 11.1 Large turbine generator bearing.

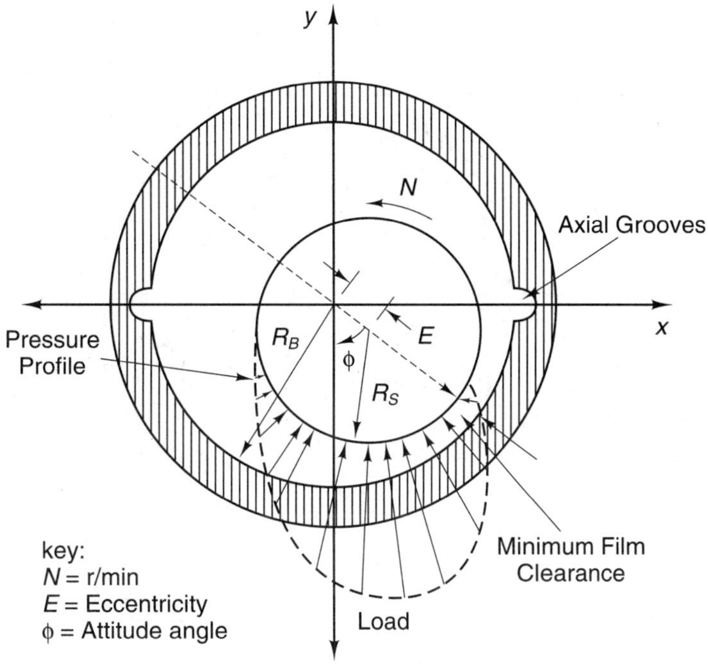

Figure 11.2 Typical two-groove cylindrical bearing.

dicating the pressure profile of this type of bearing is shown in Figure 11.2. The clearances between the shaft and bearing bore, normally in thousandths of an inch, are greatly exaggerated in the figure to demonstrate the operational principles involved. As the shaft rotates, it "wedges" the lubricating oil between the shaft and the bearing bore. This dynamic action develops the pressure profile shown, which results in an upward lifting force on the rotating shaft. The grooves ensure an adequate supply of oil.

The dynamics involved in the operation of the bearings result in a net lateral force on the bearing that moves the rotating shaft off-center somewhat. This can be seen in Figure 11.2. However, for heavily loaded shafts, this effect is relatively small and insignificant. Such is the case for most large generators and the low-pressure turbine elements because of the weights of the shafts and rotor assemblies.

For lightly loaded shafts, however, the lateral forces are more significant and cannot be neglected. This can be seen from the pressure profile of Figure 11.3. The consequence is that the shaft/bearing relationship can become unstable and result in undesirable vibrations that can reduce bearing life and affect generator component integrity and performance.

One method used to reduce instability is to design the bearing to be elliptical in shape, rather than round. A diagram of such a bearing is shown in Figure 11.4.

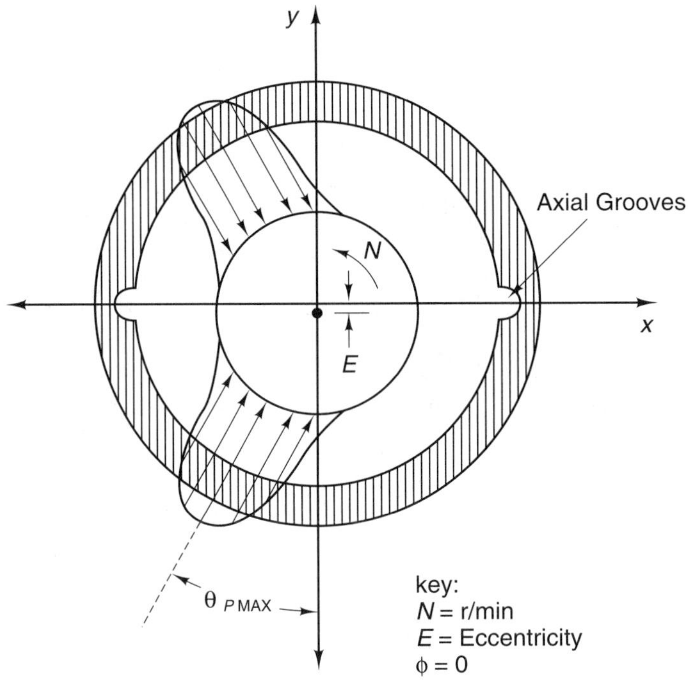

Figure 11.3 Typical lightly loaded bearing.

The top and bottom clearances of such an *elliptical bearing* are smaller than the side clearances, which results in optimizing the top and bottom high-pressure profiles, while reducing the pressure zones on the sides. This has a stabilizing effect on the shaft/bearing relationship.

For very light loads, such as high-pressure turbine elements, stability is achieved in most cases by applying *tilting pad bearings.* A three-pad bearing is shown in Figure 11.5 for illustration purposes, but the actual number of pads is usually larger. Because the pads tilt to form individual equal and opposing pressure profiles and resultant forces, this type of bearing is the most stable. However, tilting pad bearings are also the most expensive.

Because of the voltages built up in the generator shaft during operation (as discussed in Chapter 3), it is common practice to insulate the bearings on one or both ends of the generator. The exciter end bearing is invariably insulated, but some manufacturers will also routinely insulate the drive end bearings. The insulation can be either single layer or double layer. This insulation will prohibit the flow of small currents between the shaft and ground through the bearings, which can score and ultimately destroy the bearing lining, resulting in high vibrations, excessive wear, and, ultimately, a unit shutdown.

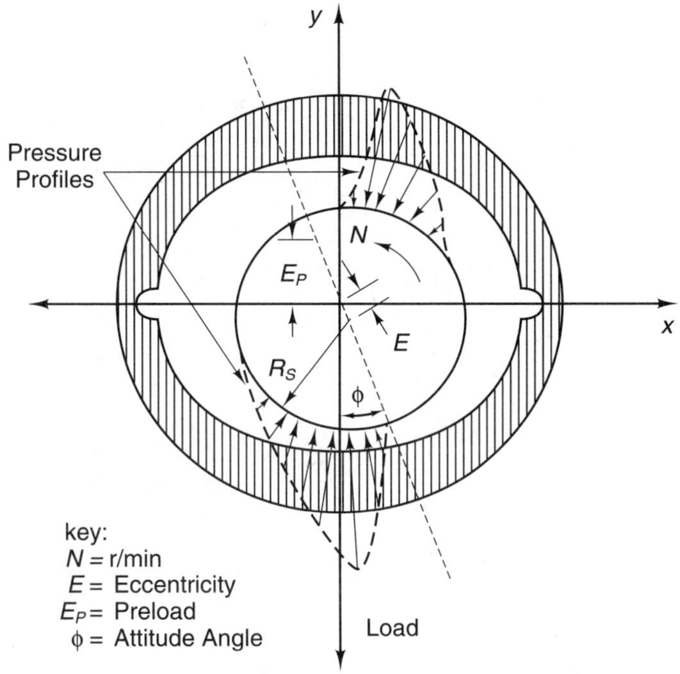

Figure 11.4 Typical elliptical bearing.

11-1.2 Lubrication Oil System

Oil is required to form the dynamic action that supports the rotating shaft on a thin film of oil. Adequate supply of this oil at the proper pressure and temperature will ensure that no damage is incurred either on the shaft journal or on the bearing.

Loss of oil to the bearing while the rotor is turning can result in loss of (*wiped*) babbitt, scored bearings, and/or damage to the rotor journals as well as to the gland seals. Moreover, as seen in Chapter 9, loss of seal oil to the gland ring when the generator is pressurized with hydrogen can lead to a leak of hydrogen from the unit, which can have serious consequences. Thus, it is critical that the lubrication oil system be operational whenever the generator shaft is rotating at any speed.

During normal operation, the lubrication oil is driven by a shaft-mounted oil pump. Electric motor-driven pumps are used to supply the oil when the unit is at less-than-full speed or as backups when the main oil pump fails. Figure 11.6 depicts one typical design of the bearing lubrication oil system.

Interlocks are generally provided by manufacturers to ensure that the rotor cannot rotate without the lubrication system in operation. A pressure switch can be used to interlock the turning gear motor and prevent its operation without the

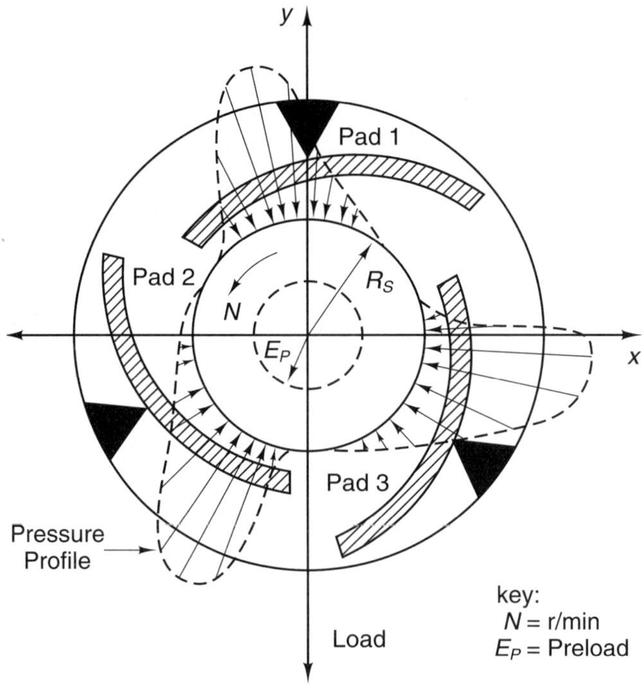

Figure 11.5 Typical three-tilting-pad bearing.

operation of the bearing lubrication oil system. Other interlocks can be used to prevent the start-up of the unit from standstill by tying the presence of steam to that of lubrication oil pressure.

The bearing lubrication system is placed in operation by activating the electric motor driven *bearing oil pump* (BOP). This pump is also used in some designs to supply bearing oil to the seal oil system through the loop seal tank. The loop seal tank is part of a common drain between the seal oil system and the generator bearings. This tank also provides a degree of isolation between the generator bearing oil and the turbine bearing oil by preventing any hydrogen in the generator from escaping into the main oil reservoir in the event of a shaft seal failure.

Vapor extractors are also used in some designs to provide a negative pressure for the drain components of the seal oil system and the generator bearings. This negative pressure helps remove hydrogen and/or oil vapors from the bearing cavity.

The main components of the bearing lubrication oil system are described in detail as follows:

Main oil pump (**MOP**): This shaft-driven pump is the main source of driving power when the unit is operating between two-thirds and full speed. It can be mounted on the turbine generator shaft extension on the turbine gov-

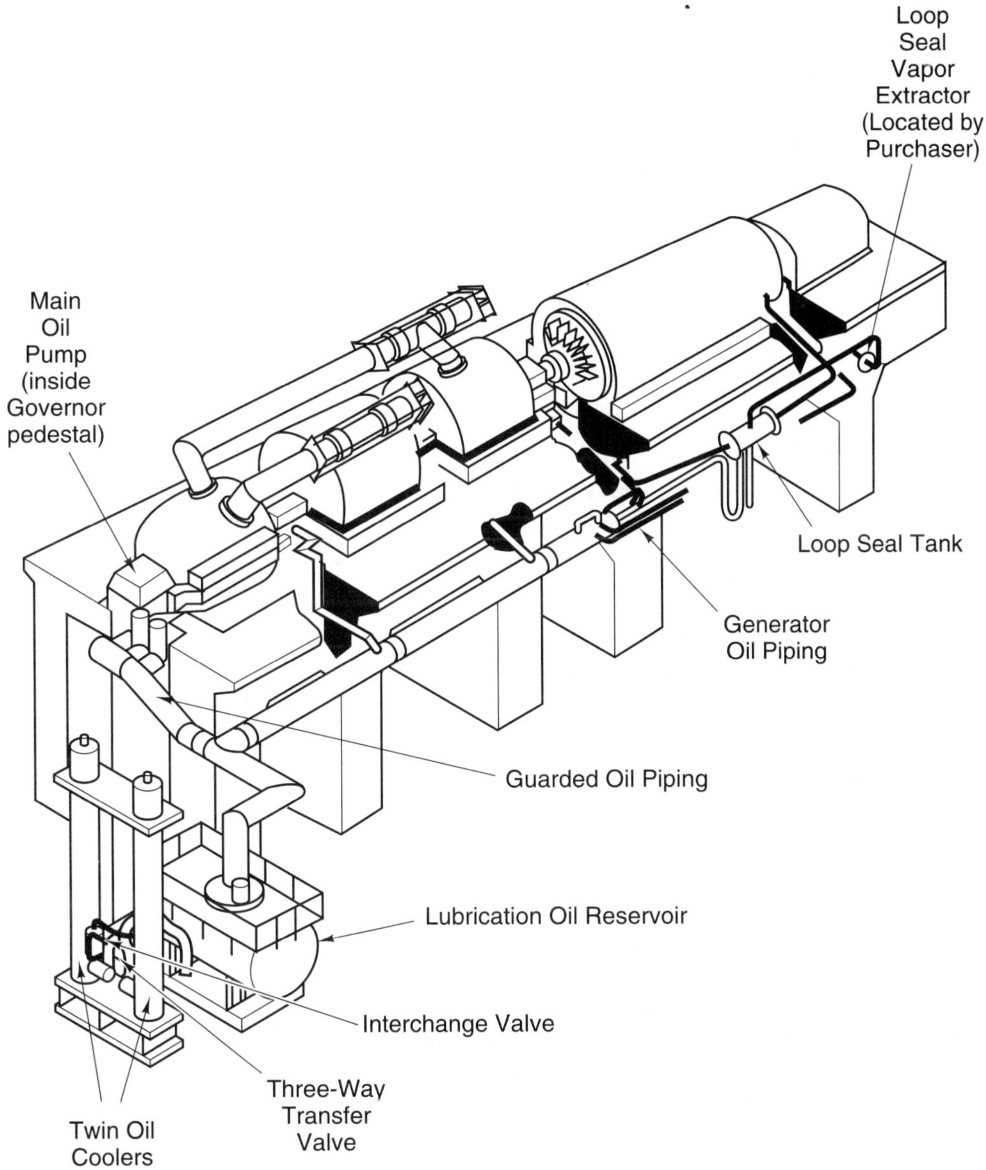

Figure 11.6 Typical bearing lubrication system.

ernor pedestal. Its typical discharge pressure is approximately 350 lbf/in^2, and in some designs, it can serve as a backup supply to the generator seal oil system, as well as to the mechanical overspeed and manual trip header.

Bearing oil pump (BOP): This AC-motor-driven pump provides lubrication to the turbine and generator bearings during turning gear and during acceleration to full speeds, as well as during deceleration below 3000 r/min. It also serves as the low-pressure seal oil backup source (the last backup available). Once the MOP reaches the appropriate speeds, the BOP is placed on standby mode, ready to be activated when the MOP fails to provide the proper bearing oil pressures.

Emergency oil pump (EOP): The emergency oil pump provides the same function as the BOP, except that it is driven from a DC source of electrical power. Its purpose is to maintain oil pressure in the case of a loss of AC power during coast-down of the unit after a turbine trip or other abnormal situation. It also serves as a backup when the MOP and BOP fail during normal generator operation, or as a backup to the BOP if the latter fails during start-up prior to the operation of the MOP.

Seal oil backup pump (SOB): This pump on the seal oil supply system is the primary backup for the air-side seal oil system described in Chapter 9. It is also used to pressurize the mechanical overspeed and manual trip header during the start-up period when the MOP is not yet available.

Lubrication oil reservoir: The reservoir is a large tank used to store the lubrication oil. During normal operation, all lubrication oil comes from this tank, and drained oil is returned to this tank through the drain lines from all the bearings in the turbine generator.

Lubrication oil cooler: The lubricating oil is heated by the friction between the shaft journal and the bearing as it is pumped through the bearings. Therefore, it must be cooled in order to be reused, and the cooler serves this purpose. As in some of the systems seen in previous chapters, there may be a single cooler or redundant coolers (partial or complete redundancy). Such coolers are typically tube-and-shell type, and the secondary heat transfer medium is plant service water. It should be noted that in most designs, during turning gear operation, the coolers are not put into operation so as to not increase the viscosity of the oil beyond the specification. This is done by shutting off water flow to the coolers.

Oil ejector: The oil ejector is a jet pump mounted at the outlet of the MOP and submerged in the lubrication oil reservoir. It is used to increase the flow of oil as it is extracted from the main oil reservoir. As the high-pressure oil from the MOP passes through its nozzle, its speed increases, causing oil from the reservoir to be drawn into a pickup chamber through holes in the mounting. Flow and pressure characteristics of the ejector are determined by its nozzle hole type and its diffuser throat size.

Vapor extractors: One or more vapor extractors are typically installed in the lubrication oil system piping in order to remove gases and oil vapors from the lubrication and seal oil systems. They can also serve to keep these gases from discharging into the turbine generator room. They must be in service continuously as long as oil circulation is active or whenever the generator is charged with hydrogen.

Oil strainer: Dirt can be a source of major problems in the bearing system. Thus, it is important to remove foreign particles in the returning oil from contaminating the oil reservoir.

Lifting pumps: Lifting pumps are commonly used in large (heavy) rotors to reduce the size of the turning gear system. These pumps are designed to inject high pressure oil between the bearing and the journal to lift the shaft slightly. This will reduce the starting torque requirements on the turning gear motor. Furthermore, the increased clearance allows larger sized contaminants in the bearing oil to pass through without damaging the shaft. Such large contaminants are likely to be present during start-up of the unit when the bearing lubrication system is started after some downtime. The lifting pumps are normally taken out of service after the unit has been turning for a period of time.

11-1.3 Lubricating Oil Quality

Foreign materials in the bearing lubrication oil can result in a bearing wipe or in journal scoring. Thus, bearing damage caused by such contamination of the bearing lubricating oil can be a significant problem. Furthermore, water entrained in the oil can be a cause of rusting in the ferrous metal parts with which it comes into contact. For this reason, manufacturers clearly specify the various characteristics of the oil, as well as maximum operating oil contamination levels. The strainer supplied with the bearing oil lubrication system by most manufacturers traps only the larger contaminants in the oil. The main oil tank, because of the natural settling ability of the contaminants in the low flow velocity environment of the tank, is also capable of controlling the contaminants in the oil to some degree. Likewise, water may settle at the bottom of the tank. However, small-sized contaminants and the large proportion of the water content are generally taken out by oil purification equipment, which is not typically furnished by the turbine manufacturer. Most generating unit operators will add oil purification equipment to maintain the purity of the oil to the manufacturer's specifications.

It should be noted that most problems with foreign material in the oil take place after an extended outage. Introduction of these contaminants may result from rags left in bearing oil lines, dirt introduced into the oil tank by cleanup work overhead, water introduced into oil during cleaning operations, weld splatter and machining chips introduced during piping modifications, rust and water introduced during transfer of the oil to and from contaminated storage facilities, as well as many other causes. Most manufacturers recommend special procedures during the outage to keep the contaminants out and/or extensive cleaning operations of the oil after the extended outage.

11.2 SENSORS AND MONITORS

Numerous sensors are installed in and around the bearing lubrication oil system to measure the oil pressure, the oil temperature, the bearing babbitt metal temperature, and the oil level in the reservoir. Only the sensors that directly affect the operation of the generator will be discussed here. Neither the interlocks between the various pumps nor the interfaces with the various turbine systems will be addressed in order to limit the discussion within the appropriate scope. Furthermore, because the turbine-end and collector/exciter-end bearings are nearly similar in their treatment, the sensors for only one of these will be discussed, with the assumption that the other one is almost identical.

11-2.1 Discharge Pressures of the Various Pumps

The pump discharge pressure is used to determine if the various motor-driven pumps are in operation. These are used by the BOP, the EOP, and the SOB pumps. Pressure switches (device number 63) are normally used for this purpose.

11-2.2 Bearing Oil Pressure

The bearing header pressure must be maintained at a certain pressure (15–20 lbf/in^2, depending on the design). If this pressure decreases to a preset level, the monitors will alarm on low pressure. Pressure switches (63) are normally used to detect low pressure. These switches may also be used to activate backup devices when the pressure drops to an unacceptably low level. One such switch can be used to start the BOP and SOB pumps when bearing oil header pressure is low. Another can be used to activate the EOP pump when the other two are on and the oil pressure remains at a low level.

11-2.3 Vapor Extractor (Negative) Pressure

The vapor extractor pressure monitor will indicate failure of the vapor extractor. A pressure switch (63) is typically used to carry this out.

11-2.4 Reservoir Level

The level of the oil tank is monitored closely for high and low levels. Level switches (71) are typically used for this measurement.

11-2.5 Lubrication Oil Cooler Outlet Temperature

The temperature of the oil as it exits the cooler must be controlled carefully to avoid various problems with the bearings as well as with the pumps. A thermocouple is typically used for this function.

11-2.6 Lubrication Oil Cooler Inlet Temperature

The temperature to the inlet of the coolers must also be monitored. One major reason is that cooling water in the cooler secondary is not turned on until the temperature of the oil reaches a certain level. This is to avoid low viscosity oil during start-up when there is relatively little heat generated at the bearings. A thermocouple is typically used for this function.

11-2.7 Bearing Oil Drain Temperature

The bearing oil drain temperature is an important measurement because it can provide information about the condition of the bearings. Thermocouples are typically used for this function.

11-2.8 Bearing Metal Temperature

The bearing metal temperature measurement provides a more direct indication of the condition of the bearing (babbitt). Thermocouples are typically used for this function.

11-2.9 Cooler Service Water Inlet and Outlet Temperatures

As with the coolers in other systems, monitoring inlet and outlet temperature can provide the ability to detect problems in the cooler. These are not typically supplied by the manufacturer, but the unit operator should consider their installation.

11-2.10 Lubrication Oil Cooler Differential Pressure

Monitoring of the lubrication oil cooler differential pressure can indicate whether or not there is flow in the coolers, and therefore, in the system. This monitor, however, is typically supplied by the turbine manufacturer, not the generator manufacturer.

11.3 DIAGNOSTICS OF THE BEARING LUBRICATION OIL SYSTEM

The main objective of the monitors in the lubrication oil system is to ensure that the oil is supplied with the appropriate pressure and temperature. Thus, the malfunctions that can affect the operation of the system will be broken down into these two categories. A third problem will be covered under "Miscellaneous Problems."

11-3.1 Problem Area #1: Reduction or Loss of Bearing Oil Pressure

There are basically three causes for a reduction or loss of bearing oil pressure:

1. Loss of pumping ability
2. Blocked bearing oil lines
3. Leakage of large amounts of oil

These will be covered in detail in the following sections. Figure 11.7 depicts the causality diagram.

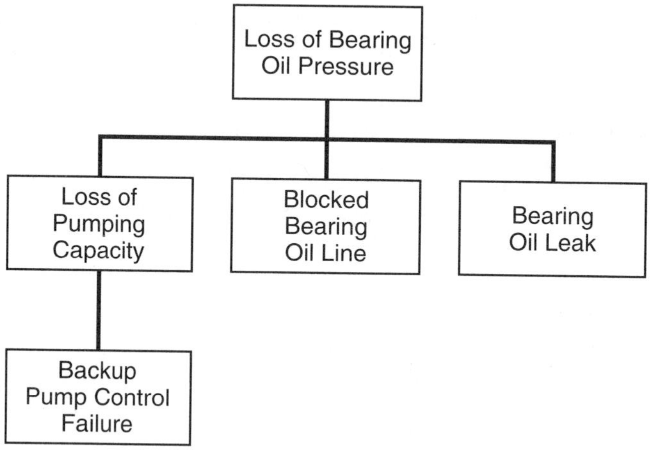

Figure 11.7 Causality diagram for loss of bearing oil pressure problem.

Loss of Pumping Capacity Because of Backup Pump Control Failure

Loss of pumping capacity could occur during normal operation, during start-up or coast-down, or during turning gear operation. If all backups are constantly tested to ensure their availability, however, loss of pumping capacity is not likely. These situations will be treated individually.

During normal operation, the MOP is the prime source of driving power, and the BOP and EOP act as backups, in that order. It is not likely that all three pumps will fail nearly simultaneously. One possibility is the failure of the sensors and interlocks that control the operation of the backup pumps. Diagnostic Box 11.1 makes the assumption that backup pump failure is due to a failure in the control system.

■ DIAGNOSTIC BOX 11.1 ■

PROBLEM: BACKUP PUMP CONTROL FAILURE DURING NORMAL OPERATION

SENSOR INDICATION:

- **Bearing oil pressure** will be low. This is the main indication of the problem.
- **Discharge pressure** of the backup pump(s) will be low or zero. This means that one or more of the backup pumps will not be in operation.

CORRECTIVE ACTION: Activate the backup pumps manually immediately, if possible. If not possible, remove the unit from service. Consult the manufacturer for specific recommendations.

During unit start-up or coast-down, the BOP is the main source of driving power for the bearing oil, with the DC-powered EOP serving as a backup. As long as there is proper voltage in the station DC system and the pump as well as the pressure-sensing devices are tested regularly, there is no reason to believe that both pumps will fail nearly simultaneously. Quite often, however, in a sudden unit trip, AC power can be lost to the station auxiliaries, leaving the DC-driven EOP as the only means of pumping the bearing oil. If the pressure-sensing devices fail or if there is insufficient charge in the station battery, the EOP may not be able to operate satisfactorily. In that case, there is little to be done to avoid damage. Thus, it is important that the DC system be properly maintained and regularly tested. Otherwise, there is little difference between this situation and the one before in how to detect it and what to do about it.

During turning gear operation, the AC-driven bearing oil pump provides the main source of oil. In most designs, the DC backup oil pump provides backup in case of failure of the main source of oil.

Blocked Bearing Oil Lines

Blockage in the bearing oil lines can impede the flow of oil to the bearings. This can take place in the main header to all bearings or in the individual lines to the bearings. Because flow is not measured anywhere throughout the system, except possibly in the cooler (through a cooler differential pressure transmitter), this is not an easy problem to detect. If the blockage is in the cooler, the pressure drop across it will be higher than normal. If it is in the main header line, it is possible that the cooler differential pressure may be lower than normal.

If the blockage is in the individual lines to one or more bearings, the only way to detect it is through temperature measurements. Because the oil performs a cooling function for the bearing, higher bearing oil drain temperatures may be indicative of this problem. It may be impossible to tell the difference, however, between this and a problem with the bearing itself.

■ DIAGNOSTIC BOX 11.2 ■

PROBLEM: BLOCKED BEARING OIL LINES

SENSOR INDICATION:

■ **Bearing oil pressure** may be low if the blockage is in the main line and if the pressure measurement is taken after the blockage. Otherwise, this may not be deviant at all.

■ **Differential pressure (drop)** of the oil through the cooler may be higher than normal if the blockage is within the cooler itself. On the other hand, it may be lower than normal if the blockage is in the main line or header, but not in the cooler itself. If blockage is in one or more of the lines to individual bearings, then the differential pressure may be normal.

■ **Bearing oil drain temperature** may be higher than normal in only some bearings if the blockage is in the individual lines or in all of them if the blockage is in the main line or header.

CORRECTIVE ACTION: If the temperatures exceed the manufacturer's limit, the unit should be brought off-line to find and repair the blockage. Consult the manufacturer for specific recommendations.

Bearing Oil Leak

Large leaks of bearing lubrication oil can result in a deficiency of lubrication oil throughout the system. This may be manifested by a drop in pressure in the header or by a rise in bearing oil drain temperature.

DIAGNOSTIC BOX 11.3

PROBLEM: **BEARING OIL LEAK**

SENSOR INDICATION:

- **Oil reservoir level** may be lower than normal. This may be the most direct indication, although not necessarily the earliest.
- **Bearing oil pressure** may be low. This may be the main indication of the problem.
- **Differential pressure (drop)** of the oil across the cooler may be lower than normal if the leak is sufficiently large to reduce flow through the cooler. This is only true, however, if the leak is located upstream of the cooler. If the leak is located downstream, the differential pressure may be normal, or even higher than normal, as more oil is pumped in an effort to maintain pressure at the bearings.
- **Temperature of the oil** at the bearing drain from all of the bearings may be higher than normal. This is a result of the lower-than-normal flow of oil to the bearings as a result of the leak.

CORRECTIVE ACTION: Repair the leak as soon as possible. Remove the unit from service if it is not possible to repair and the temperatures exceed safe levels. Consult the manufacturer for specific recommendations.

11-3.2 Problem Area #2: Improper Bearing Oil Temperatures

It is imperative that the temperature of the bearing lubrication oil be maintained within the proper range. Excessively high temperatures will result in low-viscosity oil, which will, in turn, result in less friction between low-viscosity oil and the journal to build up the oil wedge. The consequence of this is a thinner and less extensive oil film, which could cause overheating of the bearing metal and the film itself, resulting in wiped babbitt, total loss of babbitt, and damage to journals and seals.

Alternatively, excessively low temperatures and resulting high oil viscosities, particularly at start-up, can overload the motor-driven pumps. This can also cause high vibrations caused by oil whip.

There can be various causes for the high and low temperature problems. These are as follows:

- Fouled cooler
- High inlet plant service water temperature
- Inadequate plant service water flow
- Plant service water flow to cooler valved out

The causality diagram for this problem area is shown in Figure 11.8.

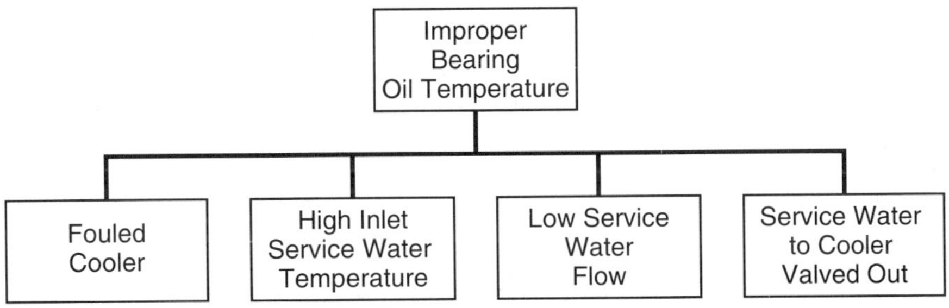

Figure 11.8 Causality diagram for oil temperature problems.

Fouled Cooler

Cooler fouling represents a reduction in the heat transfer capability of the cooler caused by corrosion or fouling of the tube walls. It does not represent the presence of an obstacle to the flow of plant service water in the cooler secondary. This is covered in a separate section. Fouling can result in less heat being transferred from the hot oil to the water, thus causing a higher cooler outlet temperature for the lubrication oil. This is generally not a problem during turning gear operation.

DIAGNOSTIC BOX 11.4

PROBLEM: COOLER FOULING

SENSOR INDICATION:

- **The lubrication oil cooler outlet temperature** (thermocouple or thermometer) will read a *high* temperature. This is the key symptom for identifying this problem.
- **The temperature rise** of the cooling water, as calculated by the difference between the outlet (plant service) water thermocouple and the inlet water

thermocouple, will be lower than normal because of the reduced amount of heat being removed from the oil across the fouled tube walls.

■ **If a cooler inlet oil thermocouple is used, the temperature drop** of the oil should be lower than normal because of the reduced amount of heat being removed from the oil across the fouled tube walls.

■ **The cooling (plant service) water inlet temperature** reading is normal.

■ **The cooler differential pressure transmitter** may indicate a higher-than-normal reading caused by the increased oil flow throughout the system as a result of the lower viscosity.

CORRECTIVE ACTION: Increase the plant service water flow to bring down the temperature of the oil (if possible). Clean out the cooler tubes during the next maintenance outage. If the oil temperature is unacceptably high, shut the unit down for immediate repair of the coolers. Consult the manufacturer for specific recommendations.

High Inlet Temperature of the Plant Service Water

In many situations, the source of cooling water is shallow lakes, where the summertime temperature of the water will rise significantly above the inlet water temperature rating of the cooler. Little can be done about this problem in the short term. The long-term solution, of course, is to replace the coolers with adequate ones. This problem assumes that the unit is in normal operation, start-up, or shut-down, and not on turning gear.

■ **DIAGNOSTIC BOX 11.5** ■

PROBLEM: HIGH INLET TEMPERATURE OF PLANT SERVICE WATER

SENSOR INDICATION:

■ **The lubrication oil cooler outlet temperature** (thermocouple or thermometer) will read a *high* temperature. This is a key symptom for identifying this problem.

■ **The cooling water inlet temperature** reading is high, another key indicator of this problem.

■ **The temperature rise** of the cooling (plant service) water, as calculated by the difference between the outlet water thermocouple and the inlet water thermocouple, may be near normal or somewhat low.

■ **If the cooler inlet oil temperature is measured, the temperature drop** of the oil should be near normal.

■ **The cooler differential pressure transmitter** may read higher than normal. This is caused by the higher flow of lubrication oil through the cooler.

CORRECTIVE ACTION: Increase the cooling water flow (if possible) to bring down the temperature of the oil. If the flow is already maximized, consider spraying water over the outside of the cooler shell as an emergency measure (if possible). If the lubrication oil temperatures are higher than the limit, take the unit out of service. Consider replacing the coolers with others that have a higher inlet water maximum temperature. Consult the manufacturer for specific recommendations.

Inadequate Flow of Plant Service Water

Occasionally, the flow of water to the cooler will become partially blocked by some foreign object. Alternatively, a leak of the secondary water may be caused by a broken pipe. In either case, the flow of water to the cooler will be reduced. This will cause a deficiency in the cooling capability of the cooler, which would likely lead to high temperature oil. Improper reassembly of the lubrication oil cooler after a maintenance outage can also result in reduced water flow. If an automatic water flow regulator is used, this may not surface as a problem until the flow is maximized and the cooling deficiency in the cooler surfaces. As has been noted previously, flow of water is generally shut off during turning gear operation.

■ **DIAGNOSTIC BOX 11.6** ■

PROBLEM: INADEQUATE PLANT SERVICE WATER FLOW

SENSOR INDICATION:

■ **The lubrication oil cooler outlet temperature** (thermocouple or thermometer) may read a *high* temperature. This is a key symptom for identifying this problem.

■ **The temperature rise** of the cooling water, as calculated by the difference between the outlet water temperature and the inlet water temperature, will be higher than normal because of the reduced cooling water flow.

■ **If the cooler inlet oil temperature is measured, the temperature drop** of the oil should be lower than normal.

■ **The service water inlet temperature** reading is normal.

■ **The cooler differential pressure transmitter** may read higher than normal. This is caused by the higher flow of seal oil.

CORRECTIVE ACTION: Increase the cooling water flow to bring down the temperature of the oil (if possible). Find the blockage or leak and correct the problem. If the oil temperature is unacceptably high and the problem cannot be fixed quickly, then shut the unit down for cleanup and repair. Consult the manufacturer for specific recommendations.

Plant Service Water to Cooler Valved Out

Interruption of water flow to the cooler because of a closed valve is a fairly common occurrence, especially when bringing a unit on-line after a maintenance outage. Fortunately, it is easy to correct, once it has been detected.

■ DIAGNOSTIC BOX 11.7 ■

PROBLEM: SERVICE WATER TO COOLER VALVED OUT

SENSOR INDICATION:

■ **The lubrication oil cooler outlet temperature** (thermocouple or thermometer) will read a *high* temperature. This is a key symptom for identifying this problem.

■ **Temperature rise** of the cooling (plant service) water, as calculated by the difference between the outlet water thermocouple and the inlet water thermocouple, will be near zero. This is another key symptom for this condition.

■ **If the cooler inlet oil temperature is measured, the temperature drop** of the oil may be significantly lower than normal.

■ **Service water inlet temperature** reading is normal. This would eliminate the possibility of high inlet water temperature being the problem.

■ **Cooler differential pressure transmitter** may read higher than normal. This is caused by the higher flow of lubrication oil through the cooler.

CORRECTIVE ACTION: Open the cooler water valves slowly. Beware of thermal shock to the cooler. Consult the manufacturer for specific recommendations in case possible damage to the unit has been incurred. Put this item in the start-up procedure checklist.

Excessive Plant Service Water Flow to the Cooler

Although excessive water flow is not a problem, per se, the significance of this condition is that low-temperature lubrication oil (high viscosity) can result in damage to the motor-driven pumps and/or in high shaft vibration at the bearing caused by oil whip. In some designs, plant service water to the cooler secondary is not turned on until the temperature of the oil increases to a specific temperature. However, failure of the control mechanism may result in water inappropriately flowing through the cooler secondary.

DIAGNOSTIC BOX 11.8

PROBLEM: EXCESSIVE PLANT SERVICE WATER TO COOLER

SENSOR INDICATION:

- **The lubrication oil cooler outlet temperature** (thermocouple or thermometer) will read a *low* temperature. This is a key symptom for identifying this problem.
- **Temperature rise** of the cooling (plant service) water, as calculated by the difference between the outlet water temperature and the inlet water temperature, will be higher than normal. This is another key symptom for this condition.
- **If the cooler inlet oil temperature is measured, the temperature drop** of the oil should be higher than normal.
- **Cooling water inlet temperature** reading may be lower than normal. The reason for the low temperature oil may be that the source of plant service water is too cold or its flow is too high for its value.

CORRECTIVE ACTION: Throttle the cooler water valves progressively until the temperature of the oil is normal. Consult the manufacturer for specific recommendations.

11-3.3 Miscellaneous Problems

This section covers one specific problem that is not covered elsewhere and can result in serious consequences. It is not a cause of failure as such, but, rather, the consequences of other failures. It is the case of hydrogen accumulation in the exciter housing caused by a loss of hydrogen seals in the generator.

In generators designed with rotating exciters, the exciter bearing is lubricated by the same system as the other bearings in the turbine and generator. As seen in Chapter 9, this is also the source of seal oil for the shaft seals. Failure of all primary as well as backup pumping capacity in the seal oil system (an unlikely case) will result in a loss of seal in the generator. Such a major loss of seal can result in hydrogen escaping through the bearing cavity through the bearing oil drain. Because the drain is not filled to capacity, the gas can escape over the oil. If the leak through the bearing oil drain is not severe, a vapor extractor may be able to remove it. However, a severe leak may result in some of the hydrogen making its way to the exciter bearing, where it may escape through the bearing and accumulate in the upper parts of the inside of the exciter housing. Because the purity of the hydrogen may now be such that it is in the flammable range, any sparks may cause an explosion. Needless to say, this represents a dangerous situation.

This situation can be prevented by venting the exciter housing continuously during normal operation. The manufacturer should be contacted for their recommendations.

SUMMARY

This chapter covered the basics of the bearing lubrication oil system. The purpose of the system is to supply lubricating oil to the bearings at the proper pressure and temperature. The lubrication oil system is typically part of the turbine system. However, only the parts that affect the generator and/or exciter/collector bearings were discussed in this chapter.

CHAPTER 12

Generator Protective Relaying

12.1 INTRODUCTION

The previous chapters describe various potential problems in a generator system that can be detected through monitoring and diagnostics. For the most part, once the problem has been identified, there is typically sufficient time for an operator to implement a corrective course of action that will avoid or minimize the damage to the unit or the system. There are some abnormal conditions in a generator, however, for which the time between detection and action is so short as to preclude intervention by a human operator. For these problems, automatic protection of the equipment needs to be designed and implemented to remove the unit from service within the allowed time limits without operator intervention. Protective relays are used in these circumstances; this chapter will describe the techniques involved.

The concepts employed in generator protection are, basically, the same as those in any protective relaying application. *Reliability* and *security* are of equal importance. The science (some call it an art) of protective relaying requires that the system always operate to isolate a circuit or circuit component when a fault or other intolerable condition exists (reliability), but to maintain the circuit or component in service when a fault is not in the protected circuit or component or the condition does not jeopardize the equipment (security). It is sometimes easier to achieve the former objective than the latter. It is in meeting both objectives that the ingenuity of the protection engineer is tested.

In the discussions of various aspects of generator protection that will follow, a unit-connected generator that is part of a large power system is assumed. It is also assumed that the reader has become familiar with the contents of Chapter 2. Where appropriate, considerations regarding other types of units will be addressed.

12.2 INSTRUMENT TRANSFORMERS

Voltage transformers (VTs) and *current transformers* (CTs) are the devices that reduce the high power system voltage and current to the 120-volt and 5-ampere level in order for these system parameters to be sensed in a safe and inexpensive manner. It is essential that the protection engineer have an understanding of instrument transformer principles and performance. A brief review of instrument transformer principles and those aspects of their performance characteristics that affect protective relaying will be given in this section.

12-2.1 Polarity

The polarity of transformer windings is an indication of the sense of the windings relative to each other. On a drawing, the polarity will be indicated by a dot or an "x." On the actual transformers they will be indicated by a colored mark of some kind.

Refer to Figure 12.1. The voltage from the polarity marked terminal to the nonpolarity marked terminal of the secondary winding is essentially in phase with the voltage from the polarity marked terminal to the nonpolarity marked terminal of the primary winding. In other words, if the transformer had an equal number of turns on both windings and only the terminals were available, one would not know if there were a transformer in the enclosure or not.

The voltage and current relationships in the transformer of Figure 12.1 are as follows:

$$I_2 = n * I_1 \qquad (12.1)$$

$$E_{CD} = n * E_{AB} \qquad (12.2)$$

Similarly, the current coming out of the polarity marked terminal of the secondary winding is essentially in phase with the current entering the polarity marked terminal of the primary winding. The polarity rule for voltages is important in VT applications, whereas the rule for currents is important in CT applications.

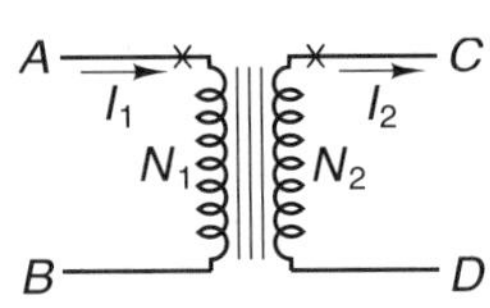

Figure 12.1 Transformer polarity.

In power and distribution transformer applications, the terms *additive* and *subtractive* polarity are used. These have little meaning here as long as the above rules about polarity marks are used. However, if one were to connect terminals *A* and *C* of the transformer of Figure 12.1 and apply a voltage to terminals *A* and *B*, the voltage between terminals *B* and *D* would be the voltage of *A* to *B* minus the voltage *C* to *D*. The polarity would thus be defined as being subtractive. If the polarity mark of the transformer secondary were on terminal *D* instead of *C*, the transformer polarity would be additive.

12.2.2 Performance

The equivalent circuit for a transformer is shown in Figure 12.2. In this figure, Z_p and Z_s are the primary and secondary impedances, respectively. Z_m is the magnetizing impedance of the transformer. In the case of instrument transformers, the load connected to the secondary winding is called the *burden*. The burdens of VTs are connected in parallel as shown in Figure 12.3. The burdens of CTs are connected in series as shown in Figure 12.4.

Because the function of a VT is to provide an accurate measure of the primary system voltage, the exciting current drawn by the transformer is not important, but

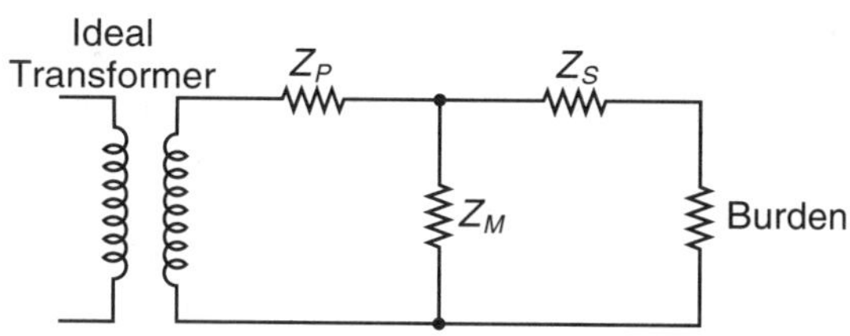

Figure 12.2 Transformer equivalent circuit.

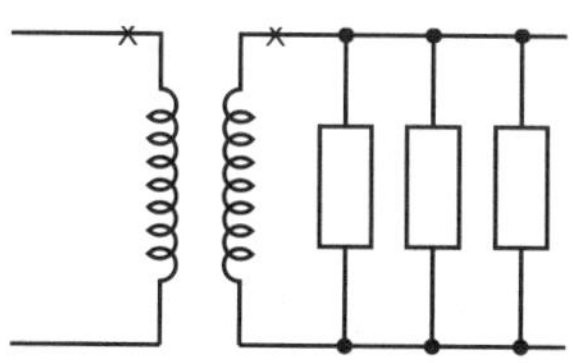

Figure 12.3 Voltage transformer burdens.

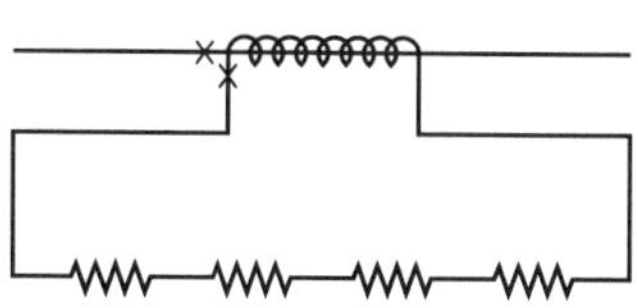

Figure 12.4 Current transformer burdens.

the internal impedance drop must be very low or negligible. In contrast, the CT must provide an accurate measurement of the primary current, so the internal impedance drop is not as important as the amount of exciting current drawn. Any exciting current drawn will be subtracted from the current reproduced in the secondary winding and will represent a ratio error.

In general, the accuracy of VTs is not as significant a factor in relaying as it is in metering applications and, therefore, nothing further will be said about this. On the other hand, CT accuracy is important in protective relaying. Industry standards define metering accuracy classes and relaying accuracy classes for instrument transformers.

Figure 12.5 is a typical excitation curve for a CT. It shows the exciting current required, in terms of the secondary winding current, for corresponding values of voltage induced in the secondary winding. Although the burdens of a CT are normally quite low, the voltage developed across them during fault conditions can be significant. This is particularly true in large substations or power plant switchyards where the distances between the CTs on circuit breakers or power transformers and the relays can be hundreds or even thousands of feet. The voltage the CT must support without intolerable error is the product of the secondary current times the burden impedance plus the internal impedance drop of the transformer.

The relaying accuracy class for a CT specifies the amount of secondary terminal voltage the CT must support without exceeding a certain error when the secondary current is 100 amperes. For example, a C800 (formerly 10L800) accuracy class means that the CT must provide a secondary potential of 800 volts when the secondary current is 100 amperes, without exceeding an error of 10%.

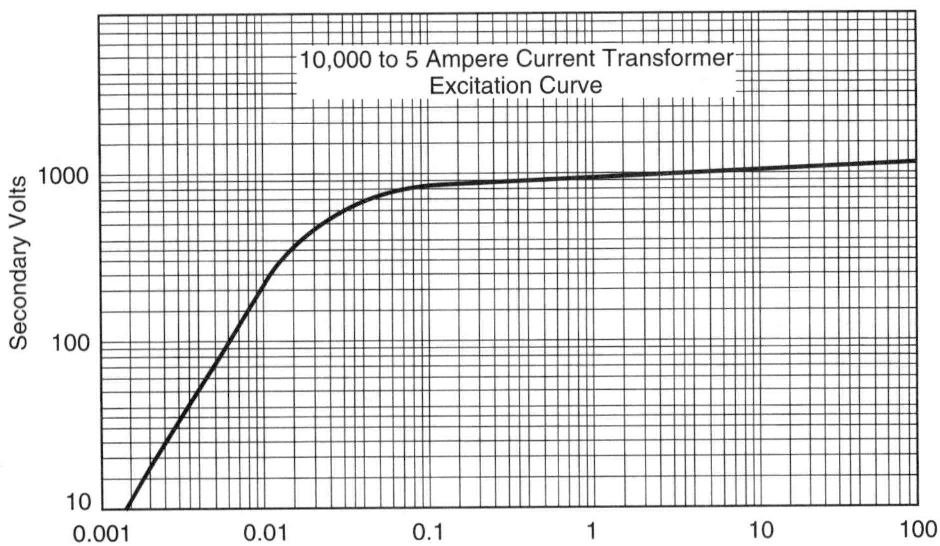

Figure 12.5 Current transformer excitation curve.

In the case of multiratio CTs, the accuracy class refers to the highest current tap. In such instances, the induced voltage for a given value of exciting magnetomotive force (mmf) is proportional to the tap used; that is, a 1200/5 multiratio transformer that could support 800 volts on the 1200 amperes tap would support only 400 volts on the 600 amperes tap for the same flux density. However, the flux density is a function of the mmf, which is, in turn, proportional to the exciting ampere turns.

Refer to Figure 12.6, which shows the excitation curves for a C800 accuracy class multiratio bushing type CT with a maximum ratio of 2000/5 amperes. Because there is only one primary turn, the tapped secondary winding will have a total of 400 turns. One ampere of exciting current in the secondary winding on the 2000/5 tap will produce enough flux to induce 980 volts. This corresponds to 400 ampere-turns of excitation. The voltage per turn is 980/400 or 2.45 volts per turn. On the 1000/5 tap, the same level of flux density will require the same number of exciting ampere turns or two amperes of exciting current. The corresponding induced voltage in the secondary winding will be $200 \times 2.45 = 490$ volts. Because exciting current is an error current, the percent error is twice as much on the 1000/5 tap as that for the 2000/5 tap with only one-half the secondary voltage.

The CTs used on large generators, circuit breakers, and power transformers are *bushing type* CTs. A bushing type CT has a toroidally wound magnetic core on which a secondary or output winding is wound. The primary winding is the conductor or bushing through the CT. This means that they all have only one primary turn. The secondary winding is distributed around the core to provide uniform magnetic coupling with the core. This is particularly important for multiratio CTs.

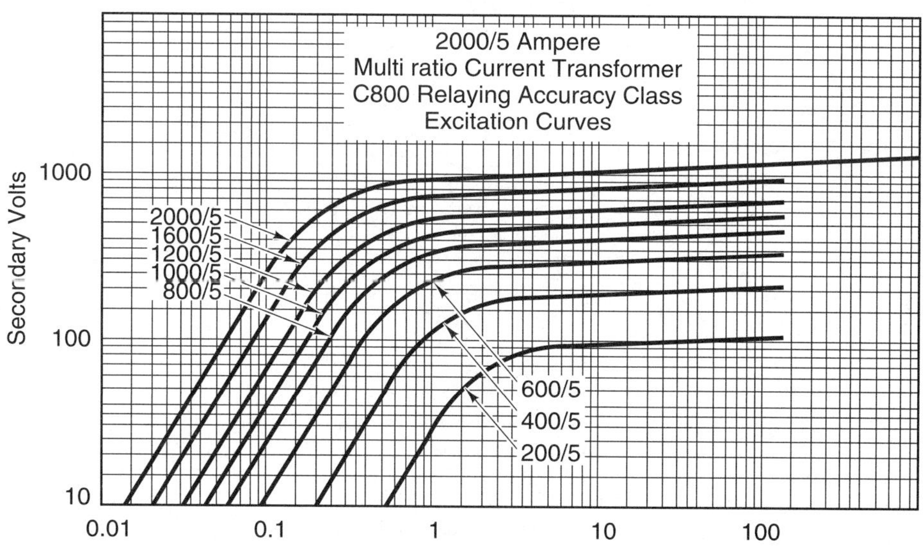

Figure 12.6 Multiratio current transformer excitation curve.

The CTs on the generator line and neutral bushings are single ratio units, but those on the high-voltage bushings of the generator step-up (GSU) and high-voltage circuit breakers are usually multiratio units.

In making a CT application, the maximum fault current that will flow through the CT must be determined and the total CT secondary burden calculated. The CT saturation curve(s) can then be used to determine the exciting current required to support the burden and CT impedance voltage.

In addition to analyzing the AC performance of CTs, it is sometimes necessary to determine whether or not the direct component of fault current will result in incorrect relay operation. Refer to Baldwin, [1980] for information on this aspect of relaying. Suffice to say here that direct current saturation of CTs can affect the performance of some types of relay systems.

12.3 STATOR FAULT PROTECTION

Various types of faults within the generator windings and in equipment connected to the generator terminals require that protection for the generator be implemented. These faults and the devices used to detect them are described in this section. The Institute of Electrical and Electronic Engineers (IEEE) has developed a list of identifying numbers for electrical devices, [IEEE, C37.102–1987]. A partial listing of these device numbers is given in Table 12.1 below.

Table 12.1 GENERATOR PROTECTIVE RELAY DEVICE NUMBERS

2	Timer
21	Phase Distance Relay
24	Volts-Per-Hertz Relay
25	Synchronizing or Synchronism Check Device
27	Undervoltage Relay
32	Antimotoring Relay
40	Loss-of-Excitation (Loss-of-Field) Relay
46	Negative-Sequence Relay
51/50	Overcurrent Relay
52	Circuit Breaker
59	Overvoltage Relay
63	Gas Pressure Relay
64	Ground Protective Relay
67	Directional Overcurrent Relay
76	Excitation Alarm or Protection Relay (DC overcurrent)
78	Out-of-Step Relay
81	Frequency Relay
86G	Generator Lockout Relay
87G	Generator Differential Relay
87N3	Third-Harmonic Differential Relay
87T	Transformer or Unit Differential Relay

12.3.1 Phase Fault Protection

Faults between the phase windings of a generator can be detected by current-type *differential* relays (87G). These relays compare the currents from the secondary windings of CTs in the neutral end of each phase with currents from CTs in the corresponding phase or line terminals. Figure 12.7 shows a basic current differential circuit. In a typical electromechanical differential relay, currents in its *restraint* (R) windings cause contact opening torque. Current in the *operating* (OP) winding causes contact closing torque. The relative strengths of the operating and restraint windings are designed to provide a percentage differential characteristic.

The current in the operating winding is the difference between the two CT output currents. It is apparent that the two CTs should have the same ratios and same performance characteristics. Ideally, if the two CTs were perfectly matched, a simple overcurrent relay could be used for the operating circuit and no restraint windings would be required. In fact, that is what was used in the early differential relay systems. It was soon learned that it was necessary to add the restraint windings to provide more security against false operation and to provide a *percentage characteristic.* The percentage characteristic of a relay is defined as the unbalance or operate circuit current as a percentage of the smallest restraint circuit current. Typically, generator differential relays have either a 10% or a *variable percentage* characteristic. The variable percentage characteristic means that the relay is very sensitive, that it has a low percentage characteristic at low values of current, but is less sensitive at high current levels. This makes the relay sensitive to low internal faults, but insensitive to heavy external faults.

If the CTs supplying the differential relays are well-matched and have good performance, a fault external to the protected zone will result in little or no current in the operating circuit of the relay, and the high value of circulating current will provide restraint. On the other hand, a fault within the protected zone will result in current in the operating circuit. This operating circuit current will result in a sufficient amount of torque to overcome the restraint torque, and contact closure or relay operation will occur.

Prudence dictates that the differential relays be supplied by CTs dedicated for that function; that is, no other devices should be connected to those CTs. The

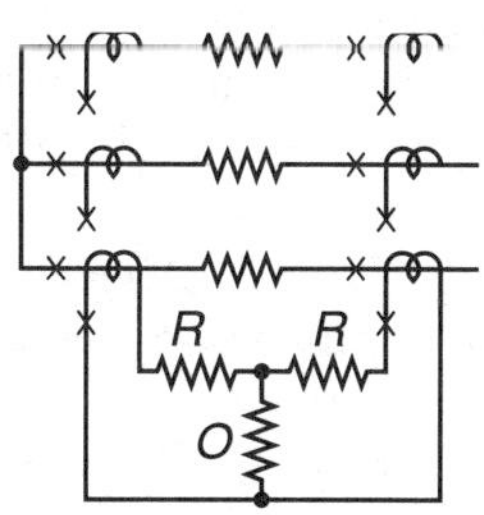

Figure 12.7 Basic differential relay.

CT circuit should be grounded at one location and only one location. Multiple grounds on the CT circuits may permit ground grid currents to flow in the relay operating coil circuit and cause incorrect relay operation, or the operating circuit could be effectively shorted out and prevent operation of the relay for a fault within the protected zone.

Transformer protection must be considered in the protection of unit-connected generators. Therefore, some comments about transformer differential relays (87T) are necessary at this point. Because a transformer has different input and output voltages, it should be obvious that the currents in the primary and secondary circuits are not equal. In addition, the GSU is connected in delta on the generator side and grounded wye on the high-voltage side. This connection results in a phase shift between the primary and secondary line voltages and currents.

To compensate for the phase shift in the power transformer, the CTs on the delta side are connected in wye, and those on the wye side are connected in delta. The available ratios of CTs, however, may not match the transformation ratio of the power transformer. Therefore, transformer differential relays must have tapped windings to accommodate the mismatch between the CT ratios. Typically these relays will accommodate a two-to-one mismatch in ratios. Transformer differential relays are also less sensitive than generator differential relays, with 50% characteristic being common.

When a transformer is energized at rated voltage, there can be a *magnetizing inrush* current that can exceed the rated current of the unit. This magnetizing inrush current may contain a direct component and harmonic currents. The inrush is a random phenomenon and depends on the magnetic state of the transformer core at the time of energizing and the point on the energizing voltage wave. Modern transformer differential relays are much more complex than the simple differential relays described earlier and have harmonic restraint circuits to preclude false operation on inrush currents. In general, these relays restrain or prevent operation on the second harmonic or the second and higher harmonic currents.

In the case of a unit-connected generator, the voltage on the step-up and auxiliaries transformers is increased gradually as the generator excitation is raised and no magnetizing inrush occurs; however, a similar phenomenon can occur during a load loss or rejection when there is a sudden jump in the generator terminal voltage.

Figure 12.8 shows a unit-connected generator and ideal differential protection zones. Actually, many utilities do not provide a separate zone for the GSU, but depend on the overall or unit differential relay as primary protection for that component. Also, many utilities do not provide differential protection for the unit auxiliaries transformers (UATs). The discussion in the following paragraphs should cause protection engineers to consider the application of differential relays on the UATs.

The UAT and secondary circuit to the main low-voltage breaker are an integral part of the unit, a fact that is sometimes overlooked. In some cases, the secondary circuit of this transformer does not have the same level of integrity as the main generator circuits. The generator main leads are usually isolated phase bus-

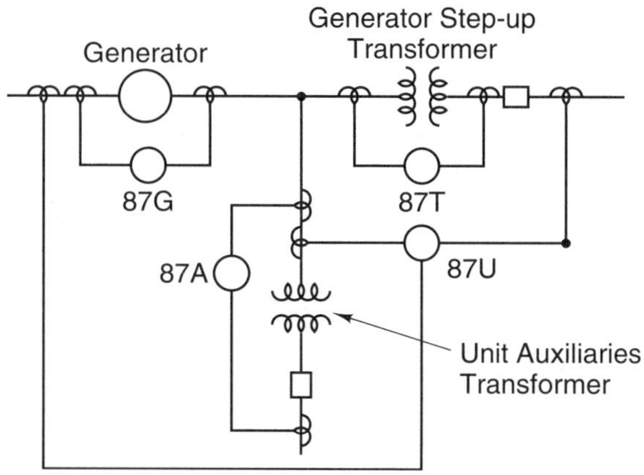

Figure 12.8 Generating unit differential zones.

work. The secondary leads of the unit auxiliary transformer may be cable or nonsegregated phase buswork.

When a fault occurs on the power system side of the unit high-voltage breaker(s) or on the load side of a unit auxiliaries circuit breaker, it is cleared as soon as the circuit breaker is opened. However, a fault that occurs on the unit side of either the main unit circuit breaker(s) in the plant switchyard or the UAT secondary circuit breaker is not immediately cleared by opening of the unit circuit breakers. The fault current will continue to flow even after the unit and generator excitation are tripped because the magnetic flux trapped in the generator will cause the generator to continue to supply current to the fault, which is still electrically connected to the unit. The rate at which the magnetic field and fault current will decay is a function of the generator characteristics and the impedance of the circuit between the generator and the fault.

A fault on the low-voltage side of the UAT represents a fairly low magnitude fault to the generator, because the transformer rating is small compared to that of the generator. However, the time integral of fault current through the transformer can cause thermal and mechanical distress to the latter. If a fault in the UAT or secondary is not detected promptly, the amount of distress is aggravated. Frequently an overcurrent relay is used to detect faults on this part of the unit. Accordingly, the time delay of this relay must be set to provide selectivity with the low-voltage relays in order to avoid a unit trip for a fault on the auxiliaries system load circuits. This may result in an unacceptable time delay for the fault described previously. A high-speed differential relay could minimize the time to trip the unit and excitation.

Large generators have an enclosure around the bushings at the neutral end of the phase windings. This enclosure should be electrically insulated from the generator frame except at one location. If this enclosure is grounded to the generator frame at more than one point or if other grounded conductors touch the enclosure, an incorrect differential relay operation may occur during heavy loads or during faults external to the generating unit. The reason for this incorrect operation is that the multiple grounds on the neutral lead enclosure permit stray magnetic flux from one or more of the neutral CTs to induce current in the enclosure. The effect is to rob the CT secondary winding of some current. In this case, the output of that neutral CT will not match the output of the corresponding phase CT. Typically, generator differential relays will operate on as little as 0.14 ampere, and the mismatch caused by the above could result in relay operation. The accepted practice is to have the differential relays (87G & 87T) connected to trip the generator circuit breaker(s) and the generator excitation by a lockout relay (86G).

12.3.2 Ground Fault Protection

Protection must be provided for ground faults that occur in the generator windings or in equipment connected to the generator terminals. Generator grounding and ground protection will be covered in the following sections.

Generator Neutral Grounding

The neutrals of most unit-connected generators are grounded by the distribution transformer method. In this scheme, shown in Figure 12.9, the primary winding of a transformer in the distribution-size class is connected between the generator neutral and station ground. A resistor (or in some cases a reactor) is connected across the transformer secondary winding.

Because a generator must be capable of operating at 105% of rated terminal voltage, the grounding transformer primary voltage rating should be at least 105% of the generator rated ground-to-neutral voltage. The resistor value is selected so

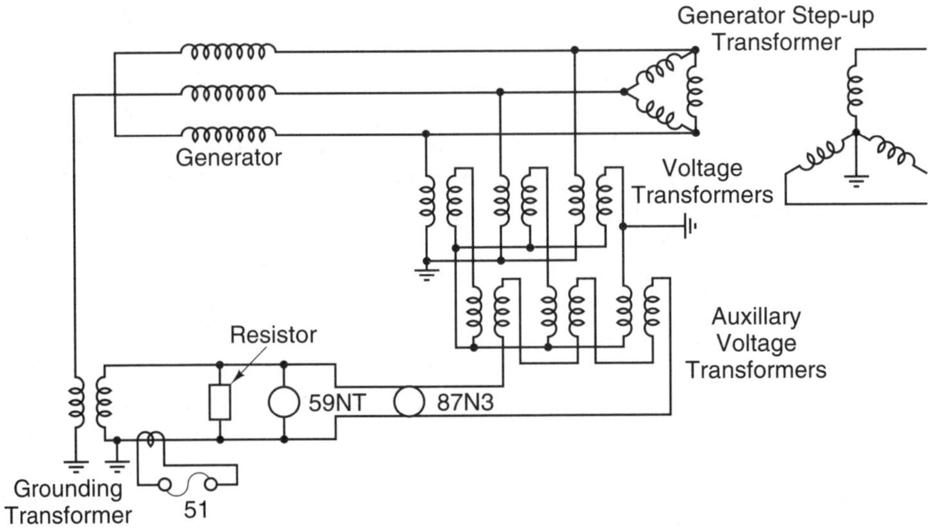

Figure 12.9 Unit-connected generator grounding.

that during a single-phase-to-ground fault, the kilowatt loss in the grounding resistor will be equal to the normal three-phase capacitive kilovars of the system, consisting of the generator and equipment connected directly to its terminals; that is,

$$R_g = \frac{X_{co}}{3} \tag{12.3}$$

where R_g is the equivalent resistance as seen from the high-voltage terminals of the grounding transformer, and X_{co} is the capacitance to ground per phase of the generator and connected system. X_{co} includes the capacitance of the generator winding, isolated phase bus or cable, main and auxiliaries transformers primary windings, and any surge capacitors connected to the generator terminals.

The actual resistance of the grounding resistor will be the equivalent resistance divided by the square of the turns ratio of the grounding transformer,

$$R_{act} = \frac{R_g}{n^2} \tag{12.4}$$

where n is the grounding transformer turns ratio.

High-resistance grounding in line with the above criteria usually results in ground fault currents between 5 and 15 amperes. For units in which the capacitance is so low that the calculated value of ground fault current is less than 5 amperes, the resistor value normally is reduced to cause at least 5 amperes to flow for a ground fault at the generator terminals. This is to ensure blowing of a VT fuse in the event of a failure in the VT.

A few generator installations employ *resonant neutral grounding*. In this case, the grounding resistor is replaced by a reactor as shown in Figure 12.10. As with resistor grounding, the equivalent reactance value as seen at the high-voltage terminals of the grounding transformer should be the same as that of a resistor.

$$X_g = \frac{X_{co}}{3} \tag{12.5}$$

As in the case of the resistor, the actual reactance is the equivalent reactance divided by the square of the turns ratio of the grounding transformer.

The system is called a *resonant system* because the inductive component of ground fault current is canceled out by the capacitive component when the system is properly tuned. If there were no losses in the system, the ground fault current would be zero. As it is, the value is usually a fraction of an ampere for a ground fault at the generator terminals. In this type of grounding, it is necessary to provide taps on both the transformer and reactor to permit tuning of the system after it is installed.

On generators with higher voltages, the primary windings of the VTs connected to the generator leads are usually connected in grounded wye. Because these transformers have such a high impedance, they are not considered to constitute a system ground. If the secondary circuits of these transformers are connected in wye, as they usually are, it is suggested that the secondary neutral not be

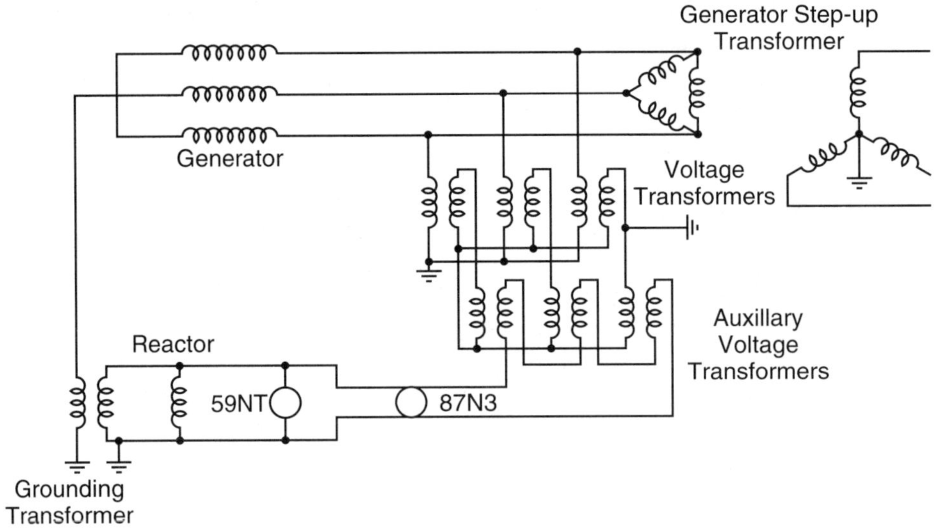

Figure 12.10 Resonant neutral grounding.

grounded, but one phase conductor or terminal should be grounded. This will be discussed in the next section.

Protection

With high-resistance grounding of the type described, the ground fault current is too low to be detected by even the most sensitive differential relays. It is therefore necessary to provide some form of sensitive ground fault protection.

In most installations, the existence of a ground fault in the generator is detected by a sensitive voltage relay (59N) connected across the grounding resistor or reactor, as shown in Figure 12.9. The relay detects the shift in the neutral voltage during the fault. Of course, the amount of neutral shift will depend on the impedance in the fault and the location in the winding where the fault occurs. A zero-impedance fault at the generator terminals will result in a complete shift of the neutral. The closer to the neutral the fault occurs, the lower the value of neutral shift and the corresponding voltage available for detection. A typical sensitivity for the voltage relay is 5.8 volts. With the grounding transformer ratio likely to be used, this will permit sensing a fault within about 5% from the neutral. Because large turbine generators are built with single turn coils, this sensitivity will usually result in detecting a fault in the neutral end coil.

The voltage generated in all generators has a third-harmonic component. The third-harmonic is a zero sequence harmonic and will, therefore, appear at the neutral. The 59N relays are built to have a much lower sensitivity to the third harmonic than to the fundamental; typically, it requires about eight times as much third-harmonic voltage to operate the relay as fundamental voltage. Some users also use an overcurrent (51) relay in the grounding transformer secondary circuit to supplement the voltage relay.

In recent years there has been a move to provide more complete ground fault protection of the generator winding than can be provided with available voltage-sensing relays. The parameter used in most of these schemes is the third-harmonic voltage of the generator. One scheme uses the absence of third-harmonic voltage at the neutral as an indication of a ground fault near the neutral of the machine. Another scheme compares the third-harmonic voltage at the neutral with the third-harmonic voltage at the terminals (87N3).

In addition to providing more complete protection, these schemes monitor the integrity of the ground circuit. If the circuit between the generator neutral and ground is open, the relay sensing the third harmonic would interpret that as a ground fault near the neutral. The monitoring of the neutral circuit integrity in this way is a valuable feature of these relays. In order to perform a high-potential test of the generator insulation, the neutral must be isolated from ground. There have been occasions in which the ground was not reconnected. These third-harmonic relays will detect such a condition.

As indicated earlier, the ground protection scheme using a voltage-sensing relay is very sensitive. If wye-wye connected VTs have both the primary and secondary neutrals grounded, a ground fault on the VT secondary circuit may cause undesirable tripping of the generating unit. If one phase conductor of the VT secondary circuit is grounded rather than the neutral, a ground fault on another phase will be reflected as a low-level phase-to-phase fault on the primary system and should not result in tripping of the unit.

The VTs may have fuses in the primary leads in series with current limiting resistors. Ironically, there are many installations in which current limiting resistors are common to more than one set of VTs, even though the VTs have individual fuses. The time delay setting of the generator ground relay(s) should be such as to permit a VT fuse to isolate a faulted transformer before the ground relay trips the unit.

The 59N ground relay(s) are normally connected to trip the generator breaker and the generator excitation by means of a lockout relay (86G). The ground relays on some older units were connected to sound an alarm rather than trip the unit. This practice is discouraged, because more serious faults have occurred as a result of ground faults that were not cleared promptly. There are some installations in which a very sensitive relay is used to alarm, and a less sensitive relay is set to trip the unit for more severe ground faults. Again, this practice is discouraged because of the possibility of more severe faults developing.

12.3.3 Turn-to-Turn Faults

As stated earlier, large turbine generators have single turn coils. Each turn, therefore, has full ground wall insulation and no turn-to-turn detection is required. Hydraulically driven generators (hydros) or generator/motors and some smaller turbine generators do have multiturn coils.

A standard differential relay will not detect a turn-to-turn fault because the current entering and leaving the winding is the same, even though there may be a large circulating current in the shorted loop. If the generator has two parallel paths per phase, a differential relay that compares the current in one parallel path with the total current for that phase may detect a turn-to-turn fault.

The protection on some hydro generators includes what is called a *split-phase scheme*. These machines usually have many parallel paths in each phase. A relay that compares the currents in one half of the parallel paths with the currents in the other half may also detect a shorted turn. Unfortunately, the hydro generators are physically large, and the air gap is not the same around the unit. The split-phase relays may operate as a result of the eccentricity of the unit unless it is compensated for.

12.4 STATOR THERMAL PROTECTION

The electrical insulation of the armature winding and the interlaminar insulation of the magnetic core can be damaged by excessive heat. It is therefore necessary to provide some means of monitoring the temperatures of the stator and conditions that could cause excessive heating of the core and coils.

12.4.1 Armature Windings

The continuous and short-time capabilities of cylindrical rotor synchronous generators are defined in ANSI Standard C50.13–1989. This standard requires that a generator be capable of operating continuously at rated kVA, power factor (PF), and any voltage not more than 5% above or below rated voltage without exceeding a safe temperature rise (Paragraph 3.1). Paragraph 6.1 of this standard states:

> 6.1 Armature Winding Short-Time Thermal Requirements. The generator armature shall be capable of operating at 130 percent of rated armature current for at least 1 min. starting from stabilized temperatures at rated conditions.
>
> NOTE (1): The permissible armature current at times up to 120 seconds, based upon the same increment of heat storage as defined in 6.1, will be:

Time (seconds)	10	30	60	120
Armature current (%)	226	154	130	116

> NOTE (2): It is recognized that armature temperatures will exceed rated load value under these conditions and, therefore, the machine construction is based upon the assumption that the number of such operations at armature currents to the limits specified in Note (1) will occur not more than two times per year.

The above table can be approximated by the following equation:

$$(I^2 - 1) * t = K \tag{12.6}$$

In this equation, I is the per unit of rated current and t is the permissible time in seconds. For the above table, the value of K is approximately 41.1.

When these limitations are plotted on log-log paper, they appear as the upper curve of Figure 12.11. Overcurrent relaying is not included in the protection of most generators. Presumably, the operator is depended upon to prevent extended oper-

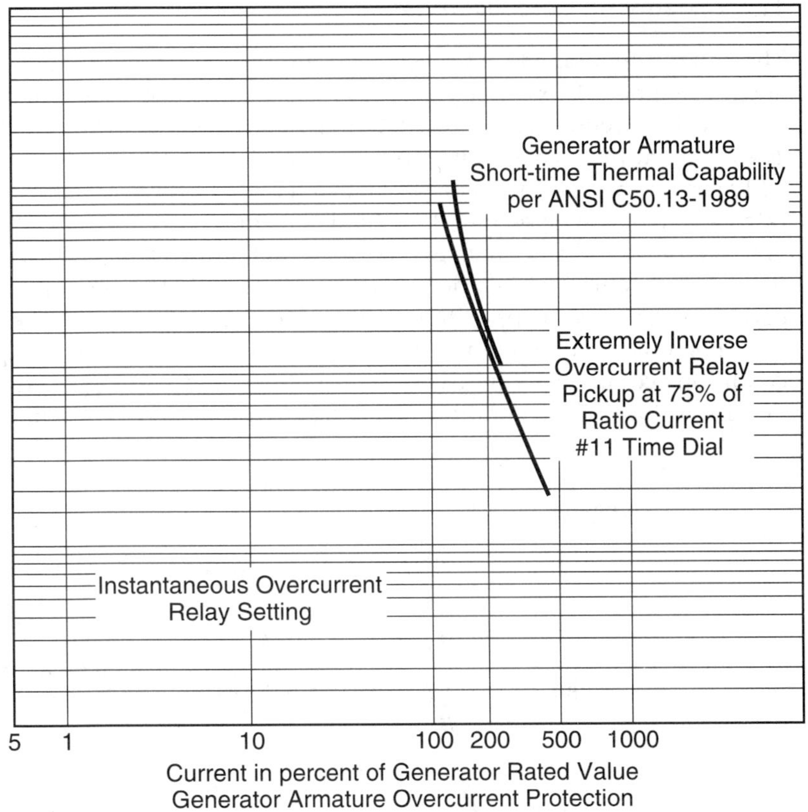

Figure 12.11 Generator armature short-time current capability.

ation at high load currents. Overcurrent protection can be provided by an extremely inverse overcurrent relay (51), which is torque-controlled by either a high dropout instantaneous overcurrent relay (50) or an undervoltage relay (27) or element. If a voltage-controlled overcurrent relay is used, particular care must be taken in setting the relay to preclude unwanted tripping of the unit if the voltage to the voltage element is lost during permissible operating conditions. An operating curve for an overcurrent relay (51) torque controlled by an instantaneous overcurrent relay (50) is shown as the lower curve of Figure 12.11. In this case, the instantaneous relay (50) is set to pick up at 115% of generator rated current. The extremely inverse relay (51) is set to pick up at below rated current, but cannot operate until the 50 relay operates. A relay such as this could be used to initiate runback of the automatic voltage regulator (AVR) and/or governor to reduce load and/or excitation on the machine if the user does not wish to trip the unit with this device. It could also be set to alarm an operator to take corrective action. Voltage-controlled overcurrent relays are sometimes used on generators as backup for the system relaying.

12.4.2 Magnetic Core Overheating

The voltage developed by a given generator is a function of the speed (frequency) and the magnetic flux density in the stator (armature) core, as seen in Chapter 2 and discussed in Section 6.2.5, but will be expanded upon in the following paragraphs. The magnetic flux is maintained by the current in the rotating field winding. Figure 12.12 is a typical saturation or magnetization curve for a large turbine generator at rated speed. It can be seen that the field current required to maintain rated voltage at rated load and PF is much greater than that required at no-load. This is because the armature current has a demagnetizing effect; that is, the mmf of the armature current is partially in opposition to that of the DC in the rotating field winding.

It should be noted that at low load or no-load, the current required for rated conditions could result in a very high voltage and corresponding magnetic flux density in the core. Operation with excessive flux densities will cause abnormal

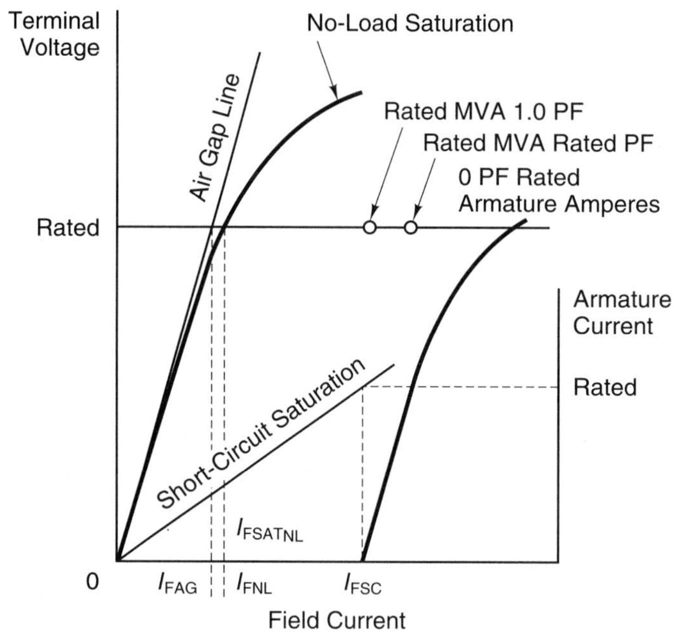

I_{FAG} = Generator field current required to produce rated voltage on the air gap line.
I_{FNL} = Generator field current required to produce rated voltage at no load.
I_{FSC} = Generator field current required to produce rated armature current with generator terminals short-circuited.

Figure 12.12 Synchronous generator saturation curves.

flux distribution in the core, which, in turn, can result in high levels of eddy currents in some core areas. These currents in the core can cause local overheating and possible arcing within the core. Once an area of the core is damaged, even normal levels of flux densities can cause further damage.

Because the voltage is a function of the product of the flux density and frequency, a device that monitors the ratio of voltage to frequency, or *volts/hertz,* will monitor the flux density of the core. The manufacturers of large turbine generators have established operating limits in terms of volt/hertz capability. Power transformer manufacturers have also established similar limits.

Instantaneous and inverse time volt/hertz relays (59/81) are available. If instantaneous relays are used, a multilevel scheme with time delay relays should be employed. On a unit-connected generator, the primary voltage rating of the GSU is frequently lower than that of the generator. In this case, the relay settings should be based on which of the two components is more limiting.

There have been a number of events in which a generator has been subjected to an excessive flux condition because there was a discontinuity in the VT circuit. Drawout VTs are vulnerable to this condition. In such a case, the volt/hertz relay may not be able to sense an excessive condition if it is supplied by the VT with the circuit discontinuity. For this reason, the volt/hertz relays should be supplied from more than one phase and/or from different transformers.

Some of the excessive volt/hertz events have occurred at no-load when the units were being placed in service. It is therefore desirable to have some form of inhibit feature in the excitation control equipment to prevent an operator from exceeding the field current required for 105% of rated voltage at no-load interlocked with the unit circuit breaker(s).

Some users disable the volt/hertz protection when the unit is on-line on the assumption that the system will hold the volt/hertz to a safe level. This is discouraged, because a load loss that occurs without the unit circuit breaker opening could cause a high volt/hertz condition. Volt/hertz protection should be applied in accordance with the limitations provided by the manufacturer of the generator and should be connected to trip the unit circuit breaker and generator excitation. (See Section 6.2.5.)

The previous discussion has addressed the potential for damage to a generator from excessive magnetic flux. In some extreme cases of underexcited operation of the generator, the end portions of the core can be subjected to overheating. The reasons for this will not be presented here, but the generator capability curve defines the underexcited limits for the machine as well as for other conditions. It is the responsibility of the operator to operate within these limits. The excitation control equipment (AVR) is equipped with a minimum excitation limiter (see Chapter 7), but this device is normally set to preclude a level of excitation low enough to jeopardize the stability of the unit. It is not intended to prevent exceeding the underexcited capability of the generator, although some users have attempted to have it do this.

12.5 ROTOR THERMAL PROTECTION

It is also important to protect the rotor from thermal effects. These are covered in the following sections. Chapter 5 has additional detail, from a rotor construction standpoint.

12.5.1 Field Winding

The short-time overexcitation for field windings of cylindrical rotor synchronous generators is specified in ANSI Standard C50.13–1989. Paragraph 6.2 of that standard states:

> 6.2 Field Winding Short-Time Thermal Requirements. The generator field winding shall be capable of operating at field voltage of 125 percent of rated-load field voltage for at least 1 min starting from stabilized temperatures at rated conditions.
>
> NOTE (1): The permissible field voltages at times up to 120 s, based upon the increment of heat storage as defined in Section 6.2, will be

Time (seconds)	10	30	60	120
Field voltage (%)	208	146	125	112

> NOTE (2): It is recognized that field winding temperatures under these conditions will exceed rated-load values and, therefore, the machine construction is based upon the assumption that the number of such operations at field voltages to the limits specified in Note (1) will occur not more than two times per year.

The preceding table can be approximated by the following equation:

$$(V_f{}^2 - 1) * t = K \tag{12.7}$$

In this equation, V_f is per unit of rated field voltage, and t is the time in seconds. For the preceding table, K is approximately 33. Note that the standard requires that the field be capable of 125% of rated voltage for one minute. This is based on the fact that most excitation systems will provide an exciter ceiling voltage of at least 125% of rated value. High-response excitation systems will inherently have a high ceiling voltage associated with them. The ceiling may be above 150% of rated field voltage.

The excitation of modern generators is controlled by continuously acting AVR (see Chapter 7). These regulators are high-gain devices that can provide full *boost* or full *buck* output for very slight variations in generator terminal voltage. Reliable operation of the power system frequently dictates the performance requirements of these regulators. However, any system condition that causes a sustained reduced voltage at the generator terminals can represent a hazard to the generator field winding. This is particularly true for generators with high performance excitation systems.

The excitation control system should include overexcitation limiting and protection features. Usually, an *overexcitation limiter* within the AVR is set to prevent the limits specified above from being reached. A backup device referred to as *overexcitation protection* is usually set to trip the generator and excitation in the event that the AVR is out of service or the limiter is ineffective in limiting the field current. The latter may be the result of a failure in the AVR.

12.5.2 Generator Rotor Body

In order to understand the need for some form of protection of the rotor body, it is necessary to review some aspects of unbalanced fault conditions and their effect on a generator.

By the method of symmetrical components, an unbalanced set of vectors or phasors can be resolved into three symmetrical sets of vectors or phasors. These sets are known as the *positive-sequence*, the *negative-sequence* and the *zero-sequence* phasors. The symmetrical components of the unbalanced currents of Figure 12.13 are shown (not to scale) in Figure 12.14. For normal operating conditions on a power system, only the positive-sequence voltages and currents are of concern.

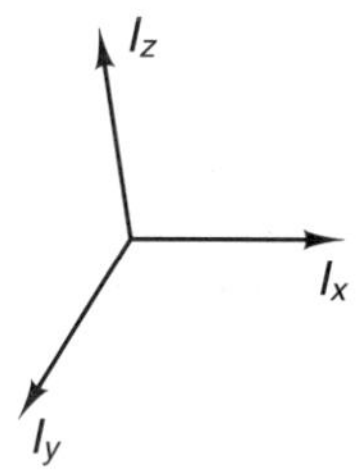

Figure 12.13 Unbalanced current phasors.

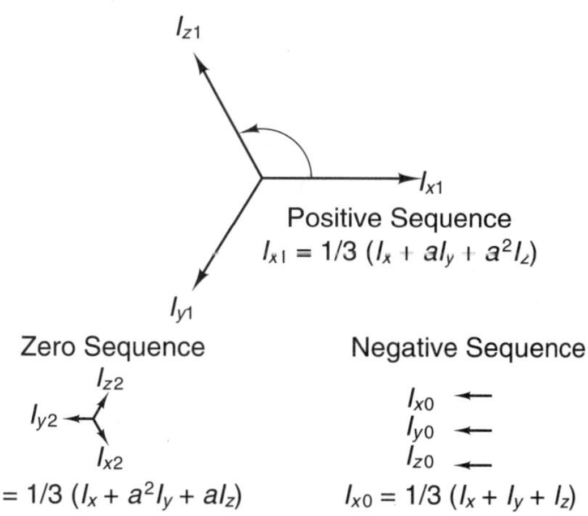

Figure 12.14 Symmetrical components of unbalanced currents.

Most faults (short circuits) on a transmission system are single-phase-to-ground faults. With the delta-wye connection of the GSU of the unit-connected generator, these faults are seen by the generator as phase-to-phase faults. In that case, there are both positive- and negative-sequence currents flowing in the armature windings.

During normal operation, the mmf of the positive-sequence currents is locked in synchronism with that of the rotating mmf established by the generator field current. The mmf of any negative-sequence armature currents rotates in the direction opposite to that of the rotor as shown in Figure 12.15. Because both the positive- and negative-sequence currents are at rated frequency, the negative-sequence mmf induces currents in the rotor body that are of twice the rated or fundamental frequency, i.e., 120 Hz in the United States. These induced currents can cause heating of the rotor body, wedges, and retaining rings, and, if protracted, damage to the rotor.

Again, ANSI Standard C50.13–1989 specifies both the short and continuous capabilities for generators in terms of the negative-sequence currents. Paragraphs 6.3 and 6.5 of this standard are reproduced below:

> 6.3 Rotor Short-Time Thermal Requirements for Unbalanced Faults. The generator rotor shall be capable of withstanding, without damage, unbalanced short circuits or other unbalanced conditions on the system or at the armature terminals resulting in values of I_2^2t as listed below.

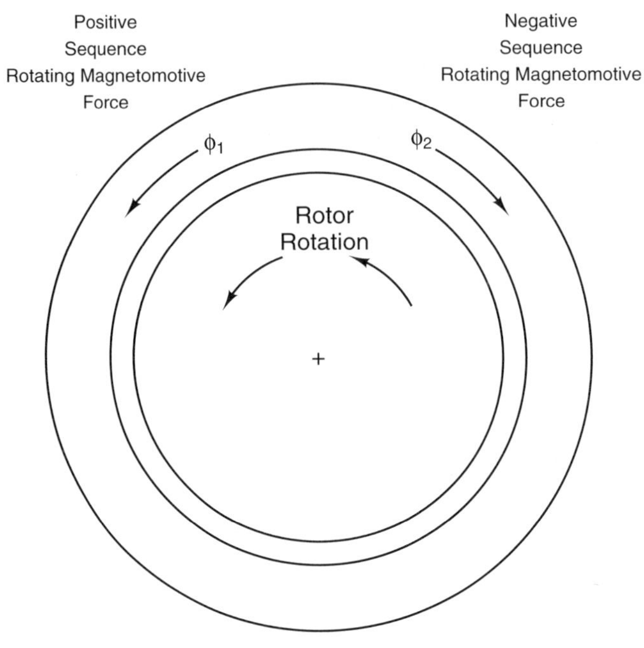

Figure 12.15 Rotating magnetic fields with unbalanced armature currents.

Type of Cylindrical Rotor Synchronous Generator	Minimum Generator Short-Time Capability Expressed in terms of I_2^2t (note)
Indirectly Cooled	30
Directly Cooled up to 800 MVA	10

I_2^2t in the preceding table is the integrated product of the square of the generator negative-phase-sequence current (I_2), expressed in per unit stator current at rated kilovolt-amperes and duration of the fault in seconds (t).

The generator unbalanced fault capability expressed in terms of I_2^2t applies for times up to 120 s, based on a constant increment of heat storage and negligible heat dissipation.

In the above criteria, the generator shall be capable of withstanding the thermal effect of unbalanced faults at the machine terminals, including the decaying effects of

(1) Field current, where protection is provided by causing field current reduction, such as with an exciter field circuit breaker or equivalent.

(2) Direct-current component of the stator current.

NOTE: Generators subjected to faults between the preceding values of I_2^2t and 200 percent of these values may suffer varying degrees of damage: for faults in excess of 200 percent of these limits, serious damage may be expected.

6.5 Continuous Unbalance Requirements. A generator shall be capable of withstanding, without injury, the effects of a continuous current unbalance corresponding to a negative-phase-sequence current of the following values, providing the rated KVA are not exceeded, and the maximum current does not exceed 105 percent of rated in any phase. Negative-phase-sequence current is expressed in percent of rated stator current.

Type of Generator	Permissible I_2 (percent)
Cylindrical rotor indirectly cooled	10
Directly cooled to 960 MVA	8
961 to 1200 MVA	6
1201 to 1500 MVA	5

These values also express the negative phase sequence current capability at reduced generator kilovolt-ampere capabilities, in percent of the stator current corresponding to the reduced capability.

One can determine the negative-sequence component of three-phase currents by using the law of cosines and the equations for symmetrical components. (These were covered in Section 5.2.4.) A generator can be protected from exceeding safe limits of rotor heating from negative-sequence-induced currents by either electromechanical or solid-state negative-sequence relays (46). The first negative-sequence relay made by one manufacturer consisted of an induction disc overcurrent element supplied by a negative-sequence current segregating network, or filter. The negative-sequence filter is supplied with all three phase currents. By means of suitably connected and scaled mutual reactors and resistors, it develops an output voltage or current proportional to the negative-sequence component of the phase currents.

Some of the older negative-sequence relays are not capable of protecting the generator from sustained negative-sequence currents between the continuous capability of the machine and about 45% of rated current. The newer solid-state relays have the capability of protecting the machine for all possibly harmful levels of such currents and can also alarm the machine operator when the maximum allowable continuous level is reached or exceeded. They also have a reset characteristic that can approximate the rotor cooling to "remember" previous unbalanced faults within a short period of time.

Industry practice is to connect the negative-sequence relay to provide an alarm for lower levels of negative-sequence currents and to trip the generator and excitation when the preset level of I_2^2t has been reached.

12.5.3 Rotor Winding Fault Protection

Because the field voltage of most generators is 750 volts or less, the insulation is not subjected to unduly high electrical stresses. The winding, however, is subjected to mechanical stress because of the high centrifugal forces and thermal expansions and contractions.

The two types of electrical failures that can occur are turn-to-turn short circuits and short circuits to the rotor body. Short-circuited turns could result in thermal and/or magnetic unbalance of the rotor with consequential vibration of the rotor. In generators that have the field terminals available, the resistance of the winding can be monitored by measuring the field voltage and current. With suitable instrumentation, this field resistance measurement is an indication of the field winding temperature and can be plotted by a recording device. A reduction in field resistance for a given operating condition is an indication of shorted turns. In rotating rectifier type of excitation systems (see Chapter 7), in which the field terminals are not continuously available, shorted turns must be found with some form of test.

In the past, a single ground on the field winding was not considered cause for immediate shutdown of a unit. The existence of a second ground, however, would represent shorted turns with the possibility of thermal unbalance, vibration, and possible forging damage. More recently, there have been incidents that suggest that a field winding ground may be the result of some previous damage, such as arcing, which caused forging damage. This has prompted some manufacturers to recommend tripping the unit as soon as the first field ground is detected. While it is important to determine where the ground is located, the generator field winding is connected to the exciter output circuit; thus, most ground detection schemes cannot indicate the location of the ground. It is usually necessary to isolate the generator field from the excitation source to locate the ground.

On generators where the field terminals are available, the field insulation to ground can be continuously monitored by ground protective relays (64), which have been available for many years. One such scheme is shown in Figure 12.16. A brief description of its operation will be given here.

Device 64 is a sensitive microvoltmeter. Resistors R_1 and R_2 are fixed or linear resistors, the values of which are determined by the rated field voltage. Resistors R_1, R_2, and the nonlinear resistor, R_3, constitute a voltage divider circuit. If a ground fault occurs in the field circuit, the field circuit will also be a voltage divider network. The voltage across element 64 will be that corresponding to the difference in voltage at the ground in the field and that at the junction of R_2 and R_3. There will always be a point in the field circuit at which the voltage is the same as that of the junction of R_2, R_3 and element 64. This is a null point, and a ground fault at that point will not be detected. During normal excursions of field voltage caused by operation of the AVR, this null point will be moved because of the voltage/resistance characteristic of the nonlinear resistor. The null point can also be moved by manual operation of the push button to short out R_1. This scheme is very sensitive and can detect insulation resistance to ground in the order of one megohm at the terminals of the field at 600 volts.

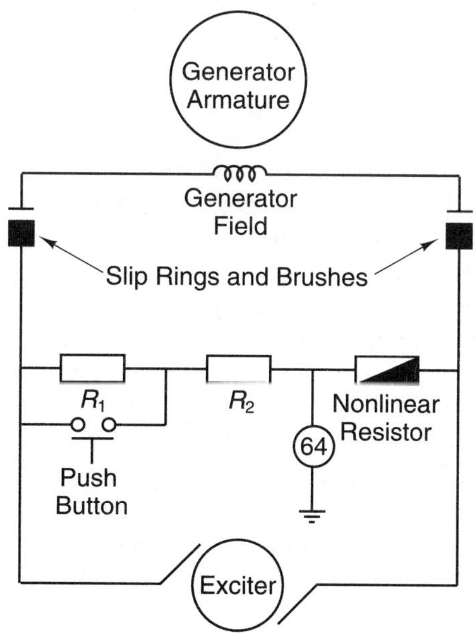

Figure 12.16 Generator field ground detection.

There is another scheme that inserts a direct voltage between one side of the field circuit and ground. This scheme is described in Section 4.4.1 of IEEE, C37.102–1987.

On rotating rectifier systems in which the field terminals are not available for continuous monitoring for grounds, the insulation to ground is usually checked, either manually or automatically, by a circuit in the excitation control equipment, utilizing brushes that are momentarily connected to slip rings connected to the output terminals of the brushless exciter. The circuitry of one scheme is similar to that of Figure 12.16. When the detection operation is initiated, there is a brush contact verifier feature to ensure that the brushes are in contact with the collector rings. During the check for grounds, the polarity of the brushes is reversed to ensure that a null point is not missed. Prudence dictates that as soon as a field ground fault is detected, the unit should be shut down in an orderly manner.

12.6 ABNORMAL OPERATING CONDITIONS

There exist various abnormal operating conditions in the generator and on the power system from which the generator or other equipment must be protected. These are described in the following sections.

12.6.1 Loss of Excitation

Loss of the direct current excitation of a generator field results in the generator running as an induction generator. It can still deliver some real power to the power system, but the speed of the unit will be above that corresponding to rated frequency. The excitation for the generator must be supplied by the other generators on the power system. The amount of kilovars drawn from the power system will depend on how heavily loaded the generator was when excitation was lost. The time it takes for the transition from synchronous to induction operation also depends on the load level and the characteristics of the generator and the power system. The reactive power drawn by the generator causes a voltage drop across the GSU.

Because the internal voltage of the generator is considerably lower than normal, the real power output of the generator is greatly reduced. This, and the fact that the generator under these conditions usually represents a severe reactive drain on the power system, dictates that the generator should be isolated from the power system. Failure to do so could jeopardize the stability of the other units on the system. In addition, sustained operation as an induction generator could cause the armature current limits of Section 12.4.1 to be exceeded.

A loss of excitation condition can be detected by a relay (40) that monitors the phasor voltage and current at the generator terminals. The ratio of this phasor voltage and phasor current represents an apparent impedance. This apparent impedance can be shown on the impedance or R–X diagram of Figure 12.17. While delivering rated load at rated PF, the apparent impedance at the machine terminals is indicated by point A of the diagram. In this case, the generator is supplying real power as well as inductive reactive power, or vars, to the power system. That is why the point A is in the first quadrant of the diagram. If the generator were ab-

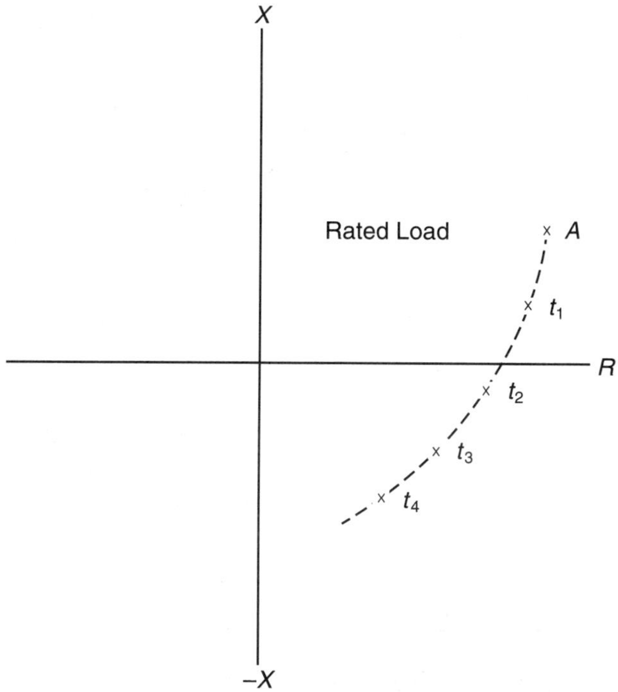

Figure 12.17 Change of the apparent impedance of a generator following loss of excitation.

sorbing vars, (that is, operating in the underexcited mode), the apparent impedance would be in the fourth quadrant of the diagram. When the generator has lost the source of DC excitation, the excitation must be supplied by the power system.

Figure 12.17 also shows how the apparent impedance of a generator may change with time after loss of DC excitation. In this example, it is assumed that the unit was operating at rated conditions prior to the loss of field. The locus of the apparent impedance is not always the same, depending on the load and system characteristics; however, the apparent impedance always ends up in the third or fourth quadrants of the diagram.

Figure 12.18 shows some typical settings for an *impedance* or *mho type* loss-of-field relay (40). Some relays do not permit inclusion of the origin of the *R–X* diagram in the setting, whereas others do. If a relay is set to include the origin as shown on the larger circle of Figure 12.18, a directional element must be used to prevent relay operation for some permissible operating conditions. These relays also have a voltage element that permits tripping by the relay only after the voltage is reduced to a preset level.

In recent years, the two-zone scheme of protection shown in Figure 12.18 has become popular. This scheme utilizes two loss-of-field relays and one or two external time delay relays. A sensitively set relay, shown by the larger circle, trips the unit after a time delay, up to one minute. This relay is intended to detect partial

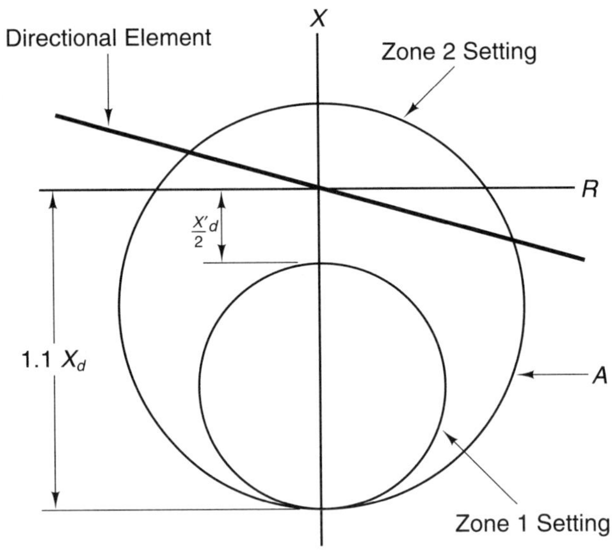

Figure 12.18 Loss-of-field relay application.

loss of field at heavy loads, or complete loss of field at very low loads on the unit. A less sensitive relay, represented by the smaller circle, trips in a much shorter time, with a maximum time delay of one second. All relay settings must be delayed long enough for the locus of the apparent impedance to move out of the operating circle for unit swings that would not result in loss of synchronism. One manufacturer's relay has a built-in time delay of 0.1 second, while that of another has a built-in delay of 0.25 second.

The loss-of-field relay(s) are usually connected to trip the unit as well as the source of excitation. The unit should not be restored to service until the cause of the problem is discovered and corrected.

12.6.2 Loss of Synchronism

As discussed in Chapter 2, there is a limit to the amount of power that can be transmitted across a transmission system or portion thereof. In spite of all the studies that are made to verify that a power system will be stable, there have been many cases of generators losing synchronism. While no attempt will be made here to address the many aspects of system stability, it is recommended that relays to detect a loss of synchronism should be applied on major units. When a unit is out of step with other units on the power system, it will experience severe shaft torque pulsations that can build up to damaging level. If the generator is tripped during the first slip cycle, these torques may not build up to dangerous levels.

There are a number of out-of-step protective relay (78) schemes available that will accomplish the action recommended above. They involve impedance type relays and logic and time delay circuits. Some users depend on the loss-of-field relay to detect an out-of-step condition. It should be emphasized that while a loss-of-field relay may, in some cases, detect the out-of-step condition, in most cases, the impedance locus will not pass into the relay operating characteristic. In the two-zone loss-of-field relay scheme, the time delays involved may be too long to prevent damage, even if the relays do detect the condition. Therefore, the loss-of-field relays should *not* be expected to provide out-of-step protection!

12.6.3 Overvoltage

In general, overvoltage protection (59) is not required on steam turbine generators if adequate volts/hertz protection is provided. Even with a full-load rejection, the overspeed is usually less than 20% above rated speed. On the other hand, hydro generators may reach 140% of rated speed. If the unit has a direct-connected exciter or static excitation supplied from the machine terminals, the voltage (without a regulator in service) is proportional to the square of the speed. In an installation such as this, it is prudent to provide overvoltage protection. Because some instantaneous AC voltage relays are responsive to volts/hertz, a DC voltage relay supplied by a full-wave rectifier connected to the generator VTs could be used for this function.

12.6.4 Abnormal Frequency

The primary concern associated with operating a loaded steam turbine generating unit at other than rated frequency (speed) is the avoidance of resonant frequencies of the long, tuned, low-pressure blades in the turbine. Operation of these blades, under load, at a speed that results in coincidence with the blading natural frequency in a lower mode and a multiple of the actual turbine running speed frequency can lead to fatigue damage of the blades. Fatigue damage is a cumulative effect, and most manufacturers give off-frequency limits in terms of cumulative times of operation in specified frequency ranges. During start-up, all turbines experience multiple resonance points in the tuned blades as the units are brought up to rated speed. In this case, the resonance points occur at low steam flow, and the stress levels are not considered damaging. During start-up, hold speeds corresponding to resonant frequencies should be avoided. A secondary concern regarding off-frequency operation is that running the generator below rated speed can result in damage to the hydrogen seals and/or in wiped bearings caused by higher vibrations.

The times at which generating units are required to operate at abnormal frequencies while loaded are infrequent. Even so, no single off-frequency event should approach the limits defined by the turbine manufacturer. Ideally, a device that keeps an account of the accumulated time of operation in specified frequency bands could provide a means of alarming and tripping. This could be accomplished by a computer algorithm.

Underfrequency protection schemes (81) [Warner, 1976] have been proposed to the industry. Most utility companies belong to an area pool and/or a reliability group. These organizations have defined certain requirements for generating station operation during severe system disturbances. These requirements are based on maximizing the reliability of the interconnected system. They may conflict with some forms of underfrequency protection proposals; therefore, it is suggested that an underfrequency protective system that is compatible with the off-frequency guidelines or limits imposed by the operating agreements of the interconnected groups or reliability organizations be employed. When it is suspected that the off-frequency limits of a unit have been exceeded, the turbine manufacturer should be consulted for recommended action.

12.6.5 Motoring

Motoring, as discussed here, refers to the condition in which the steam flow to the turbine is interrupted, but the generator is still electrically connected to the power system.

As long as the excitation is maintained on the generator and the unit is in synchronism, there is no hazard whatsoever to the generator. Once again, however, the hazard is to the low-pressure blading in the turbine. During normal operation, the steam flow through this blading acts as a cooling medium. During motoring, when there is no steam flow, there can be a buildup of heat in this blading. Most turbine manufacturers allow motoring for not more than a period of one minute.

The antimotoring protection (32) applied on a given unit should be in line with the turbine manufacturer's recommendation. Over the years, the manufacturer's recommended antimotoring schemes have changed because of changes in the governing and turbine control systems. One manufacturer's recommended scheme uses a pressure transducer, which indicates loss of steam flow across the first stage of the turbine, combined with a sensitive reverse power relay and auto-stop oil pressure to initiate tripping of the unit.

Typically, the real power drawn by the generator while motoring as described above will be less than 3% of the unit kilowatt rating. In many cases, it is much less. It also depends on whether or not vacuum is maintained on the condenser. The reverse power relay must, therefore, be quite sensitive. It is desirable that it be able to operate reliably at 0.5% of rated power into the generator. Combustion turbines that are connected to a compressor would draw a much higher value of motoring power because of the compressor load. In this case, a much less sensitive reverse power relay could be applied.

There have been incidents of severe overspeed of turbine generator units that resulted in modification of some antimotoring recommendations. It is generally agreed that an extreme overspeed condition represents a more serious hazard than a motoring condition. This attitude has resulted in modern antimotoring schemes being designed to avoid an overspeed condition. In other words, the scheme should not operate until there is positive indication that power flow is into the generator.

Many utilities use the antimotoring protection during a normal shutdown of a unit. In doing so, the operator initiates closing of the valves to interrupt the steam supply to the turbine. The generator breaker(s) is then tripped by the antimotoring protective devices when there is power flow into the generator. There have been cases in which the generator excitation was tripped, but the generator breaker remained closed. This results in the generator operating as an induction motor. In this case, the generator is not in synchronism with the other units on the power system and must draw excitation current from the power system. Although the current drawn from the power system is not as high as it would be if the unit had been at high load level, the unit should be isolated in some way. There will also be slip frequency currents induced in the rotor. If vacuum is not maintained on the turbine, both the armature current and the induced rotor currents will be greater because of the increased slip caused by the windage losses in the turbine.

In one case, the generator breaker was an independent pole-operated device, and one or two of the poles remained closed. Unlike the motoring situation described above, this does represent a hazard to the generator rotor because of the currents that can be induced in the rotor body by the slip frequency and negative-sequence stator currents.

12.6.6 Inadvertent Voltage Application

Application of system voltage to the terminals of a generator while it is at standstill or on turning gear is one of the most serious hazards to which the machine can be subjected. Unfortunately, this has happened to many units, resulting in severe damage to some of them. Depending on the switching device through which voltage is applied, the voltage may be a balanced or unbalanced set of three-phase voltages or a single-phase voltage. The event causes a severe electrical and mechanical shock to the machine, which results in high mechanical stresses on the armature winding and on the rotor and shaft system of the unit.

Application of balanced three-phase voltages is usually the result of closing a circuit breaker. Application of single-phase or unbalanced voltages is usually the result of a disconnect switch closing or flashover of one or more poles of a circuit breaker. These latter events will be discussed later.

When three-phase voltage, either balanced or unbalanced, is applied to the generator at near standstill, the generator attempts to accelerate as an induction motor with the turbine as a mechanical load. In addition to the high mechanical shock, the stator currents induce high currents in the rotor body. If the voltages are not balanced, there will be both positive- and negative-phase-sequence current components in the armature current. In this case, the rotor currents will contain both slip frequency components and components that are at twice the rated frequency minus the slip. As the unit accelerates, these currents will excite a multitude of torsional frequencies in the shafts. Unless the machine is immediately isolated from the power system, severe damage may occur. There have been a number of units that have been severely damaged because they were not isolated promptly. There have been many units that have been inadvertently energized, but suffered little or no damage because they were isolated quickly. In general, those units that were isolated within one second suffered no damage. However, this does not guarantee the same results for future incidents. The manufacturer should be consulted when such an incident takes place.

Application of balanced three-phase voltage may be detected with some of the relays in a normal complement of protective relaying. At standstill, the impedance that the generator presents to the system is the subtransient reactance. If the subtransient reactance falls within the operating characteristic of the loss-of-field relay, it will be detected by that device. In the two-zone scheme of loss-of-field protection described earlier, the relay that would detect this condition must have a short time delay (less than one second) associated with it. A sensitive directional power relay, such as might be used for antimotoring protection, could detect the condition, but the time delays used in antimotoring protection are too long to prevent damage.

On a unit-connected generator, the GSU is connected in delta on the generator side and grounded wye on the system side. If one pole of a high-voltage switch is closed, line-to-ground voltage will be applied to one phase of the GSU. When an isolating switch is accidentally closed, one pole will usually flashover before the other two because of the slow speed of switch closing and because the poles are not as precisely adjusted as are those of a circuit breaker. This will result in full line-to-line voltage applied between two terminals of the generator and half line-to-line voltage on the third phase. There will be both positive- and negative-phase-sequence currents in the stator that will induce currents in the rotor. Because the loss-of-field relays are not really polyphase relays, they may or may not detect the application of single-phase voltage, depending on the phase relationships between the voltages and currents supplied to the relay during the event.

There have been incidents of flashover of one or more poles of the high-voltage circuit breaker when a unit was being placed into service, but before synchronizing, or when the circuit breaker was tripped while removing a unit from service. During this time, the unit circuit breaker poles will periodically see voltages from zero to twice the normal line-to-ground voltage across the open contacts. If the insulation strength of the open contact gap is inadequate to prevent flashover, the generator will be electrically connected to the power system through the failed pole or poles. Although the shock to the unit is much less that it would be if the unit were at standstill, there is still potential for significant damage to the unit.

In one incident, a unit was being removed from service, and one pole of the open unit circuit breaker flashed over. The remote line circuit breaker was tripped by backup relays, but one phase conductor of a long section of extra high-voltage line was left connected to the unit through the failed circuit breaker pole. The capacitance of the line was sufficient to delay the decay of generator voltage, even though the generator excitation had been tripped. In this case, the positive-sequence component of the line charging current maintained the magnetic flux of the generator, but the negative-sequence component caused double-frequency currents to be induced in the rotor.

There have also been cases of a shut-down generator being energized through the UAT, which was back fed from the start-up transformer. In this case, the impedance of the two auxiliaries transformers in series limits the current to the generator to a much lower value than that from the high-voltage system, and the hazard to the unit is considerably reduced. However, the condition should be detected and the unit isolated as promptly as possible, because failure to do so may also cause failure of one or both of the UATs.

One scheme proposed uses three very inverse directional overcurrent relays (67) connected to operate for current into the unit. They should be connected to CTs in the generator leads, and potential for the directional units should be taken from the generator VTs. The overcurrent setting of the relays should be as low as possible without exceeding the continuous rating of the relay at that setting during full-load operation of the generator. If that setting or the relay location will not detect energizing the generator through the UAT, additional relays that will do so should be applied in a suitable location in the circuits. Figure 12.19 is a typical generator capability curve showing the area of operation of this scheme.

It should be pointed out that it is the practice of some companies to remove the fuses of the VTs or withdraw drawout VTs during a shutdown for reasons of personnel safety. Because the directional elements of these relays require voltage for operation, the scheme would be disabled in this case and could not provide protection.

Another scheme utilizes an instantaneous overcurrent relay (50) supervised by a frequency relay (81), which disables the overcurrent relay when the unit is near rated speed. This scheme will not detect the flashover of a circuit breaker or switch when the unit is being removed from service or just before synchronizing as described earlier. In addition, special care must be taken when selecting the type of frequency relay to be used if the VTs will be disconnected during a shutdown. The frequency relay must be one that interprets the absence of voltage as a low frequency. Some relays have a minimum voltage below which they will not operate.

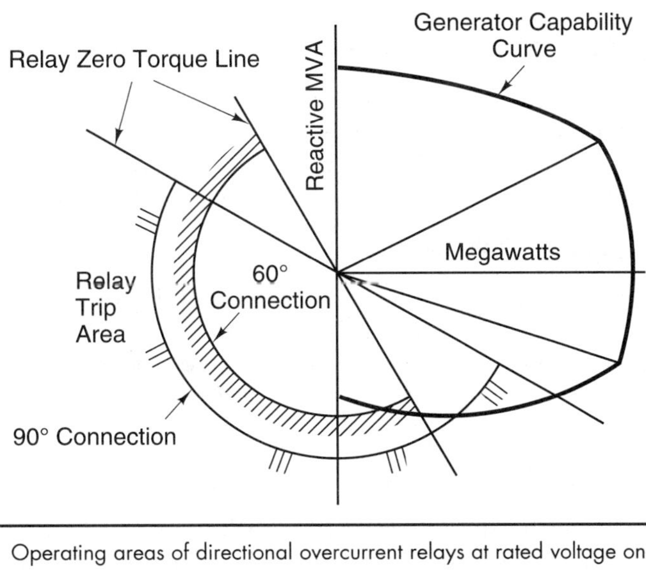

Figure 12.19 Operating areas of directional overcurrent relays at rated voltage on a typical generator capability curve.

A special relay for protecting generators during shutdown was developed by one relay manufacturer. This relay seems to overcome the shortcomings of the two schemes described above. It is called a *dead machine* relay.

Because this event has occurred in one form or another so many times, it is recommended that some form of reliable protection, specifically for this function and compatible with company maintenance practices, be provided for every generator. In addition, interlocking of switches and circuit breakers to prevent the event from happening should be employed along with suitable training, not only of unit and system operators, but also maintenance personnel whose actions could result in one of the above described events.

12.6.7 Subsynchronous Resonance

As discussed in Chapter 2, series capacitor banks are sometimes used to increase the power transfer capability of a transmission line, usually to enhance system stability. These banks have the potential for causing a subsynchronous resonance (SSR) problem, which could damage a generating unit. In addition to providing corrective measures to dampen out subsynchronous oscillations, should they occur, the generator protection should include protective relays to detect these oscillations and isolate the unit if they are sustained.

12.7 EXCITATION SYSTEM PROTECTION

The excitation system for a generator must also be protected for faults or conditions that represent a hazard to either the generator main field or the exciter. This protection is discussed in Chapter 7.

12.7.1 Exciter Thermal Capability

In a well-designed generating unit, the exciter thermal capability will equal that of the generator main field. The field protection described in Section 12.4.1 should provide adequate protection for the exciter as well as the generator main field. This should hold true for either a static or a rotating exciter.

12.7.2 Exciter Field Ground

The field of a rotating exciter is normally supplied by a power amplifier in the voltage regulator. Because the field terminals are available, it is possible to provide a circuit and devices to detect a ground fault on the exciter field. This is an optional feature with some equipment manufacturers.

12.8 BACKUP PROTECTION

As in all devices, there is a need for backup protection in the generator. This section describes this important aspect of generator protection.

12.8.1 Phase Faults

The complement of protective relays on a generating unit may include relays to isolate the unit from the power system for transmission system faults that should be cleared by the system protection, but are not. One common form of back-up on a unit-connected generator is provided by a *distance* or impedance type relay (21). The voltage for this relay should be taken from VTs on the generator terminals, and the current should be supplied by CTs in the generator neutral. The directional sensing of the relay should be such that the relay "looks" out into the system. A typical setting or reach would be 90% of the longest line out of the station. This scheme is shown in Figure 20.

A distance type relay (21) in the switchyard, looking back toward the unit and set to reach part way into the GSU, is another form of backup protection. In this case, the relay, in part, is backup for the GSU protection, but it may also be primary protection for the leads between the GSU and the switchyard. These leads are sometimes quite long, in the order of 0.25 mile or so.

A pilot wire relay system is sometimes used as protection for the leads between the GSU and the switchyard circuit breaker. Both of these lead protection schemes are shown in Figure 12.21.

Another form of backup, usually used on non-unit-connected generators, is a voltage-controlled overcurrent relay. This relay has an induction disc overcurrent element that is torque-controlled by an instantaneous undervoltage element. The voltage element is usually set at about 80% of rated voltage, and the overcurrent element should be set to prevent the short-time thermal capabilities described in Section 12.3.1 from being exceeded. One disadvantage to this type of relay is that if the potential is lost, the relay loses its restraint. This can result in an undesired trip of a loaded generator.

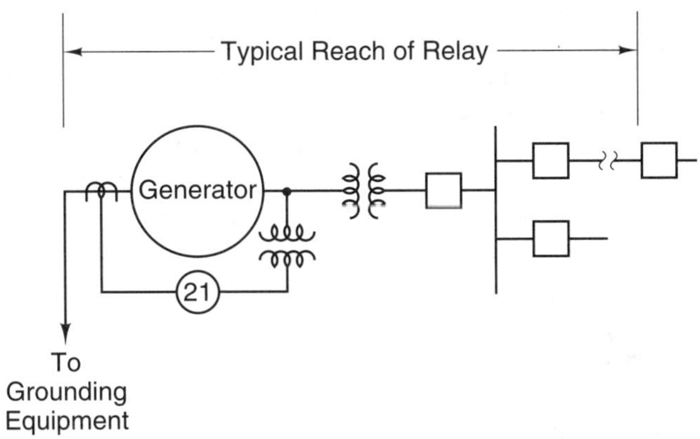

Figure 12.20 Generator backup distance relay.

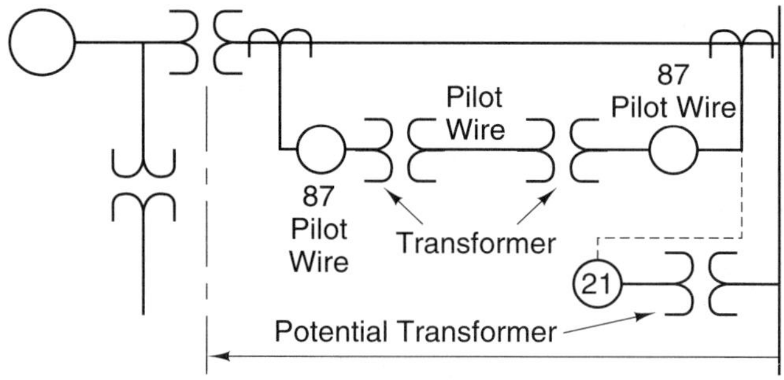

Figure 12.21 Transformer lead protection.

12.8.2 Ground Faults

Ground fault backup protection for the generator and equipment connected to its terminals may be provided by an overcurrent relay connected in the secondary of the grounding transformer as described in Section 12.2.2. Ground fault backup for transmission system faults may be provided by an overcurrent relay supplied by a CT in the GSU high-voltage neutral circuit. This device may be a directional overcurrent relay.

12.9 SYNCHRONISM CHECK DEVICES

Synchronism Check Devices: Many generating units have automatic synchronizing or synchronism check relays (25) to either control or supervise the connection of a generator to the power system. These are more in the realm of control equipment than protection, but they are intended to prevent damage to a generator by closing the circuit breaker when there is too great a difference between the generator voltage, frequency, or phase angle and that of the power system. In some cases the 25 device gives the operator a "window" in which he can close the circuit breaker. In other cases, the device itself may energize the circuit breaker closing circuit.

Moreover, some older generators employ an overcurrent protection scheme that monitors the current magnitudes in the three phases coming out of the generator to protect the system against the large oscillations resulting from an out-of-phase synchronization. However, this does little to protect the generator itself.

SUMMARY

This chapter discusses the various relaying schemes used to protect the generator against faults whose occurrence is too sudden to permit a diagnosis and orderly corrective action by the unit operators. These are quite important to ensure the overall protection of the generator.

REFERENCES

[ANSI, C50.13–1989] "Requirements for Cylindrical Rotor Synchronous Generators," American National Standard C50.13–1989.

[Baldwin, 1980] Baldwin, M. S., W. A. Elmore, J. J. Bonk, "Improved Turbine Generator Protection for Increased Plant Reliability," *IEEE Transactions,* PAS, Vol. 99, No. 3, May/June 1980, pp. 982–989.

[IEEE, 76CH] "Transient Response of Current Transformers," IEEE Publications 76 CH 1130–4 PWR.

[IEEE, C37.102–1987] "IEEE Guide for AC Generator Protection," IEEE STANDARD C37.102–1987.

BIBLIOGRAPHY

[IEEE, 1989] "Inadvertent Energizing Protection of Synchronous Generators," Committee Report, *IEEE Transactions on Power Delivery,* Vol. 4, No. 2, April 1989, pp. 965–977.

[IEEE, C37.2–1991] IEEE Standard C37.2–1991, "IEEE Standard Electrical Power Systems Device Function Numbers," 1991.

[Westinghouse, #B1235] *Applied Protective Relaying,* Westinghouse Publication #B7235 (now Asea-Brown-Boveri Corp.).

[Warner, 1976] Warner, R. E., T. L. Dillman, and M. S. Baldwin, "Off-Frequency Turbine-Generator Unit Operation," *Proceedings of the 38th American Power Conference,* Chicago, Ill., April 20–22, 1976.

CHAPTER 13

Automated Diagnostic Systems

13.1 INTRODUCTION

The objective of this chapter is to complete the discussion of the subject of generator diagnosis by introducing the concept of automated diagnosis by computer. One of the enabling technologies for this effort has been the emergence of *artificial intelligence* as an application-ready technology. This technology will be discussed in this chapter.

13.2 EVOLUTION OF PLANT MONITORING SYSTEMS

Monitoring of plant sensors has progressed from the early days of periodically sending someone out to read local gauges and manually record their values on paper to automatically monitoring the values of key parameters through a computer and activating an alarm whenever a limit is exceeded.

The earliest automated monitoring systems consisted of switches that illuminated an alarm window in an annunciator panel in the control room when a predetermined limit had been exceeded. Analog electrical parameters, such as voltages and currents, were also monitored in the control room and recorded on strip charts. Data histories were maintained mostly through hand-recorded logs, strip charts, and the operator's memory. Diagnosis was performed by operators or engineers knowledgeable and experienced in the operation and troubleshooting of generators. In general, this process suffered from insufficient data on which to base a diagnosis.

With the advent of computers, the monitors have become more powerful, capable of monitoring parameters that could not be monitored in the past. Additionally, plant computers have significantly enhanced the ability of the plant operators to acquire and maintain a history of many sensors simultaneously through the use of database management systems (DBMS). Such histories also include a record of all alarms and the time at which they came on and/or were cleared.

Diagnoses, however, are still being made by the same type of experienced individuals, either unit operators or plant/headquarter engineers. In contrast with its earlier versions, the diagnostic process at this time suffers from too much information. Even experienced engineers and operators cannot efficiently interpret the large volume of information being recorded by the data acquisition system. This can be referred to as *overmonitoring* the generating unit. Such a data overload on the part of the operator makes it difficult for the operator to focus on the critical indications that provide clues to incipient failures. Key indications are, therefore, often lost in the confusion.

In summary, while the data acquisition and alarm process has improved significantly over the past two decades, to the point that humans can no longer keep up with it, the process of diagnosing the actions of a generator is still at the same stage it was 20 or more years ago. It is clear, therefore, that any improvements in monitoring and diagnostics must come from the automation of the diagnostic process. If the knowledge of diagnostic experts could be collected, represented, and manipulated by a computer, this goal could be accomplished. Surprisingly, however, this seemingly simple task has resisted automation for many years. The following section describes the difficulties involved as well as the technology that has the promise to overcome these difficulties.

13.3 DIAGNOSTIC KNOWLEDGE

Computers have traditionally been excellent at performing computational tasks that are well defined and leave no ambiguity as to what step to take next. Certainly, traditional mathematical and logical computations fit such a description very well. Manipulation of data that is exact in nature (the data as well as the manipulation) also falls under this classification. The common thread of these elements is the ability to write a precise *algorithm* for the execution of the task. Naturally, they are referred to as algorithmic processes. Computers are capable of carrying out such tasks much faster and more effectively than a human.

But there exist other tasks that are not as well defined or clearly expressed. Such tasks cannot be easily represented through an algorithm. For example, any five-year-old child can recognize a tree when she sees one, regardless of whether the tree is an oak, a pine, or a palm, or whether it is day or night, or if it is partially hidden behind a building. Yet, to write an exact algorithm that allows a computer to process a digital image and pick out a tree within it is quite difficult. Furthermore, if the tree is partially hidden, in the shade, or part of a forest, then the difficulty level increases many-fold.

Humans depend on what is referred to as *heuristic knowledge* in order to carry out mental tasks that cannot be easily described through algorithms. This knowledge is often learned through experience. It allows humans to perform some tasks more efficiently than computers, even though computers have more memory and their processing speed is significantly faster than that of the human brain. These abilities partly make up what is considered to be human intelligence.

Other tasks besides image recognition that are generally performed much more efficiently by a human than by a computer are the following:

- Planning
- Scheduling
- Classification
- Design
- Natural language communication
- Situational awareness

These seemingly disparate tasks have in common the fact that a decision must be made based on a large number of inputs whose values can range widely. Heuristic knowledge allows humans to cut to the core and only pay attention to a limited number of inputs by recalling certain similar cases in the their past experience. Such knowledge can be quite difficult to acquire and represent in the traditional computer programming languages and techniques.

The diagnostic process is, in many ways, also representative of such heuristic thinking. Diagnosticians are also faced with a large number of inputs, most of which are not significant for diagnosing any one type of incipient fault. They also depend heavily on prior experience. The more times that a diagnostician has seen a particular type of problem, the more likely he is to identify it correctly when it surfaces again in the future.

The solution to the problem, therefore, lies in developing and using a new means of representing and exercising the heuristic knowledge inherent in the decision-making abilities of diagnostic experts. The field of artificial intelligence, (in particular, one of its branches called *expert systems*) is an ongoing effort to provide these new techniques. This is described in the following section.

13.4 ARTIFICIAL INTELLIGENCE AND EXPERT SYSTEMS

Artificial Intelligence (AI) has been loosely defined as computational techniques that allow computers to perform tasks that normally require human intelligence. AI has also been defined as "computational tasks that humans perform better than computers" [Rich, 1991].

Because no universally accepted definition of human intelligence exists, it can be quite difficult to apply that standard to AI. Thus, the second definition is preferred. The tasks described in the previous sections are some of those that humans perform better than computers at the present time. It can be expected that when some of these become effectively solvable through a computer, they will cease to be considered part of the AI realm.

Expert systems, in particular, deal with knowledge-intensive tasks, such as diagnostics. They have the ability to capture the heuristic knowledge that experts use to perform their diagnostic function. Expert systems are composed of three parts:

- The *inference engine* is a generic mechanism that exercises knowledge in order to reach a conclusion based on the combination of that knowledge and the applicable data being presented. The inference engine represents the underlying reasoning process that is typically common to many types of diagnostic tasks, whether it be generator diagnosis, automobile diagnosis, or medical diagnosis. It is based on the concept that the thinking process of a diagnostician is the same; only the domain in which the diagnostician deals is different. Thus, a particular inference engine will generally be capable of manipulating knowledge from different domains.
- The *knowledge base* contains the domain-specific knowledge. It is the knowledge that makes the system applicable to generator diagnosis or to diagnosis of heart disease. This separation of the domain-specific knowledge from the generic manipulation engine makes the development of expert systems relatively easy, when compared to conventional software systems.
- The *auxiliary features* of an expert system can be quite significant. Although they include many features that are only significant to the expert system developer, one very important one to a user is the ability to explain the reasoning that the expert system followed in order to arrive at a particular conclusion. This is called the *explanation feature* and is quite popular with users who are unsure of the system's ability to produce a correct answer.

What sets expert systems apart from conventional computer programs, therefore, is their ability to represent and use heuristic knowledge. But this knowledge must be represented in a way that can be easily manipulated by the inference engine. This is the subject of *knowledge representation,* and it is an important one because of the conflicting requirements that the knowledge be represented expressively (i.e., useful to the computer as well as to the developer) and concisely (i.e., the knowledge not be voluminous).

Expert systems typically employ one or both of the following means of knowledge representation: *rules* and *frames.* These are explained in the next sections.

13-4.1 Rules

The knowledge used by many people is conditional in nature. Each bit of conditional knowledge is composed of a *premise* and an *action.* The premise presents a set of conditions or tests that must be satisfied before the action is carried out. The action need not be a real action; it could simply be inferring a new bit of information that was derived through the use of the knowledge. Such representation of conditional knowledge is referred to as *rules.*

But conventional programs, such as FORTRAN and BASIC, also support conditional statements. Why, then, can they not be used for this purpose? The answer is that while rules and conditional statements appear on the surface to be the same, they are not. Their major difference is how they are employed. In algorithmic programming, a conditional statement must dictate which path the computation follows, based on one or a series of tests. Following more than one path simultaneously is not typically possible in conventional systems, because a computer must be told exactly what to do next. The fuzzy nature of heuristic knowledge makes this impossible. Thus, Figure 13.1 shows how algorithmic knowledge can be represented in a decision tree. Notice how each diamond asks a question that must be answered with a yes or no. However, the expert system, shown in Figure 13.2, does not require strict response to an answer because there are many ways that the system can branch out. This latter style of computing is referred to as *declarative programming.* Figure 13.2 shows a knowledge base, which is represented as a network. The *nodes* (round or rectangular boxes) are the variables or states of the system and the *arcs* (lines between the boxes) are the rules that determine the value of the states. The knowledge network, called an *inference network,* provides the conditional relationships from the input *sensors* (the circles at the top of the figure) to the output *conclusions* to be reached (the squares at the bottom of the diagram). If a particular path is blocked by the absence of a piece of data, other rules will be executed that also affect the same conclusion. Figure 13.3 shows two examples of rules.

The simple rule network of Figure 13.3 can be pictorially represented as one sensor input, the "Outlet water temperature," one conclusion (diagnosis), "Blockage of the water in the coil #52," and one intermediate node, "Outlet water tem-

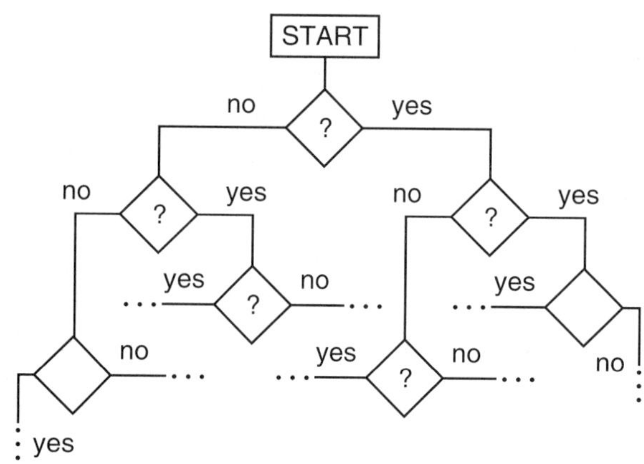

Figure 13.1 Conditional structure of conventional languages.

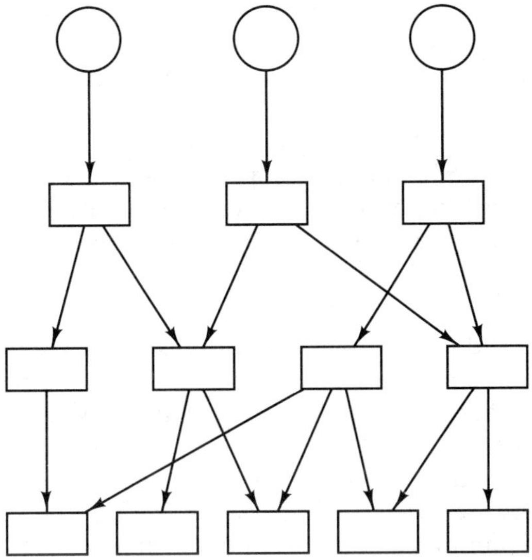

Figure 13.2 Knowledge network for an expert system.

perature is higher than normal," RULE-001 and RULE-002 provide the conditional relationships that link these nodes.

There exist two general types of expert systems that are based on rules: *forward-chaining* and *backward-chaining*. A forward-chaining system (also called *data-driven*) looks at incoming data and tries to derive a conclusion based on such data. This is especially useful in automated monitoring and diagnostic systems in which there is data coming in continuously that is to be compared to alarm levels and used in general diagnosis.

Backward chaining, on the other hand, does not look at the data first, but looks at the possible conclusions that could conceivably be made and attempts to justify each possible conclusion with data in existence. The nature of the backward-chaining inference engine is such that only the data that are necessary

RULE-001	
Premise:	IF the outlet water temperature > 90 Deg. C,
Action:	THEN Outlet water temperature = higher than normal.
RULE-002	
Premise:	IF outlet water temperature of coil #52 is higher than normal,
Action:	THEN There is a blockage of water in coil #52, CF = 0.85.

Figure 13.3 Examples of rules in an expert system.

for verification of a particular conclusion are requested from the user. Thus, this type of system is particularly applicable to diagnostics when the data must be entered by a user by hand into the keyboard. This type of reasoning is also called *goal-driven.*

Although the majority of expert systems are of the rule-based type, systems based on other techniques, such as frames, are becoming more common because developers have begun to encounter the inefficiencies of rule-based representations as they are applied to larger and more complex domains.

13-4.2 Frames

Frames are used to represent structured knowledge that can be difficult to represent with rules alone. It appears to be more passive because there is little intrinsic "motion" as there is in rules. However, it can also be used to represent relationships between ideas or actions, such as cause-and-effect relationships. Thus, they can also be effectively used in diagnostic knowledge representation.

A frame is a structure that assigns an object or a concept with attributes and values. For example, if the concept of a generator is to be represented, it could be done with frames in the fashion shown in Figure 13.4 below.

13-4.3 Advantages and Disadvantages of Expert Systems

Expert systems have the following advantages:

- They make possible the availability of human-like reasoning around the clock.
- They can handle incomplete information.
- They are easier to code than conventional systems because of their unique separation of domain-specific knowledge from the generic-reasoning mechanism.

GENERATOR	
a-kind-of:	electrical device
an-element of:	power plant
drive:	steam turbine
type:	alternating current
NANTICOKE-1	
a-kind-of:	generator
poles:	2
terminal voltage:	20,000
rated output:	800 MW
excitation-type:	rotating

Figure 13.4 Frames describing a specific generator.

- Their automated nature makes it possible to obtain consistent answers to the same problem across different locations or at different times.
- They can handle the uncertainty inherent in human problem-solving in the real world.

They also have some disadvantages that must be understood so as to not excessively depend on them:

- They are brittle in that they don't know when they don't know. They will always try to provide an answer, whether the question is within their limit of expertise or not. An inexperienced user may improperly assume that a correct answer is being provided when, in fact, it is not. This can be largely mitigated through the use of uncertainty indications, because with this feature they can provide a quantitative measure of their confidence in the diagnosis. Presumably, if the system is developed correctly, the response to a problem outside their domain will have a low certainty factor.
- They cannot handle common sense unless it is explicitly programmed in, which can be quite extensive.
- They cannot learn by themselves. This is one key aspect that separates them from humans.

There are two main systems in existence that are capable of diagnosing large electrical generators: GEMS by EPRI/General Electric and GenAID by Westinghouse. Although expert systems to do diagnostics for components of generators exist, these are the only two systems that can do a comprehensive job of diagnosing all aspects of the generator. These expert systems will be discussed in the following two sections.

13.5 THE GENERATOR EXPERT MONITORING SYSTEM (GEMS)

The GEMS system was developed by Ontario Hydro and SRI International under EPRI and Niagara Mohawk Power Corp. sponsorship. Development work began in 1987 by Ontario Hydro and SRI International. A prototype system has been operating in Ontario Hydro's Nanticoke power station and at Niagara Mohawk's Oswego plant. A commercial version of the system is being offered by The General Electric Company.

GEMS is a comprehensive generator diagnostic system. It covers all significant subsystems of the generator, such as the generator stator winding and core, the rotor winding and forging, the excitation system, the seal oil system, the hydrogen auxiliary system, and the stator cooling water system (for water-cooled generators). It focuses quite a bit of attention on the generator stator windings, because that is where most serious failures are known to occur.

GEMS is a local on-site system, in which all computation is performed within a local computer in the plant. Data are obtained either directly from sensors in the machine or through a data link from the plant computer, if one exists. GEMS provides an excellent window-based user interface and can trend sensor points as well as variables calculated from various sensor values (such as PF, average temperatures, etc.). GEMS can, in many cases, discover sensor inaccuracies as well as sensor failures.

GEMS can manipulate uncertainty, whether from the knowledge itself or from the sensor readings. It represents the knowledge using a knowledge network called *Bayesian Belief Networks,* which represent the uncertainty internally in the links between states. The belief networks are implemented in frames.

13.6 THE GENERATOR ARTIFICIAL INTELLIGENCE DIAGNOSTIC SYSTEM (GenAID)

The GenAID system was developed by Westinghouse Electric Corp. Development work began in 1980, and the first prototype was fielded in 1984 at the Texas Electric Utilities Co. Martin Lake Station Units 1, 2, and 3. Commercial installation of the system exists in nearly 20 units at this time.

One significant difference between GenAID and GEMS is that whereas GEMS performs all diagnoses locally at the plant site, the diagnostic function of GenAID is located at a central site where diagnosis is performed for all units connected to the system. The GenAID expert system receives the data through a remote link from data acquisition computers at the plant sites, carries out the diagnostic functions, and sends the conclusions back to the plant control room through the same link.

GenAID can also manipulate uncertainty, although it does it in a different manner. GenAID represents the knowledge using an inference network of rules and uses the *Confidence Factor* (CF) formalism to define, calculate, and propagate uncertainty throughout the network. Confidence factors were originally developed as part of the landmark MYCIN project in expert system diagnostics [Buchanan, 1984], and represent a more generalized version of Bayesian probability.

SUMMARY

This chapter covered the automation of generator diagnostics. It described the concept of expert systems, upon which automated diagnostics is largely based. Furthermore, this chapter discussed the two leading generator automated diagnostic systems, GEMS and GenAID.

In conclusion, there is ample room for improving the diagnostic process for electrical generators. One way to do this is to apply the recently available expert systems technology to the process. Expert systems attempt to represent the knowledge of a diagnostic expert in a computerized system capable of carrying out diagnosis automatically. This has the effect of wading through the volume of data presently being received at the control room and focusing on the few and typically insidious indications of incipient faults on generators.

Readers interested in GEMS are directed to the General Electric Field Sales office nearest their location. Those interested in GenAID are likewise directed to contact their Westinghouse representative for more information.

REFERENCES AND FURTHER READING

The reader is referred to the following for more information on expert systems:

[Buchanan, 1984] BUCHANAN, B. G., and E. H. SHORTLIFFE, *Rule-based Expert Systems,* Addison-Wesley, Reading, Mass., 1984.

[Gonzalez, 1993] GONZALEZ, A. J., and D. D. DANKEL, *The Engineering of Knowledge-based Systems: The Theory and the Practice,* Prentice Hall, Englewood Cliffs, N.J., 1993.

[Rich, 1991] RICH, E., and K. KNIGHT, *Artificial Intelligence,* 2d Ed., McGraw Hill, New York, N.Y., 1991.

Index